Optimizing Community Infrastructure

Butterworth-Heinemann is an imprint of Elsevier
The Boulevard, Langford Lane, Kidlington, Oxford OX5 1GB, United Kingdom
50 Hampshire Street, 5th Floor, Cambridge, MA 02139, United States

Library of Congress Cataloging-in-Publication Data
A catalog record for this book is available from the Library of Congress

British Library Cataloguing-in-Publication Data
A catalogue record for this book is available from the British Library

ISBN: 978-0-12-816240-8

For information on all Butterworth-Heinemann publications visit our website at https://www.elsevier.com/books-and-journals

Publisher: Brian Romer
Acquisition Editor: Brian Romer
Editorial Project Manager: Michelle Fisher
Production Project Manager: Poulouse Joseph
Cover Designer: Alan Studholme

Typeset by TNQ Technologies

Optimizing Community Infrastructure

Resilience in the Face of Shocks and Stresses

Edited by

RYAN M. COLKER, B.A., J.D.
Vice President, Innovation
International Code Council
Executive Director
Alliance for National & Community Resilience
Washington, DC, United States

ELSEVIER
Butterworth-Heinemann
An imprint of Elsevier

List of Contributors

Natalie Ambrosio, BSc
Editor
Four Twenty Seven
Berkeley, CA, United States

Allison Hoadley Anderson, FAIA, LEED AP
Principal
unabridged Architecture
Bay St Louis, MS, United States

Bilal M. Ayyub, PhD
Professor
Center for Disaster Resilience
Center for Technology and Systems Management
Department of Civil and Environmental Engineering
University of Maryland
College Park, MD, United States

Jerry P. Brashear, MBA, PhD
The Brashear Group LLC
Ashland, OR, United States

Joyce Coffee, MCP, LEED AP
President and Founder
Climate Resilience Consulting
Chicago, IL, United States

Ryan M. Colker, JD, CAE
Vice President, Innovation
Executive Director,
Alliance for National & Community Resilience
International Code Council
Washington, DC, United States

Jeff Dagle, MSEE, PE
Chief Electrical Engineer
Electricity Infrastructure Resilience
Pacific Northwest National Laboratory
Richland, WA, United States

Cindy Davis, CBO
Deputy Director of Building and Fire Regulations
Virginia Department of Housing and Community Development
Commonwealth of Virginia
Richmond, VA, United States

Jason Hartke, PhD
President
Alliance to Save Energy
Washington, DC, United States

Alice C. Hill, JD
Research Fellow
Hoover Institution
Stanford University
Washington, DC, United States

Michael E. Hooker, MBA
Executive Director
Onondaga County Water Authority
Syracuse, NY, United States

John S. Jacob, PhD
Texas Community Watershed Partners
Texas A&M AgriLife Extension Service
Houston, TX, United States

William Kakenmaster, BA
Research Assistant
Hoover Institution
Stanford University
Washington, DC, United States

Yoon Hui Kim, PhD, MPhil
Director of Advisory Services
Four Twenty Seven
Berkeley, CA, United States

Samantha A. Medlock, CFM
President
Climate Risk Advisors, LLC and
Adjunct Professor of Law
Santa Barbara & Ventura Colleges of Law
United States

Geoffrey G. Miller, PE, BCEE
Deputy Executive Director
Onondaga County Water Authority
Syracuse, NY, United States

Devesh Nirmul, CEM, CSDP, LEED AP O+M
Hannon Armstrong Sustainable Real Estate/
CounterpointeSRE
Annapolis, MD, United States

Robert G. Ottenhoff, MCRP, BA
President & CEO
Center for Disaster Philanthropy
Washington, DC, United States

M. John Plodinec, PhD
Associate Director
Resilience Technologies
Community and Regional Resilience Institute
Washington, DC, United States

Allison C. Reilly, PhD
Assistant Professor
Center for Disaster Resilience
Department of Civil and Environmental Engineering
University of Maryland
College Park, MD, United States

James Tim Ryan, CBO
Codes Administrator (ret.)
City of Overland Park, KS, United States

James Schwab, FAICP, BA, MA
Principal
Jim Schwab Consulting LLC and
Adjunct Assistant Professor
University of Iowa
School of Urban and Regional Planning
United States

John Scott, BOMA Fellow, RPA
Managing Director
Colliers International
Clearwater, FL, United States

Stacy Swann, BA, MBA, MTS
Climate Finance Advisors, BLLC
Washington, DC, United States

Vice-Chair
Board of Directors
Montgomery County Green Bank
Rockville, MD, United States

Timothy P. Taber, PE, BCEE
Vice President
Barton & Loguidice
Liverpool, NY, United States

Ziyue Wang, MEM, BA
Climate Finance Advisors, BLLC
Washington, DC, United States

Charris R.H. York, MS
Texas Community Watershed Partners
Texas A&M AgriLife Extension Service
Houston, TX, United States

Author Biographies

NATALIE AMBROSIO, BSC

Natalie Ambrosio, Editor, Four Twenty Seven, is an adaptation expert and science communicator specialized in distilling technical information into actionable insights on climate risk and resilience across sectors. At Four Twenty Seven, Natalie manages publications and communications, writing about climate change's economic impacts, strategies for investors to assess and manage risk, and the interconnected nature of climate resilience. Previously, Natalie contributed to a nationwide assessment of cities' vulnerabilities to climate change and their readiness to adapt, at the Notre Dame Global Adaptation Initiative (ND-GAIN). Natalie holds a BS in Environmental Science and a certificate in Journalism, Ethics, and Democracy from the University of Notre Dame.

ALLISON HOADLEY ANDERSON, FAIA, LEED AP

Allison Anderson founded unabridged Architecture, a firm specializing in built works that are inherently defensible against climate challenges including structures armored against natural and manmade hazards, adaptive reuse to prepare for the next century of service life, and urban strategies to accommodate water and prevent flooding. After Hurricane Katrina devastated her community, the firm's work focused attention on sustainability, adaptation, and resilience, responding with a deep understanding of the importance of place and tradition within the context of modern design. The firm was an early adopter of sustainability in architectural practice. The distinction between sustainability and resilience to climate hazards is one that she has studied in research, practice, and education. Her work in climate adaptation spans project scales, from urban infrastructure and first-responder shelters to residences that weather coastal storms. She received a Bachelor of Architecture at the University of Southern California and a Master of Architecture from the University of Texas. Allison has taught at the University of Texas and Louisiana State University and was the 2015 Favrot Visiting Chair in Architecture at Tulane University. She is a fellow of the American Institute of Architects and chairs the AIA Resilience and Adaptation Advisory Group.

BILAL M. AYYUB, PHD

Dr. Ayyub is a University of Maryland professor of Civil and Environmental Engineering, director of the Center for Technology and Systems Management, professor of Reliability Engineering, and professor of Applied Mathematics and Scientific Computation. Dr. Ayyub's main research interests are risk, resilience, uncertainty, decisions, and systems applied to civil, mechanical, infrastructure, energy, defense, and maritime fields. Dr. Ayyub is a distinguished member of the American Society of Civil Engineers and a fellow of the Structural Engineering Institute, the Society for Risk Analysis, American Society of Mechanical Engineers, and Society of Naval Architects and Marine Engineers. Dr. Ayyub completed projects for governmental and private entities, such as the National Science Foundation, Department of Defense, Hartford, Chevron, Bechtel, etc. Dr. Ayyub is the recipient of several awards and research prizes from ASCE, ASNE, ASME, ENR, the Department of the Army, etc. He has coauthored more than 650 publications including 8 textbooks and more than 15 edited books. He is also the founding Editor-in-Chief of the ASCE-ASME Journal of Risk and Uncertainty in Engineering Systems. His most recent edited book on Climate-Resilient Infrastructure published by ASCE was included in the 2017 Engineering-News Record Newsmakers.

JERRY P. BRASHEAR, MBA, PHD

September 11, 2001, induced Mr. Brashear to leave a 20+ year career in consulting and academic (University of Texas at Austin) research and development in energy policy, technology, and exploration and development risk management and to dedicate himself to risk management for critical infrastructures, especially the lifelines—water, energy, transportation, and communications. At George Mason University, he directed the 2003–05 multiuniversity assessment of the National Capital Region's infrastructures. He then served as senior fellow of the ASME Innovative Technology Institute, codirecting 6 year's development and testing of Risk Analysis and Management for Critical Asset Protection (RAMCAP), a "bottom-up" risk management process designed to maximize security and resilience, including the version tailored to the water sector. Based on that, he wrote the first draft of the ANSI/AWWA J100-10 Standard Risk and Resilience Management of Water and Wastewater Systems and has continued to serve on the Standards Committee through its 2019 update. From 2010 to the present, Mr. Brashear has extended this risk management process to other infrastructures, interdependencies among infrastructures, and regional risk management for social benefits. He holds degrees from Princeton (AB Magna cum laude), Harvard Business School (MBA), and Michigan (PhD in Urban and Regional Planning).

JOYCE COFFEE, MCP, LEED AP

Joyce Coffee is founder and president of Climate Resilience Consulting, a Certified B Corp that works with clients to create practical strategies that enhance markets and communities through adaptation to climate change. She is an accomplished organizational strategist and visionary leader with more than 25 years of domestic and international experience in the corporate, government, and nonprofit sectors implementing resilience and sustainability strategies, management systems, performance measurement, partnerships, benchmarking, and reporting.

Joyce is a senior sustainability fellow at the Global Institute of Sustainability and advises various high-level resilience groups, including the Climate KIC's City Finance Lab, the Climate Bond Initiative's Adaptation and Resilience Expert Group, EU Technical Expert Group (TEG) on Sustainable Finance, the Global Adaptation and Resilience Investment work group, the Anthropocene Alliance, Global Adaptation and Resilience Investment work group, the MIT Climate CoLab, Partnership for Resilience and Preparedness, the Climate Service, US Green Building Council's Illinois chapter, and the UNISDR building disaster scorecard.

She received a BS in biology, environmental studies, and Asian studies from Tufts University and a master's degree in city planning from the Massachusetts Institute of Technology. She is the author of the Climate Adaptation Exchange Blog.

RYAN M. COLKER, JD, CAE

Ryan M. Colker is Vice President, Innovation at the International Code Council. He also serves as Executive Director of the Alliance for National and Community Resilience (ANCR), a national coalition working to provide communities with the tools necessary to holistically assess and improve their resilience. Prior to joining ICC, Colker served as Vice President at the National Institute of Building Sciences where he led the Institute's efforts to improve the built environment through the collaboration of industry stakeholders from both the public and private sectors. At the Institute he directed the Consultative Council which develops findings and recommendations on behalf of the entire building community and served as staff director of the Council on Finance, Insurance, and Real Estate; the National Council on Building Codes and Standards; the Off-Site Construction Council and the Institute's STEM Education Program. He is a recognized expert on emerging issues within the built environment including resilience, building performance, and off-site construction and speaks and writes frequently on these subjects. Previously, he served as Manager of Government Affairs for ASHRAE and Program Director of the Renewable Natural Resources Foundation. He graduated from The George Washington University Law School, and holds a BA with honors in environmental policy from the University of Florida.

JEFF DAGLE, MSEE, PE

Jeff Dagle has worked at the Pacific Northwest National Laboratory in Richland, Washington, operated by Battelle for the US Department of Energy (DOE), since 1989 with a focus on electric power system reliability and security. Recent project highlights include leading the North American SynchroPhasor Initiative (NASPI) and serving on the DOE Grid Modernization Laboratory Consortium leadership team. Past career accomplishments include leading the data requests and management task for the 2003 blackout investigation, supporting DOE with on-site assessments following Hurricane Katrina in 2005, leading the cyber security reviews for the DOE Smart Grid Investment Grants and Smart Grid Demonstration Protections associated with the American Recovery and Reinvestment Act of 2009, and serving as a member of the National Infrastructure Advisory Council (NIAC) study group in 2010 to establish critical infrastructure resilience goals. More recently, Mr. Dagle has been invited to serve on three National Academy study committees: analytical research foundations for the next-generation electric grid; enhancing the resilience of the Nation's electric power transmission and distribution system; and modernizing the US electricity system. He received BS and MS degrees in Electrical Engineering from Washington State University in 1989 and 1994 and is a licensed professional engineer in Washington State.

CINDY L. DAVIS, CBO

Cindy L. Davis is the deputy director of the Division of Building and Fire Regulations at the VA Department of Housing and Community Development (DHCD). The Division of Building and Fire Regulations is responsible for the promulgation of the Virginia Uniform Statewide Building Code and Statewide Fire Prevention Code, as well as the VA Building Code Academy, which provides training and certification and tracks continuing education of all code officials and technical assistants. The Division also administers the Amusement Device Technical Advisory Committee, Industrialized Buildings Program, the Manufactured Housing Program, and the State Technical Review Board.

Cindy currently serves as Secretary/Treasurer for the Board of Directors of the International Code Council. In Virginia, Davis serves on the Board of Directors for Viridiant (formerly EarthCraft Virginia).

JASON HARTKE, PHD

Dr. Jason Hartke serves as the president of the Alliance to Save Energy, the nation's premier nonprofit working to advance energy efficiency and energy productivity. As president, he leads a bipartisan alliance of business, government, environmental, and consumer leaders dedicated to policies and initiatives that accelerate energy efficiency across all sectors.

Previously, Hartke led the Department of Energy's efforts to advance energy efficiency in commercial buildings, which account for nearly 20% of the nation's energy use. He also spent nearly a decade as a senior executive at the US Green Building Council, where he led advocacy and policy across all levels of government. His efforts resulted in a fourfold increase in green building policies, the passage of historic federal investment, and new federal leadership programs. Jason served in the Clinton Administration, working in the West Wing of the White House in the Office of Intergovernmental Affairs. In 2018, The Hill recognized him as "top lobbyist" for his advocacy leadership. He serves on the board of the Global Center for Climate Resilience and the Keystone Policy Center.

Jason received his PhD in public policy from George Mason University. He holds his master's degree in journalism and mass communication from the University of North Carolina at Chapel Hill.

ALICE C. HILL, JD

Alice Hill is a research fellow at the Hoover Institution at Stanford University. She previously served at the White House as special assistant to President Barack Obama and senior director for Resilience Policy on the National Security Council. Hill led the creation of national policy to mitigate catastrophic risk, including the impacts of climate change. Before joining the White House, Hill served as senior counselor to the Secretary of the Department of Homeland Security (DHS), as an ex-officio member of the Third National Climate Assessment, and as chief of the white-collar crime unit in the Los Angeles US Attorney's Office.

MICHAEL E. HOOKER, MBA

Mike Hooker has been the executive director of the Onondaga County Water Authority (OCWA) for 26 years and has been in the water industry for 42 years working in both the public and private sectors. He has an MBA from Fairleigh Dickinson University and a BS in Management from Montclair State College.

JOHN S. JACOB, PHD

Dr. John Jacob is the director of the Texas Community Watershed Partners program and professor and extension specialist with the Texas AgriLife Extension through the Department of Recreation, Parks, and Tourism Science. Jacob holds BS and MS degrees from Texas Tech University and a PhD from Texas A&M University, all in soils and natural resources. He is registered as a professional geoscientist with the State of Texas and is a professional wetland scientist. Jacob has mapped floodplains, soils, and wetlands. His research and writing focus on resilience and land health, connecting walkability, and watersheds.

Texas Community Watershed Partners provides education and outreach to local governments and citizens about the impact of land use on watershed health and water quality. The TCWP works with local communities to increase both social and physical resilience. The TCWP currently has 12 to 15 staff members with programs in sustainable urban planning, watershed management, habitat restoration, sustainable landscapes, and water quality issues.

WILLIAM KAKENMASTER, BA

William Kakenmaster serves as a research assistant at the Hoover Institute and is an MSc candidate in comparative politics at the London School of Economics and Political Science. His research concerns global environmental politics, especially climate change and security, climate change and democracy, and climate change adaptation and resilience. Kakenmaster received a BA in International Studies, summa cum laude, from American University's School of International Service in 2017.

YOON HUI KIM, PHD, MPHIL

Dr. Yoon Kim, Director of Advisory Services, Four Twenty Seven, is an adaptation expert with more than a decade of experience working with public and private sector entities in the United States and globally to assess climate risks and identify climate resilience opportunities. At Four Twenty Seven, she leads the Advisory Services and works closely with corporations, investors, and governments to assess climate risks, support the integration of adaptation into strategic planning processes, strengthen climate policy and governance, and build capacity. Yoon holds a DPhil in Development Studies from the University of Oxford.

SAMANTHA A. MEDLOCK, CFM

Samantha Medlock currently serves as President of Climate Risk Advisors and as an Adjunct Professor of Law at the Colleges of Santa Barbara and Ventura. Prior to these positions she served as the head of the North America Capital Science and Policy Practice with Willis Towers Watson, addressing large-scale risk and resilience requirements for clients across corporate, institutional, and public sectors. She has more than 20 years of experience in land use and disaster law and has testified in Congress and provided expert opinion in complex disaster litigation. Sam has served numerous roles in government as White House Senior Advisor, land use planner, and floodplain manager. Sam also served as ASFPM's policy counselor in Washington, DC, for 5 years before being detailed to the Obama White House to lead climate resilience initiatives in the Council on Environmental Quality and the Office of Management and Budget. She is a certified floodplain manager, a Juris Doctor graduate with honors of Vermont Law School, and earned a Bachelor of Science summa cum laude from Texas Woman's University. She is an adjunct professor of Law at the Santa Barbara and Ventura Colleges of Law, serves on the Advisory Committee for the Natural Hazards Center at the University of Colorado-Boulder, and is a contributor and lecturer at the University of Cambridge Institute for Sustainability Leadership.

GEOFFREY G. MILLER, PE, BCEE

Geoff Miller is the chief operating officer for the Onondaga County Water Authority (OCWA) and has been with OCWA for 13 years and has more than 30 years of water industry experience as a consultant and in the public sector. Geoff is a Board-Certified Environmental Engineer and is a licensed New York State Professional Engineer. He is a graduate of Clarkson University with a BS in Civil and Environmental Engineering and a BS in Industrial Distribution.

DEVESH NIRMUL, CEM, CSDP, LEED AP O+M

Devesh Nirmul is a leader within the energy, sustainability, and resiliency space with demonstrated experiences in government and public policy, residential and commercial real estate/property management, and sustainability-focused nonprofit organizations. He works on expanding the PACE Financing Market in Florida, helping to accelerate the uptake of sustainable and resilient building improvements for both new construction and retrofits.

He has served in a variety of sustainability leadership positions and authored chapters for BOMI's High-Performance Sustainable Building Investments Coursebook and the Financial Resources and Technology Transfer chapter of the US third National Communication to the UNFCC. He managed the launch of the Chicago Commercial Building Energy Initiative with the Environmental Defense Fund (EDF), served as the Energy and Sustainability Director at FirstService Residential (FSR), was the first Sustainability Manager for Miami-Dade County implementing $12.5M of Energy Block Grant funds and spearheading the County's Sustainability Plan and Green Buildings program, was the University of Florida's first Urban Sustainability Agent, and consulted on climate change mitigation and adaptation solutions for the US Agency of International Development (USAID).

Devesh sustains his effectiveness by maintaining globally recognized industry credentials including

LEED AP O+M, Certified Energy Manager (CEM), and Certified Sustainable Development Professional (CSDP) credentials.

ROBERT G. OTTENHOFF, MCRP, BA

Robert G. Ottenhoff, a veteran in philanthropy, nonprofit leadership, and entrepreneurship, is the inaugural president and CEO of the Center for Disaster Philanthropy. CDP seeks to transform how donors think about, respond, and give to natural disasters, moving it from reactive response, to one focused on increasing strategy and impact. CDP offers information, analysis, and reports about disasters on its website http://disasterphilanthropy.org/ and provides tools, expert analysis, and strategic guidance. *The Chronicle of Philanthropy* recognized the launch of CDP as "one of the five high points of 2012" http://bit.ly/12IhYbu.

Prior to joining CDP, Bob spent a decade as president and CEO of GuideStar, an industry leader in the use of providing high-quality data to help donors make better decisions and improve nonprofit practice. He led efforts to develop www.guidestar.org into a nationally respected, comprehensive source of reports and services on more than 1.5 million nonprofits and built partnerships with many leading corporations and foundations. While there, Bob developed a sustainable "freemium" business model, which supports free and fee-based services to more than 10 million users annually and generates most of GuideStar's operating revenues.

Before GuideStar, he had more than 25 years of management experience in public broadcasting, including nearly 10 years as chief operating officer, and acting president of the Public Broadcasting Service (PBS); serving as executive director of the New Jersey Public Broadcasting Authority; and founding WBGO-FM, in the New York-New Jersey metropolitan area.

M. JOHN PLODINEC, PHD

As part of the Community and Regional Resilience Institute (CARRI), Dr. John Plodinec is responsible for identifying and evaluating technologies that can enhance a community's resilience. His most important contributions have been development of action-oriented tools that operationalize the "Whole Community" concept, including CARRI's Community Resilience System and its Campus Resilience Enhancement System. He has also helped several communities and universities develop plans to recover from economic and natural disasters.

Dr. Plodinec also coordinated development of an action plan for management of woody biomass and debris generated by disasters in support of the federal government's Woody Biomass Working Group. This effort involved a team from seven federal agencies as well as coordination with other major stakeholders in the American forest enterprise.

Dr. Plodinec also developed CARRI's Resilient Home Program, aimed at improving the survivability of American homes to natural disasters. This built on earlier work he did while at Mississippi State University, where he led the University's efforts to develop programs related to severe weather events. As part of a joint program with the International Code Council and other partners, he has led initial development of a Community Resilience Benchmark System. He is currently assisting Northeastern University's Global Resilience Institute in development of its Resilience Enhancement System.

ALLISON C. REILLY, PHD

Dr. Allison Reilly is an assistant professor of Civil and Environmental Engineering at the University of Maryland, College Park. Her area of expertise includes risk and resilience of infrastructure systems under climate change, infrastructure maintenance, and electric power system reliability following hurricanes. She is particularly interested in characterizing the dynamics between infrastructure resilience, policy, and individual behavior.

Prior to her appointment at the University of Maryland, College Park, Dr. Reilly was a research fellow in the Department of Industrial and Operations Engineering at the University of Michigan and a postdoctoral research associate in the Department of Geography and Environmental Engineering at Johns Hopkins University. In addition, Dr. Reilly was a research analyst for the Homeland Security Studies and Analysis Institute, a federally funded research and development center in support of the Department of Homeland Security, in Arlington, VA. While there, her primary focus was on national-level infrastructure protection. Dr. Reilly holds an MS and PhD in Civil Engineering from Cornell University and a BS in Civil Engineering from Johns Hopkins University. She is a member of the Society for Risk Analysis, INFORMS Decision Analysis Society, and the American Society of Civil Engineers.

JAMES (TIM) T. RYAN, CBO

Tim Ryan served the City of Overland Park for 40 years; all in the Building Safety Division. He served that community as a field inspector, plans examiner, field supervisor, and code administrator from 1998 until 2017. He currently serves as the executive director of the International Association of Building Officials and as a national and international code consultant. Tim graduated from Pittsburg State University with a Bachelor of Science Degree in Construction Management and Technology. He is certified in 15 separate categories of building code administration and management. He is well recognized as a public speaker and is highly regarded as an instructor on several topics of building and fire codes, legal aspects, emergency response, leadership, and management. He served on the Board of Directors for the Building Officials and Code Administrators (BOCA), Intl. from 1993 through 2002, including as President of the board. He served 7 years on the Board of Directors for the International Code Council (ICC). He was appointed to the Board of Directors for the National Institute of Building Sciences in 2006 and is currently serving as one of six board members to be appointed to that Board by the President of the United States; Tim was appointed by President Obama in 2012. Tim was awarded ICC's Bob Fowler award for his outstanding leadership in furthering a safer built environment in 2012. Tim was awarded the Mortimer M. Marshal Award from the National Institute of Building Sciences in January 2018 for outstanding lifetime achievement.

JAMES SCHWAB, FAICP, BA, MA

James C. Schwab, FAICP, is the principal of Jim Schwab Consulting LLC and adjunct assistant professor in the School of Urban and Regional Planning at the University of Iowa. Until 2017, he was the manager of the Hazards Planning Center at the American Planning Association. He earned MA's in both Urban and Regional Planning and Journalism from the University of Iowa.

JOHN SCOTT, BOMA FELLOW, RPA

John K. Scott was appointed Chair of Colliers Broker Sustainability Practice Group in 2012. John has direct responsibility and oversite of the Florida Real Estate Management Services (REMS) department, where he directs a portfolio of over 285 retail, office, and industrial properties totaling more than 32 million square feet. He leads a team of more than 110 professional and support personnel specializing in real-estate management services. The Sustainability Practice Group focuses on ensuring the Colliers network has knowledge and understanding of sustainability tools and resources available to them. Under his direction and leadership, this group has gained more than 50 Colliers team members across the globe. John also is the National Lead for the Colliers | TH Real Estate Governance team.

John previously served as an Executive Committee Board Member with BOMA International and has held numerous positions at the local, national, and international level. His passion for sustainability and energy intertwined him with DOE in 2007 and led to positions as Chair of the Commercial Real Estate Energy Alliance (2010–13), Co-Chair 2015 Building Energy Summit, and Chair of the Berkeley FLEXLAB Executive Advisory Board (2013–Present). John is currently authoring multiple chapters of a BOMI educational textbook on high-performance building valuation.

STACY SWANN, BA, MBA, MTS

Stacy Swann is the CEO and founding partner of Climate Finance Advisors. Ms. Swann has more than two decades of experience in finance, fund management, and development, including with the International Finance Corporation (IFC) where she was head of IFC's Blended Finance Unit and was responsible for managing and investing more than $750 million in donor funding for climate-smart investments. While at IFC, she also supported World Bank Group efforts to work with policymakers and other financial institutions on issues related to blended finance, climate finance, climate-smart financial policies, and incorporating practical policy and investment approaches to managing climate risk. During 2014, she had a secondment with the US Department of Treasury as their Sr. Advisor on Climate Finance, supporting efforts related to the Green Climate Fund and the Paris Agreement. Prior to joining the World Bank Group, she worked in the private sector as a developer of infrastructure projects in India and Singapore.

In addition to running Climate Finance Advisors, Ms. Swann is currently Vice-Chairperson of the Board for the Montgomery County Green Bank, the United States' first county-level green bank, and she sits on the Board of the Women's Council on Energy and Environment (WCEE). She is an adjunct professor at American University's Kogod School of Business, and she has coached finalists of GoodCompany's Climate Ventures 2.0 and CPI's Climate Finance Lab and Finance for Resilience (FiRe) Awards, platforms to crowdsource and champion new ideas to accelerate finance for climate-smart, resilient investments.

TIMOTHY P. TABER, PE, BCEE

Timothy Taber is Barton and Logiudice's Asset Management Discipline Leader and has more than 24 years of water industry experience as a consultant assisting organizations with asset management and technology implementations. Tim is a Board-Certified Environmental Engineer and is a licensed New York State Professional Engineer. He is a graduate of University of Buffalo with a BS in Civil and Environmental Engineering and a graduate of Syracuse University with an MS in Engineering Management.

ZIYUE WANG, MEM, BA

Ziyue Wang served as an intern at multiple organizations engaged in environmental and finance policy including Climate Finance Advisors, ICF International, Everbright Trust, and AXA SL Financial Group. He holds a bachelor of science degree from the University of Science and Technology in Beijing and a master's in Environmental Management from Duke University.

CHARRISS R.H. YORK, MS

Charriss York is an extension program specialist at the Texas Community Watershed Partners, a program of the Texas A&M AgriLife Extension Service. She works on stormwater and watershed planning projects, and her efforts focus on increasing awareness about stormwater, nonpoint source pollution, and green infrastructure. York has been involved with projects to implement on the ground examples of stormwater best management practices in the Houston-Galveston Region for more than 10 years. Charriss holds a BS in Biology from Truman State University and an MS in Botany from Oklahoma State University.

Acknowledgements

Pulling together a book of this magnitude and breadth is no easy task. Multiple people both upfront and behind the scenes have contributed to the final result. Michelle Fisher, Poulouse Joseph and the team at Elsevier have been a pleasure to work with and provided valuable support. The chapter authors have given their time and expertise and embraced my vision of the book as a study across and between resilience disciplines and infrastructure systems. Many have provided a "peek behind the curtain" of how their colleagues are addressing (or not addressing) resilience challenges.

My wife Allison has been incredibly supportive as I spent countless hours researching, writing, and editing. My colleagues at the National Institute of Building Sciences (NIBS) and the International Code Council and Alliance for National & Community Resilience (ANCR) have been a constant source of motivation and encouragement. Their recognition of the importance of this topic has been invaluable. Henry Green, who recently retired as NIBS President, was a fantastic mentor and allowed me to undertake this project. It was through my work at NIBS that I met many of the chapter authors and developed a strong network of resilience champions. My resilience work continues at ICC and ANCR where ICC CEO Dominic Sims and Senior Vice President Sara Yerkes have wholly embraced the need for community-level resilience. The ANCR Board of Directors (Warren Edwards, Evan Reis, Tom Phillips, Gina Bocra, Harrison Newton, Amy Schmidt, Bryan Soukup, and Mike Lesnick) along with supporters from the Meridian Institute (Brad Spangler and Isabella Soparkar) and John Plodinik are on the front lines of implementing many of the strategies identified in this book, and I am proud to be working alongside these leaders to help communities realize enhanced levels of resilience.

Formulating this resilience vision and the path forward does not just materialize overnight; there were several people and experiences that brought me to today. While it is impossible to recognize them all, there are several that stand out. My seventh-grade science teacher at A.D. Henderson University School, Don Stone, inspired a love of science and the environment, launching me on the trajectory I am on today. The professors and administrators at the University of Florida (Go Gators!) who had the vision and foresight to recognize that solutions to complex environmental problems required a new cadre of polymaths with the comfort and experience to look across disciplines for solutions. Their leadership in developing the interdisciplinary College of Natural Resources and Environment provided me and others with a strong foundation for facilitating change.

The Renewable Natural Resources Foundation (RNRF) was an ideal place to start my career following law school. Its focus on identifying interdisciplinary solutions to natural resource problems further illustrated the need for holistic, coordinated, and collaborative approaches to tough challenges. It also exposed me to the world of professional and scientific societies and the valuable work they do in research, standards setting, sharing of best practices, and supporting advancement within their disciplines. I continue to engage many of the same organizations and leaders to this day. RNRF's Executive Director Bob Day has been leading this work for decades.

Doug Read, a past director of government affairs at ASHRAE, gave me the encouragement to take on the important initiatives that would support the achievement of high-performance buildings and communities. He helped me find the confidence and experience that brought me to today.

I am eternally grateful to my mentors and look forward to seeing how the next generation of resilience leaders develops and implements the solutions we desperately need.

Introduction to Infrastructure Resilience

Since the beginning of life on earth, species have struggled to survive in the face of events that challenge their existence. Dinosaurs proved vulnerable to atmospheric changes due to asteroid impacts, whereas cockroaches proved largely immune to the environment around them. As a species, humans have generally fared well in the face of hazards, but we are frequently reminded of the fragility of human life and how the decisions we make impact our ability to survive.

In 2017 and 2018, these reminders were significant. In the United States 2017 tied 2005 for the most disasters resulting in $1 billion or more of damage and far exceeded the total annual damages attributed to these disasters. These 16 disasters caused over $300 billion in damages—roughly equaling the cost of all new buildings constructed in the United States over the same year [1,2]. According to the National Oceanographic and Atmospheric Administration, 2018 saw 14 weather and climate disaster events in the United States with losses exceeding $1 billion each. These events included one drought event, eight severe storm events, two tropical cyclone events, one wildfire event, and two winter storm events (Fig. 1).

These numbers do not include disasters that do not meet the $1 billion threshold, but still have significant impacts on local communities. Additionally (and not insignificantly), these disasters result in injuries and deaths. The Camp Fire in California burned most of November 2018 leaving 85 people dead, nearly 14,000 residences destroyed and over 153,000 acres burned [3]. Overall, the 2018 events resulted in the deaths of 247 people and had significant economic effects on the areas impacted [1]. Hurricane Maria struck Puerto Rico on September 20, 2017 causing extensive damage and around 3000 deaths [4].

As the number and cost of these events increase, the ability to respond to these events becomes a challenge. The capacity of emergency management professionals at the local, state, and federal level can be overwhelmed. The availability of professionals to implement recovery plans and reconstruct communities is spread thin. The ability for governments to fund the recovery efforts is becoming unsustainable.

Communities, governments, citizens, and business owners need cost-effective solutions to lessen or even avoid the impacts these types of events bring. Making smart investment decisions in advance of a hazard—or mitigation—has been proven effective. Building off a 2005 study that found mitigation grants offered by the U.S. Federal Emergency Management Agency (FEMA) saved $4 for every $1 invested, the National Institute of Building Sciences (NIBS) has been examining multiple mitigation strategies across the public and private sector [2]. The results of the NIBS study feature prominently throughout this book as it represents a robust examination of multiple mitigation strategies and captures quantitatively the long-understood notion that "an ounce of prevention is worth a pound of cure (Fig. 2)."

Building codes have long been a solution for establishing a minimum level of protection as discussed in Chapter 12. NIBS found that the adoption of model building codes from 1990 through 2018 have produced a national benefit of $11 for every $1 invested. They have also resulted in the creation of 30,000 jobs over that time frame [2]. These benefits are not just financial; they represent avoided casualties, property damage, business interruptions, and insurance costs. Modern codes ensured that the state of Alaska sustained minimal damage, speeding recovery following the Port MacKenzie earthquake. In Florida, following Hurricane Michael, buildings built to modern code requirements fared far better than buildings built to older standards. Still, there are many places where building codes are not adopted or are woefully out of date.

Although community leaders know they must be prepared to face potential risks, they often struggle with hard decisions. Communities are complex systems. Nodes of the system are challenged daily. Chronic challenges such as homelessness and crime require daily attention. In the face of a hazard event, the functionality of the community is further stressed. The community is

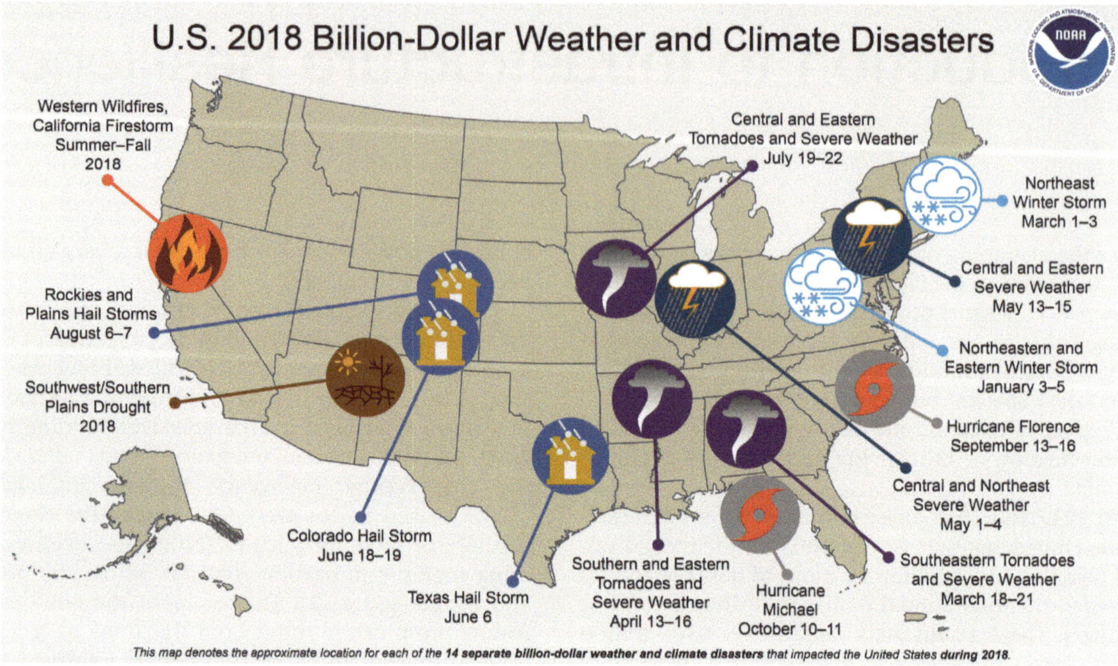

FIG. 1 U.S. 2018 billion-dollar weather and climate disasters. (Reproduced from: National Oceanic and Atmospheric Administration. [1])

National Benefit-Cost Ratio Per Peril (*BCR numbers in this study have been rounded)	Exceed common code requirements	Meet common code requirements	Utilities and transportation	Federally funded
Overall Hazard Benefit-Cost Ratio	4:1	11:1	4:1	6:1
Riverine Flood	5:1	6:1	8:1	7:1
Hurricane Surge	7:1	Not applicable	Not applicable	Too few grants
Wind	5:1	10:1	7:1	5:1
Earthquake	4:1	12:1	3:1	3:1
Wildland-Urban Interface Fire	4:1	Not applicable	Not applicable	3:1

FIG. 2 Benefit cost ratios for various mitigation measures. (Courtesy: National Institute of Building Sciences. [2])

only as strong as its weakest link. If the energy or water distribution system fails businesses cannot open. Closed businesses mean decreased tax revenue and citizens not earning incomes. Decreased tax income means less services, impacting the ability for a community to attract new businesses. This interconnectedness presents a challenge for policymakers and others charged with assuring the safety and sustainability of a community and its citizens.

Business as usual will not protect the social, economic, and environmental integrity of communities and the systems that make them work. A new approach is needed—one commonly characterized as resilience. Achieving resilience requires planning and action at multiple scales across multiple segments of the economy.

Scholars, policymakers, and practitioners have spent considerable effort over the past few years defining resilience. Although this is an important exercise for garnering interest in the topic, it does not necessarily lead to the actual achievement of resilience. The discussion later provides a bit of context on what resilience means. However, the achievement of resilience requires understanding the current landscape and the policies, strategies, and activities that facilitate achievement of resilience. The bulk of this book focuses on identifying those strategies.

DEFINING RESILIENCE

In its 2012 publication *Disaster Resilience: A National Imperative*, The National Academies defined resilience as "The ability to prepare and plan for, absorb, recover from, or more successfully adapt to actual or potential adverse events [5]." As the Academies point out, this definition is consistent with the definition used in the international disaster policy community through the United Nations International Strategy for Disaster Reduction [6] and U.S. governmental agencies including through Presidential Policy Directive 8 on national preparedness [7].

In 2014, representatives from organizations representing the design, construction, management and regulation of buildings and infrastructure came together to sign a statement recognizing the need for resilience and the importance of a cooperative approach to achieving it [8]. In the Statement, signatories recognize the National Academies' definition of resilience. As of late 2018, nearly 50 organizations have signed on to the statement [9].

This book embraces the definition established by the National Academies but recognizes that multiple academic papers have been devoted to parsing the multiple definitions of resilience [10–12]. That discussion will not be recreated here, but some important highlights are worth noting.

Many academics point to Holling's 1973 work "Resilience and Stability of Ecological Systems" as the origin of modern resilience theory [10,13]. Holling described resilience as the ability of an ecological system to continue functioning when changed, but not necessarily remaining the same. He further differentiates from "engineering resilience" that is focused on maintaining a state of equilibrium that it would revert to after a disruption.

In the case of communities and infrastructure systems, the end state following a disruption is unlikely to be the same as the original state. This is often for good reason—the original state contributed to the vulnerability in the first place, a new, potentially improved, state would be far more desirable.

It is not hard to see the parallels between an ecological system and the functions of communities. Community departments and infrastructure have specific roles and responsibilities that are interdependent and work collectively to deliver a functioning enterprise. If an element or elements of the system are disrupted an intervention is necessary (either internal or external) to get things moving to return to normal or enter a new state. The term ecosystem is often used to describe the concept of a set of interacting parts in the service of a higher goal.

In Chapter 14, Plodinec provides a visual representation of a community's level of resilience. In its steady-state condition, the community operates on a trajectory. In some cases, the community is in decline or under stress, in others it is improving. A shock (whether a disaster event or a social disruption) can significantly alter a community, triggering a need to recover. The significance of the disruption and the time needed to return to "normal" (or even a better state) represents the community's resilience (Fig. 3).

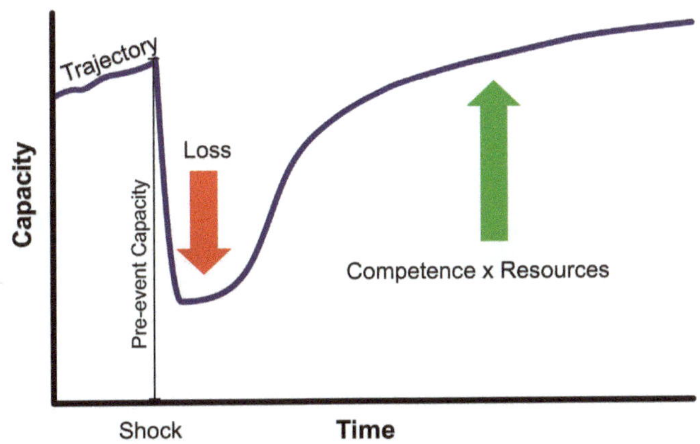

FIG. 3 Graphical representation of a community's resilience. (Adapted from UK Department for International Development: Alliance for National and Community Resilience.)

Shocks	Stresses
Infrastructure failure	Affordability
Hurricanes	Aging population
Earthquakes	Environmental degradation
Wildfires	Sea level rise
Heat waves	Growing wealth gap
Blizzard	Drought
Epidemics	Species extinction
Flooding	Aging infrastructure
Tornadoes	Population growth
Acts of terrorism	Unemployment
Civil unrest	Melting polar ice
Dam failure	Global warming
Subsidence	Food scarcity
	Increasing pollution

FIG. 4 Examples of shocks and stresses. (Reproduced from: American Institute of Architects, Designing for Resilience, 2018. http://content.aia.org/sites/default/files/2018-09/Designing_for_Resilience_V04.pdf. [14])

Although hazard events such as hurricanes, earthquakes, floods, and wildfires capture the nation's attention, they are not the only adverse events communities face. Resilience requires addressing both the chronic and the acute incidents. Put another way, communities face both shocks and stresses. Fig. 4 provides examples of the types of shocks and stresses a community may face. When a community already under stress faces a shock, the impacts can be amplified. Fig. 5 illustrates this dynamic. Truly resilient communities can adequately address both types of events.

The fate of American manufacturing hubs provides a very visible case study on the importance of focus on community stresses. Following the decline of the steel industry, cities such as Pittsburgh needed to rapidly reinvent itself. Today, it is a hub of innovation in sectors such as energy, healthcare, and technology.

Case Study: Pittsburgh Reborn

Pittsburgh's story parallels the struggle and opportunity for rebirth based on changes in both the national and local economy. In the late 19th and early 20th century, the city saw its prominence rise with the Industrial Age and the importance of steel. At the peak, Pittsburgh was producing 60% of the nation's steel or 25 million tons of steel annually. The city's infrastructure grew alongside the boom including educational and cultural resources.

Around 1950, the city's population peaked at 677,000 residents. Accompanying this industry-based prosperity were stresses on residents and the natural environment including severe air and water pollution, acid mine drainage, and polluted soils.

Following World War II, as other countries increased steel production, domestic production fell, leaving Pittsburgh without the industry that drove its prior prosperity. The population fell significantly up until recently when it leveled out at about half its prior peak. However, much of the architecture, infrastructure, and cultural offerings of the past remain. These resources accompanied by a low cost of living and a shift to new economic strategies have set Pittsburgh on a trajectory to a new level of resilience.

As a member of the 100 Resilient Cities network, Pittsburgh has undertaken steps to understand its current position, the needs of its residents, the potential shocks and stresses it could face and develop a path forward. Its resilience initiatives cross four p's: people, place, planet, and performance with multiple initiatives underway within each [16].

The definition of resilience can be applied to multiple systems on multiple scales. This allows recognition that each community function has its own resilience (often defined and achieved by disciplinary actors within those functions) and that the system as a whole also has its own level of resilience. Achieving resilience at a community level requires examination and coordination across all community functions.

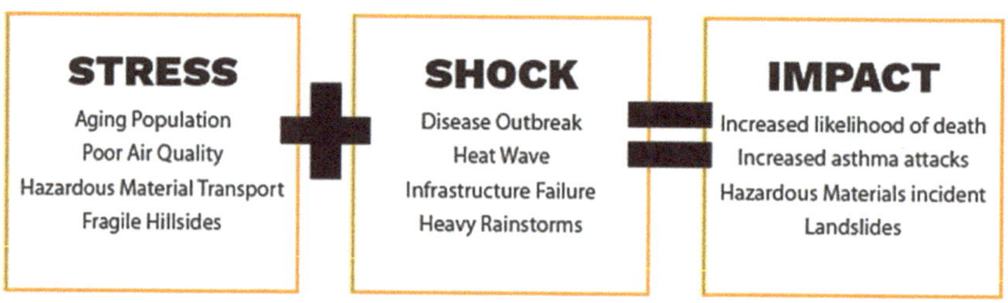

FIG. 5 Relationship between stresses, shocks, and their impacts. (Courtesy: City of Pittsburgh, ONEPGH: Pittsburgh's Resilience Strategy, 2017. http://pittsburghpa.gov/onepgh/documents/pgh_resilience_strategy.pdf. [15])

Some scholars have defined resilience as "malleable" or a "boundary object," allowing adaptability to multiple disciplines and stakeholders [10]. Vale enumerates, "The biggest upside to resilience is the opportunity to turn its flexibility to full advantage by taking seriously the actual interconnections among various domains that have embraced the same terminology [17]."

Upon consideration of 25 different definitions of "urban resilience" and considerations of the tensions among them, Meerow et al., developed their own definition recognizing these tensions. They define urban resilience as, "the ability of an urban system—and all its constituent socio-ecological and socio-technical networks across temporal and spatial scales—to maintain or rapidly return to desired functions in the face of disturbance, to adapt to change, and to quickly transform systems that limit current or future adaptive capacity [10]." Meerow et al. identifies that the concept of "urban" varies but is typically characterized as a system or network [10]. For this work, we consider the concept of "urban" broadly to include cities and communities—basically any collection of interconnected functions.

Mitigation is another term closely related to resilience. From a hazards perspective, mitigation is "the effort to reduce loss of life and property by lessening the impact of disasters [18]." This mitigation can take many forms, but is centered around undertaking measures that improve resilience. The cost effectiveness of many of these measures have been studied by NIBS and others [2]. Within the climate science community, mitigation centers on "efforts to reduce or prevent emission of greenhouse gases [19]." Adaptation in the climate change context is more closely aligned with the concept of preparations to lessen impacts. As the climate science community and the hazards community come together, an agreement on terms and the associated educational needs must be addressed.

Addressing the resilience of infrastructure in the face of evolving climate-related risks presents a particular challenge. Reilly and Ayyub present an engineering-based path forward in Chapter 13, but policymakers and resilience practitioners at large need some organizing principles and strategies to lay the groundwork. Thought leaders came together in late 2018 to identify a potential path forward. Through that effort, they identified four principles and seven strategies for climate resilient infrastructure that were ultimately endorsed by organizations including the Alliance for National & Community Resilience, American Institute of Architects, American Society of Civil Engineers, and National Council for Science and the Environment [20].

The four principles to guide development of more resilient infrastructure are as follows:

1. Be Proactive. Do what we can now with both existing knowledge and foresight. Uncertainty should not preclude action.
2. Be Fair. Consider the implications of decisions for those who are particularly vulnerable. We need to directly and consistently engage affected communities in decision-making.
3. Be Inclusive. Engage all stakeholders early and often throughout the entire process. They should include knowledge generators, knowledge users, and impacted communities. An inclusive process helps to ensure that decisions are grounded in the best available information and fit the needs and values of those affected. Inclusivity can also reduce future conflict, avoid negative unintended consequences, identify a strong pool of options, and increase support for the measures chosen.
4. Be Comprehensive. Consider the full range of risks and means to address them through planning, financing, and engineering. A holistic approach includes integrating social and ecological resilience into decisions where appropriate. Strong social dynamics and healthy, functioning ecosystems are critical to adaptive capacity—increasing communities' and regions' ability to respond effectively to both chronic stresses and extreme events.

The seven strategies for climate-resilient infrastructure are as follows:

1. Make better decisions in the face of uncertainty
2. View infrastructure systematically
3. Take an iterative, multihazard approach
4. Improve and inform cost–benefit analysis
5. Mainstream nature-based infrastructure
6. Jump-start resilience with immediate actions
7. Plan now to build back better

In the chapters that follow, individual authors may use their own, slightly different, definition for resilience. Although, in some cases, the subtle variations in definition may be important for the specific discipline or system, the general concepts remain the same.

Resilience and the U.S. Federal Government

The federal government plays a significant role in implementing national policy and influencing state and local policies that support national- and community-level resilience. They have also taken steps to protect important national infrastructure. The following are some of the steps federal agencies have taken.

Continued

PRESIDENTIAL POLICY DIRECTIVE 8: NATIONAL PREPAREDNESS

Presidential Policy Directive 8 (PPD-8): National Preparedness was instituted during the Obama administration and is "aimed at strengthening the security and resilience of the United States through systematic preparation for the threats that pose the greatest risk to the security of the Nation, including acts of terrorism, cyber attacks, pandemics, and catastrophic natural disasters [21]." The Department of Homeland Security (DHS) was charged with developing a national preparedness goal.

The national preparedness goal was defined as, "A secure and resilient nation with the capabilities required across the whole community to prevent, protect against, mitigate, respond to, and recover from the threats and hazards that pose the greatest risk [22]."

PPD-8 also includes definitions that support the national preparedness goals and are generally consistent with the goals outlined in this book. Relevant definitions are duplicated here:

- The term "national preparedness" refers to the actions taken to plan, organize, equip, train, and exercise to build and sustain the capabilities necessary to prevent, protect against, mitigate the effects of, respond to, and recover from those threats that pose the greatest risk to the security of the Nation.
- The term "resilience" refers to the ability to adapt to changing conditions and withstand and rapidly recover from disruption due to emergencies.
- The term "mitigation" refers to those capabilities necessary to reduce loss of life and property by lessening the impact of disasters. Mitigation capabilities include, but are not limited to, community-wide risk reduction projects; efforts to improve the resilience of critical infrastructure and key resource lifelines; risk reduction for specific vulnerabilities from natural hazards or acts of terrorism; and initiatives to reduce future risks after a disaster has occurred.

DISASTER RECOVERY REFORM ACT OF 2018

The U.S. Congress in the Disaster Recovery Reform Act of 2018 (DRRA) charged FEMA with issuing a final rulemaking defining resilient and resilience. The DRRA also incorporated important resilience-related reforms to the Robert T. Stafford Disaster Relief and Emergency Assistance Act (the Act governing disaster response and recovery). DRRA established an avenue for ongoing funding of predisaster mitigation grants that contribute to community resilience. It also authorized FEMA to assist states in the adoption and enforcement of building codes. Public assistance funding can now be used to restore and replace damaged structures to the latest editions of codes and standards. It is too early to tell exactly how these updates will be applied and how FEMA will ultimately define resilience. Outside the DRRA requirements, FEMA has actively embraced the importance of resilience.

FEMA REORGANIZATION AND STRATEGIC PLAN

FEMA's 2018–22 Strategic Plan sets out a path to achieving FEMA's vision of "A prepared and resilient Nation (Fig. 6)." The Plan focuses on three strategic goals intended to mobilize the whole community under a shared responsibility [23].

- *Build a Culture of Preparedness*. Resilience is the backbone of emergency management. The Nation's ability to weather storms and disasters without experiencing loss significantly reduces our risk. The most successful way to achieve disaster resilience is through preparedness, including mitigation. Strategic Goal 1 promotes the idea that everyone should be prepared when disaster strikes. To be prepared, however, we must all understand our local and community risks, reflect the diversity of those we serve, and foster partnerships that allow us to connect with a diverse Nation. People who are prepared will be able to act quickly and decisively in the face of disasters, thereby preventing death and injuries, minimizing loss of property, and allowing a more rapid and efficient recovery.
- *Ready the Nation for Catastrophic Disasters*. Catastrophic disasters, including low- and no-notice incidents, can overwhelm the government at all levels and threaten National security. They are life-altering incidents for those impacted, causing a high number of fatalities and widespread destruction. Catastrophic disasters disrupt lives and hurt our communities—economically and socially.
- *Reduce the Complexity of FEMA*. The Nation faces an evolving threat and hazard environment. FEMA must be flexible and adaptable to meet the needs of individuals and communities, and it must deliver assistance and support in as simple a manner as possible. We must innovate and leverage new technology to reduce complexity, increase efficiency, and improve outcomes.

The Strategic Plan is not just directed at the agency, but at the emergency management community and the nation as a whole. It leverages the well-established research from NIBS [2] and others that wise investments in mitigation up front can reduce the impacts of events and limit FEMA's need to deploy response and recovery resources to when it is truly needed.

In parallel with release of the Strategic Plan, FEMA undertook a reorganization to support an increased focus on achieving resilience. Headed by a deputy administrator, FEMA Resilience incorporates the Federal Insurance and Mitigation Administration, Grant Programs Directorate, National Continuity Programs, and National Preparedness Directorate.

NIST COMMUNITY RESILIENCE PROGRAM

The National Institute of Standards and Technology (NIST) has long been engaged in scientific research to assure the safety of buildings. NIST investigated and made findings following significant disasters including the World Trade Center collapse on September 11, 2001 [24]; the Station Nightclub fire [25]; and the May 2011 tornado in Joplin, Missouri [26]. NIST's work expanded to community-level efforts to improve safety and resilience in 2014 following a recognition that guidance for planning and implementing resilience measures as well as science-based tools for measuring resilience and supporting evaluation of various strategies was lacking. The Community Resilience Program has three main components: developing science-based tools and metrics to support and measure resilience at the community scale and support economic evaluation of alternative solutions to improve resilience; engaging community resilience stakeholders for input and feedback to products, such as guidance, tools, and metrics, for planning and implementing resilience measures; and learning from hazard events that impact the built environment to better understand adverse hazard impacts on the built environment and communities and developing improved tools and methods for field studies [27].

The NIST Community Resilience Planning Guide for Buildings and Infrastructure Systems was released in 2016 and provides a practical and flexible six-step approach to help communities improve their resilience by setting priorities and allocating resources to manage risks for their prevailing hazards (Fig. 7) [28]. The Economic Decision Guide for Buildings and Infrastructure Systems provides a standard economic methodology for evaluating alternative resilience options for improving community resilience [29]

DOD MISSION ASSURANCE

The Department of Defense (DoD) has embraced the concept of resilience as essential to continuous achievement of their mission. They define mission assurance in DoDD 3020.40 as, "A process to protect or ensure the continued function and resilience of capabilities and assets—including personnel, equipment, facilities, networks, information and information systems, infrastructure, and supply chains—critical to the execution of DoD mission-essential functions in any operating environment or condition [30]." Energy and water resilience have been areas of particular focus. In 2015, the U.S. Army released the Energy Security and Sustainability (ES^2) Strategy, recognizing the role of energy, water, and land as mission enablers [31]. The Strategy looks to maintain continuity of operations and maximize flexibility in system design through strategies such as microgrids and combined heat and power. Training of personnel, reducing resource demand, increasing resource efficiency, and driving innovation are additional elements of the Strategy.

Within this overall recognition of the importance of resilience planning and implementation, DoD has also specifically focused on climate-related risk. The U.S. Navy issued a Climate Change Roadmap in 2010 recognizing the national security challenge presented by climate change. The Roadmap outlines the Navy's approach to observing, predicting, and adapting to climate change [32]. In 2014, the full DoD issued its own Climate Change Roadmap, recognizing the potential impacts on the military's built and natural infrastructure and its acquisition and supply chains [33]. The DoD found potential effects ranging from an increased demand for disaster relief and humanitarian assistance to increased inundation, erosion and flooding, and changing heating and cooling demand impacting energy intensity and operating costs. The Strategic Environmental Research and Development Program and Environmental Security Technology Certification Program have begun examining the concept of nonstationarity (as described further in Chapter 13) and the needed information and research to support resilient infrastructure and installations [34].

Reinhardt and Toffel have summarized the Navy's approach to managing climate change and offer lessons for the business community [35].

RESILIENCE IS A WICKED PROBLEM

Community resilience is a modern-day version of what Rittel and Webber call a "wicked problem [36]." Achieving resilience requires a mix of social and technical solutions. When, where, and how to implement technical solutions requires understanding the impacts on both the physical and the social geography of a place. A sea wall can keep rising seas at bay for one community, but significantly impact a neighboring community. Vulnerable populations are particularly susceptible. Often, they occupy the most vulnerable land and have limited resources to respond to changing risks.

The adverse events contemplated by the Academies [5] do not just consider acute physical events (shocks) but extends to chronic physical and social events (stresses) as well—drought, loss of a major employer, an economic downturn, and poverty all impact

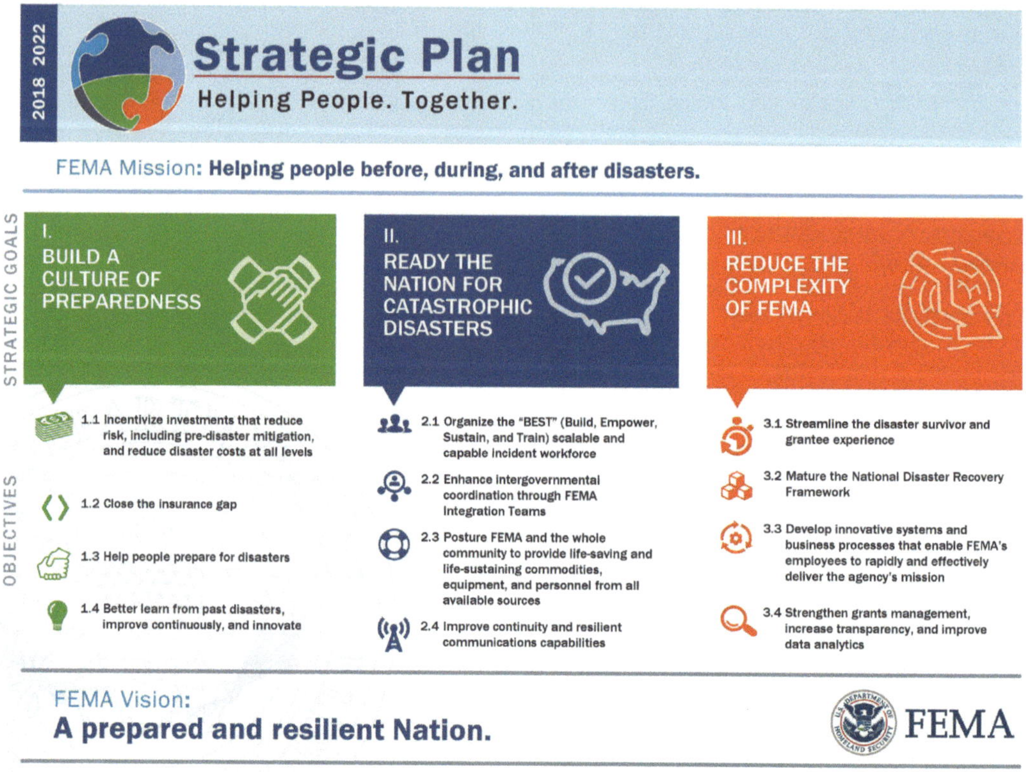

FIG. 6 FEMA 2018 to 2022 Strategic Plan. (Reproduced from: FEMA [23].)

community and individual resilience. Additionally, the risks and potential strategies to address those risks vary widely by community—again, a function of the social and physical geography of the place.

It is this combination of challenges and their potential resolution that make resilience a wicked problem. Kolko updates the concept of wicked problems for the 21st century [37]. He summarizes Rittel and Webber's concept of wicked problems succinctly: "A wicked problem is a social or cultural problem that is difficult or impossible to solve for as many as four reasons: incomplete or contradictory knowledge, the number of people and opinions involved, the large economic burden, and the interconnected nature of these problems with other problems." Although this book outlines the complexities in solving resilience challenges, hopefully it does not give the reader the impression that achieving resilience is impossible. In fact, achieving resilience is a dynamic and ongoing process as new threats emerge, new solutions and technologies develop, and the nature of society (socially, economically, and environmentally) changes.

Ritter and Webber identify 10 distinguishing properties of wicked problems [36]:

1. *There is no definitive formulation of a wicked problem.* The information needed to understand the problem depends upon one's idea for solving it. That is to say to describe a wicked problem in sufficient detail, one has to develop an exhaustive inventory of all conceivable solutions ahead of time … This property sheds some light on the usefulness of the famed "systems approach" for treating wicked problems.
2. *Wicked problems have no stopping rule.* Planning problems do not present criteria to know when "the" or "a" solution is found. The planner terminates work on a wicked problem not for reasons inherent to the logic of the problem but because of considerations external to the problem—running out of time, money, or patience.
3. *Solutions to wicked problems are not true-or-false, but good-or-bad.* Normally, many parties are equally equipped, interested, and/or entitled to judge the solutions, although none has the power to set formal decision rules to determine correctness.

FIG. 7 Six-step process to planning for community resilience. (Reproduced from: National Institute of Standards and Technology, Community Resilience Planning Guide for Buildings and Infrastructure: Volumes I and II, NIST Special Publication 1190, 2016.)

4. *There is no immediate and no ultimate test of a solution to a wicked problem.* Any solution, after being implemented, will generate waves of consequences over an extended—virtually an unbounded—period of time. The full consequences cannot be appraised until the waves of repercussions have completely run out, and we have no way of tracing all the waves through all the affected lives ahead of time or within a limited time span.
5. *Every solution to a wicked problem is a "one-shot operation"; because there is no opportunity to learn by trial and error, every attempt counts significantly.* Every implemented solution is consequential. It leaves "traces" that cannot be undone. Whenever actions are effectively irreversible and whenever the half-lives of the consequences are long, every trial counts. Moreover, every attempt to reverse a decision or to correct for the undesired consequences poses another set of wicked problems, which are in turn subject to the same dilemmas.
6. *Wicked problems do not have an enumerable (or an exhaustively describable) set of potential solutions, nor is there a well-described set of permissible operations that may be incorporated into the plan.* In such fields of ill-defined problems and hence ill-definable solutions, the set of feasible plans of action relies on realistic judgment, the capability to appraise "exotic" ideas and on the amount of trust and credibility between planner and clientele that will lead to the conclusion, "OK let's try that."
7. *Every wicked problem is essentially unique.* By "essentially unique" we mean that, despite long lists of similarities between a current problem and a previous one, there always might be an additional distinguishing property that is of overriding importance.
8. *Every wicked problem can be considered to be a symptom of another problem.* The level at which a problem is settled depends on the self-confidence of the analyst and cannot be decided on logical grounds. The higher the level of a problem's formulation, the broader and more general it becomes and the more difficult it becomes to do something about it. On the other hand, one should not try to cure symptoms and therefore one should try to settle the problem on as high a level as possible. Marginal improvement does not guarantee overall improvement.
9. *The existence of a discrepancy representing a wicked problem can be explained in numerous ways.* The choice of explanation determines the nature of the problem's resolution. Somewhat but not much

exaggerated, you might say that everybody picks that explanation of a discrepancy that fits his intentions best and that conforms to the action-prospects that are available to him. The analyst's "world view" is the strongest determining factor in explaining a discrepancy and therefore, in resolving a wicked problem.

10. *The planner has no right to be wrong.* Here the aim is not to find the truth, but to improve some characteristic of the world where people live. Planners are liable for the consequences of the actions they generate, and the effects can matter a great deal to those people that are touched by those actions.

Although the concept of wicked problems started out as a description for social or cultural problems, the complex nature of communities and the need to protect them from a panoply of risks make achievement of community resilience a 21st century wicked problem. The characteristics of this modern wicked problem are provided in the context of the initial characteristics of wicked problems.

1. *Resilience is a function of the specific needs of a community from a social, economic, and infrastructure perspective.* High-level definitions such as those provided by the National Academies and high-level benchmarks such as those developed by the Alliance for National, Community Resilience (ANCR) can provide direction and strategy, but the ultimate achievement of resilience must be determined within the individual community.
2. *A community cannot be resilient to every possible stressor.* Time, money, and/or patience will far outstrip a community's ability to develop and implement solutions.
3. *Resilience is in the eye of the beholder.* Social scientists, elected officials, engineers, planners, advocates, environmental scientists, healthcare professionals, and economists all have ideas and solutions to achieve resilience, but solutions are not binary and there is no definitive arbiter of "correctness."
4. *Resilience is only truly known following an adverse event.* We cannot plan for all possible adverse events. Solutions are often implemented based on the most likely or the most impactful event, sometimes precluding action in response to other possible events. Future losses avoided is a very difficult metric to define and is not easily incorporated into economic and policy decision-making.
5. *Solutions identified to achieve resilience may preclude the implementation of other potential solutions.* Infrastructure is long-lived, and its planning and implementation often take significant investments of time and financial resources.
6. *Communities are socially and physically unique as are the potential solutions to their challenges.* There is no defined set of interventions that will address all the needs within a community.
7. *Communities and risks are dynamic.* Daily decisions made at the community level influence its capacity to respond to an adverse event. Evolving risks such as those posed by climate change may change the effectiveness of yesterday's solutions.
8. *A community's vulnerability and an effective approach to achieve resilience is based on thousands of past decisions.* Decisions on land use, building codes, social services, economic development, education, and other community functions all define the community.
9. *With a holistic approach to resilience, the exact combination of solutions leading to resilience is infinite and cannot be defined by a single discipline.*
10. *Resilience is the manifestation of assuring that human rights around health, safety, and welfare are not ignored.* Every member of society—particularly those with expertise or influence—have an obligation to contribute to their community's resilience.

The easy problems before society—the ones that could be solved by optimization within a single discipline or system—have largely been solved. It is now time to take the leap to address the wicked problems—the ones that require interdisciplinary problem solving, integrative processes and society-wide goal setting. These challenges are only going to get harder: an aging population, shrinking government budgets, and changing risks are all looming on the horizon. Hopefully, the diversity of disciplines and experts represented in this book advance understanding and knowledge, illuminate the importance of interdisciplinary solutions and help overcome the characteristics that make achievement of resilience a wicked problem.

A SYSTEMS APPROACH TO RESILIENCE

As pointed out earlier, a systems approach may be particularly useful in addressing wicked problems. Meadows defines a system as "an interconnected set of elements that is coherently organized in a way that achieves something [38]." She goes further to describe the concept of hierarchical systems and that highly functional systems balance the welfare, freedoms, and responsibilities of the subsystems and total system. Cabrera and Cabrera describe systems thinking as an attempt to "better align how we think with how the real world works. The real world works in systems—complex networks of many interacting variables. Often nonlinear, complex, and unpredictable, real-world systems seldom correspond with our desire for simplistic, hierarchical, and linear explanations. Systems thinking is the field of study that attempts to understand how to think better about real-world systems and the real-world problems we face [39]."

In the resilience context, the planet can be considered the ultimate system, with two parallel and interconnected sets of hierarchical subsystems—natural systems and social and geopolitical systems. Natural systems branch from the macrolevel (e.g., global climate) to the microlevel (e.g., individual watersheds and even the highly local anatomy of shorelines). Social and geopolitical systems range from international organizations such as the United Nations to the national, state, local and even neighborhood scale. Within these geopolitical constructs, social systems such as education systems, healthcare organizations, and cultural institutions aim to address important needs—ideally in a coordinated fashion.

In some places, human constructs have located in vulnerable natural environments due to a variety of reasons (e.g., access to important resources, economic opportunity, or "affordability"). In some places, natural systems (sometimes due to human disruptions) have changed, bringing new risks (e.g., sea-level rise due to climate change in places such as Miami or the construction of a levee in a community upstream). The resilience of these systems individually and collectively is what allows humans to survive and even thrive.

Meadows points out that the behavior of a system cannot be known just by knowing the elements of the system. "Because resilience may not be obvious without a whole-system view, people often sacrifice resilience for sustainability, or for productivity, or for some other more immediately recognizable system property … Remember that hierarchies exist to serve the bottom layers, not the top. Don't maximize parts of systems or subsystems while ignoring the whole [38]."

Cabrera and Cabrera offer an optimistic thought on the intersection of wicked problems and systems thinking, "Systems thinking gives us hope that the collective efforts of many to understand their little part of the world can come together to better understand the world as a whole. It gives us hope that we are going to be able to solve our most wicked problems [39]."

INFRASTRUCTURE AS A COMMUNITY SYSTEM

ANCR has identified 19 functions that represent the social, organizational, and infrastructural components of communities (Fig. 8). Each of these functions have their own level of resilience, but the resilience of the community relies on the resilience of the whole. Without a high-level understanding of the system, resources may be overinvested in one subsystem, foregoing potential investment in other subsystems. If buildings are made to withstand the risks they are likely to face, but the electrical or water infrastructure is left vulnerable, the community is unlikely to fully recover following a hazard event. Without energy or water, buildings become uninhabitable. If the transportation network breaks down, businesses become inaccessible. A community's focus on infrastructure is necessary for achieving resilience, but it is not sufficient. Infrastructure is just one system in a larger ecosystem.

Meerow et al. developed a conceptual schematic of urban systems, recognizing key subsystems of socio-economic dynamics, infrastructure and form, material and energy flows, and governance networks [10]. See Fig. 9. The schematic clearly illustrates the interconnectedness across these subsystems and the need to view community-level challenges holistically. Focus on just one subsystem can lead to unintended consequences in others. However, it is unrealistic to attempt to capture the interactions across all subsystems in one book.

FIG. 8 Community functions. (Source: Alliance for National and Community Resilience.)

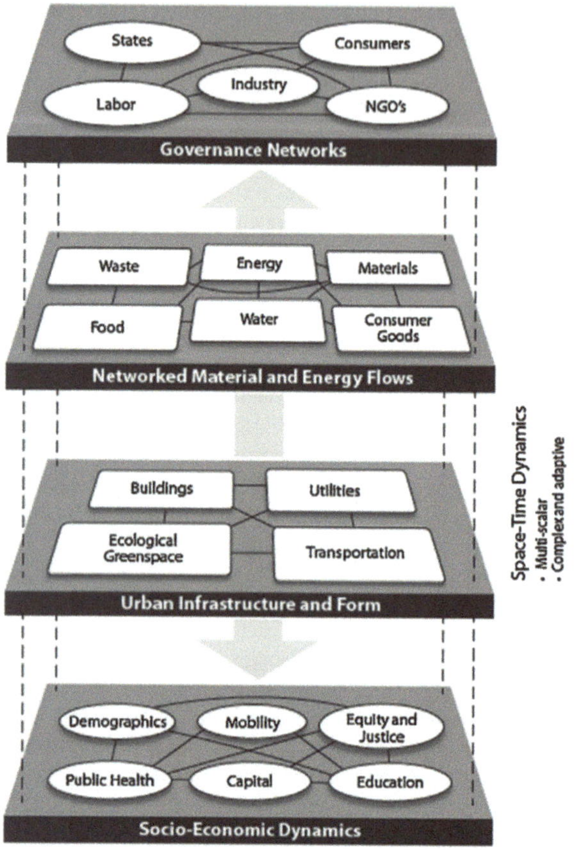

FIG. 9 Conceptual schematic of a community system. (Courtesy of: S. Meerow, J.P. Newell, M. Stults, Defining urban resilience: a review, Landscape and Urban Planning 147 (2016).)

This book uses the elements of infrastructure and form as its entry point. Although material and energy flows are not addressed specifically, the inferred implication is that infrastructure is the means for delivering these essential services. Without an effective transportation system, goods cannot move; energy and water utility structures underpin the delivery of electricity, natural gas, and potable water and the treatment of wastewater; and buildings provide the facilities to allow these systems to work. Here, the resilient aspects of infrastructure are presented from the perspective of representatives within governance networks including states, industry, and nongovernmental organizations. The book also touches on one essential component of socioeconomic dynamics, capital. Without capital, the ability to effect change is severely limited.

ABOUT THIS BOOK

This book features contributions from subject matter experts and thought leaders representing multiple disciplines and community functions. Although presented as individual chapters and organized into related themes, this book is intended to provide the reader with a broader understanding of resilience as a holistic strategy to deliver safe, strong, secure, and sustainable communities. Short introductions to each theme provide context on how each of these themes contribute to the big picture. This format is designed to build understanding of how various disciplines contribute to solutions and how solutions are developed at the intersections of disciplines.

Often, practitioners (and policymakers) will point to the availability of financial resources as a barrier to resilience investments. Chapters 6 and 7 focus on the demand and supply side of capital and how the financial markets are beginning to recognize the importance of resilient infrastructure.

Although not initially designed to specifically address the resilience challenges provided by climate change, many chapter authors (rightly so) provided significant focus on how climate change is altering risks and vulnerabilities. Some disciplines and infrastructure systems are making progress in this regard—others are proceeding slowly. Reilly and Ayyub explore methods for addressing emerging risks in Chapter 13.

While, this book outlines the importance of a holistic approach to achieving community resilience, it largely focuses on the physical and infrastructure aspects of community resilience. It acknowledges, but does not address in depth, the parallel requirements of social systems that support community resilience. Chapter 15 touches on some of these important intersections. Policies, practices, and programs that support community health, education, commerce, culture, and community are equally important. In many cases, these functions cannot be resilient without a supporting infrastructure of buildings, utilities, and transportation networks. As understanding and interdisciplinary collaboration grows, hopefully future works will examine the important intersections between physical and social community functions to support delivery of resilient communities.

Meadows eloquently sets a path forward for solving systems level problems, "Seeing systems whole requires more than being 'interdisciplinary,' if that word means, as it usually does, putting together people from different disciplines and letting them talk past each other. Interdisciplinary communication works only if there is a real problem to be solved, and if the representatives from the various disciplines are more committed to solving the problem than to being academically correct. They will have to go into learning mode. They will have to admit ignorance and be willing to be taught, by each other and by the system [24]."

This book is intended to be an initial step in helping representatives of various disciplines to be taught by each other and by the system to support achievement of a resilient system-of-systems. To shift resilience thinking to be holistic and recognize the complexities that create vulnerabilities.

REFERENCES

[1] NOAA National Centers for Environmental Information (NCEI), U.S. Billion-Dollar Weather and Climate Disasters, 2019. https://www.ncdc.noaa.gov/billions/.

[2] Multihazard Mitigation Council, Natural Hazard Mitigation Saves: 2018 Interim Report, Principal Investigator K. Porter, co-Principal Investigators C. Scawthorn, C. Huyck, Investigators: R. Eguchi, Z. Hu, A. Reeder, P. Schneider, Director, MMC, National Institute of Building Sciences, Washington, D.C. https://www.nibs.org/resource/resmgr/mmc/NIBS_MSv2-2018_Interim-Repor.pdf.

[3] K. Lam, Death toll drops to 85 at Camp Fire; 11 people remain missing, USA Today (December 3, 2018).

[4] S. Fink, Nearly a year after Hurricane Maria, Puerto Rico revises death toll to 2,975, New York Times (August 28, 2018).

[5] The National Academies, Disaster Resilience: A National Imperative, The National Academies Press, Washington, DC, 2012.

[6] United Nations International Strategy for Disaster Reduction, Terminology, 2011. https://www.unisdr.org/we/inform/terminology.

[7] The White House, Presidential Policy Directive-8: National Preparedness, 2011. https://www.dhs.gov/xlibrary/assets/presidential-policy-directive-8-national-preparedness.pdf.

[8] K. Weeks, Building organizations commit to promoting resilient design, Architect Magazine (May 14, 2014).

[9] Industry Statement on Resilience, http://aiad8.prod.acquia-sites.com/sites/default/files/2018-04/Resilience_Statement_2018-0410.pdf.

[10] S. Meerow, J.P. Newell, M. Stults, Defining urban resilience: a review, Landscape and Urban Planning 147 (2016).

[11] Y.Y. Haimes, On the definition of resilience in systems, Risk Analysis 29 (4) (April 2009).

[12] A. Rose, Defining and measuring economic resilience to disasters, Disaster Prevention and Management 13 (4) (2004).

[13] C.S. Holling, Resilience and stability of ecological systems, Annual Review of Ecology and Systematics 4 (1973).

[14] American Institute of Architects, Designing for Resilience, 2018. http://content.aia.org/sites/default/files/2018-09/Designing_for_Resilience_V04.pdf.

[15] City of Pittsburgh, ONEPGH: Pittsburgh's Resilience Strategy, 2017. http://pittsburghpa.gov/onepgh/documents/pgh_resilience_strategy.pdf.

[16] City of Pittsburgh, Resilient Pittsburgh, 2016. http://apps.pittsburghpa.gov/redtail/images/542_PRA_2016.pdf.

[17] L.J. Vale, The politics of resilient cities: whose resilience and whose city? Building Research & Information 42 (2) (2014).

[18] FEMA, What is Mitigation? https://www.fema.gov/what-mitigation.

[19] United Nations Environment Program, Mitigation. https://www.unenvironment.org/explore-topics/climate-change/what-we-do/mitigation.

[20] Hoover Institution, Ready for Tomorrow: Seven Strategies for Climate-Resilient Infrastructure, Stanford University, April 2019.

[21] B. Obama, Presidential Policy Directive/PPD-8: National Preparedness, March 30, 2011. https://www.dhs.gov/presidential-policy-directive-8-national-preparedness.

[22] U.S. Department of Homeland Security, National Preparedness Goal, second ed., September 2015. https://www.fema.gov/media-library/assets/documents/25959.

[23] Federal Emergency Management Agency, 2018-2022 Strategic Plan, 2018. https://www.fema.gov/strategic-plan.

[24] National Institute of Standards and Technology, Federal Building and Fire Safety Investigation of the World Trade Center Disaster: Final Report of the National Construction Safety Team on the Collapses of the World Trade Center Tower, September 2005. https://www.nist.gov/engineering-laboratory/final-reports-nist-world-trade-center-disaster-investigation.

[25] National Institute of Standards and Technology, Report of the Technical Investigation of the Station Nightclub Fire, June 2005. https://www.nist.gov/publications/report-technical-investigation-station-nightclub-fire-nist-ncstar-2-volume-1.

[26] National Institute of Standards and Technology, Final Report, National Institute of Standards and Technology (NIST) Technical Investigation of the May 22, 2011, Tornado in Joplin, Missouri. Mary 2014, https://www.nist.gov/publications/final-report-national-institute-standards-and-technology-nist-technical-investigation.

[27] National Institute of Standards and Technology, Community Resilience Program. http://www.nist.gov/programs-projects/community-resilience-program.

[28] National Institute of Standards and Technology, Community Resilience Planning Guide for Buildings and Infrastructure: Volumes I and II, NIST Special Publication 1190, 2016.

[29] National Institute of Standards and Technology, Community Resilience Economic Decision Guide for Buildings and Infrastructure Systems, NIST Special Publication 1197, 2015.

[30] Office of the Under Secretary of Defense for Policy, DOD Directive 3020.40: Mission Assurance, November 29, 2016.

[31] U.S. Army, Energy Security & Sustainability (ES^2) Strategy, March 2015. https://www.army.mil/e2/c/downloads/394128.pdf.

[32] U.S. Navy, Task Force on Climate Change/Oceanographer of the Navy. U.S. Navy Climate Change Roadmap, April 2010. https://www.navy.mil/navydata/documents/ccr.pdf.

[33] Department of Defense, Climate Change Adaptation Roadmap, 2014. https://www.acq.osd.mil/eie/downloads/CCARprint_wForward_e.pdf.

[34] ESTCP, Workshop Report: Nonstationary Weather Patterns and Extreme Events: Informing Design and Planning for Long-Lived Infrastructure. ESTCP Project RC-201591, November 2017. https://www.serdp-estcp.org/News-and-Events/Blog/Nonstationary-Weather-Patterns-and-Extreme-Events-Workshop-Report.

[35] F.L. Reinhardt, M.W. Toffel, Managing Climate Change: Lessons from the U.S. Navy. Harvard Business Review, July-August 2017. https://hbr.org/2017/07/managing-climate-change.

[36] H. Rittel, M. Webber, Dilemmas in a general theory of planning, Policy Sciences (1973).

[37] J. Kolko, Wicked Problems: Problems Worth Solving, Austin Center for Design, 2012.

[38] D. Meadows, Thinking in Systems, Sustainability Institute, 2008.

[39] D. Cabrera, L. Cabrera, Systems Thinking Made Simple: New Hope for Solving Wicked Problems, Odyssean Press, 2015.

Contents

PART V
BUILDINGS

PART VI
POLICIES & PRACTICES

PART VII
CONCLUSION

Introduction

WHY RESILIENCE?

Hardly a place on earth exists that is not vulnerable to some hazard. Whether a natural hazard or a man-made one, shocks and stresses impact communities. As outlined in the introduction to this book, in the United States, the number and impact of hazard events is increasing. This trend is happening globally as well [1]. While the magnitude of the numbers is staggering, action to reduce the impacts of future events does not always follow.

Since the 2005 National Institute of Building Sciences (NIBS) study on the value of Federal Emergency Management Agency (FEMA) mitigation investments, decision-makers had the information to make a compelling argument for hazard mitigation. Yet, funding for predisaster mitigation remained minor compared with the funding provided postdisaster. Until recently, the Pre-Disaster Hazard Mitigation Grant Program at the FEMA paled in comparison with disaster recovery—associated programs. According to the Pew Charitable Trusts, just 6% of FEMA's mitigation program funding between fiscal year 2007 and 2016 went to predisaster mitigation [2] (see Fig. 1).

At a state and local level, many communities have not kept pace with implementing mitigation strategies like up-to-date building codes despite considerable research on their effectiveness (see the Introduction to Part V and Chapter 12 for a discussion on the role of building codes).

Recent devastation caused by Hurricanes Dorian, Michael, Irma, Maria and Harvey; the Camp and Tubbs Fires; and other recent events provide examples of the level of devastation communities can face. Postevent evaluations of community infrastructure show that prudent policy decisions and investments can lessen the impacts of such hazard events. The magnitude 7.0 earthquake that rattled Anchorage, Alaska, on November 30, 2018, had the potential to cause significant damage, yet only a handful of structural fires developed, and no one died. The limited damage was attributed to the proactive implementation of building codes [3].

Despite the growing body of knowledge and the harrowing visuals of destruction from recent events, resilience investments are just now coming to the fore. The recent passage of the Disaster Recovery Reform Act (DRRA) with its increased emphasis on mitigation, the release of the National Mitigation Investment Strategy (NMIS) by the Mitigation Framework Leadership

The National Mitigation Investment Strategy

The Post-Katrina Emergency Management Reform Act of 2006 established the Mitigation Framework Leadership Group (MitFLG) to organize mitigation efforts across the federal government. Composed of federal, state, local, tribal, and territorial public sector representatives, the MitFLG integrates federal efforts to deliver the mitigation core capabilities in the National Mitigation Framework and assesses the effectiveness of these capabilities across the United States. Following Hurricane Sandy, the Government Accountability Office (GAO) found that federal agency coordination around mitigation investments was lacking, which lead to the reduced effectiveness of investments. GAO encouraged the MitFLG to develop a coordinated investment strategy.

The MitFLG released an NMIS in August 2019. The NMIS is intended to serve as a single, cross-agency strategy aimed at advancing mitigation investments to support whole community resilience. The strategy focuses on three goals:

- **Goal 1: Show how mitigation investments reduce risk.** The whole community will build a shared understanding of mitigation investment and its value. Specifically, the whole community will understand how effective mitigation investments can protect people, homes, neighborhoods, cultural and historic resources, ecosystems, and lifelines (for example, communications, energy, transportation, and water). The federal government and its nonfederal partners

Continued

will create a shared vocabulary and common measures to communicate information about risk and find opportunities to educate, hire, train, and develop a base of qualified mitigation professionals.

- **Goal 2: Coordinate mitigation investments to reduce risk.** The whole community will coordinate mitigation investments through shared risk information, reinforced strategies for risk reduction, and easier access to existing funding. Such coordination will help the whole community justify mitigation investments and choose the most cost-effective and reasonable actions.
- **Goal 3: Make mitigation investment standard practice.** The whole community will factor mitigation into investment decisions, especially for buildings and infrastructure. The federal government and its nonfederal partners will use and expand financial products and approaches for mitigation investment—including funding, incentives, and financial risk transfer opportunities. The federal government and its nonfederal partners also will make mitigation standard professional practice critical to safeguarding lifelines, services, and national safety and security.

Under each high-level goal are a series of recommendations to support its achievement. The MitFLG intends to coordinate implementation of the goals and recommendations across all components of the community [4].

Group [4], the availability of new mitigation funding through the Community Development Block Grant Disaster Recovery Program (CDBG-DR) [5], and FEMA's 2018–22 Strategic Plan [6] all form a growing wave of activity that is coalescing around increased focus on resilience.

However, to be truly impactful, resilience and hazard mitigation must move beyond the responsibility of emergency managers to become embedded in planning and decision-making for all professionals impacting a community. From individual businesses and structures to community-wide programs, every decision has ripple effects. The chapters in this part make the case that the time has come for action on implementing holistic resilience strategies.

As elegantly explained by Hill and Kakenmaster in Chapter 1, communities rely on infrastructure, an infrastructure that is increasingly challenged by both natural and human systems. Climate change provides a relatively new threat environment that is already stressing infrastructure. Governments, infrastructure owners and managers, and designers must prepare for these evolving threats and ensure investments in infrastructure capture the risks across its life cycle. Hill and Kakenmaster provide timely examples of resilience challenges faced by a community and the proactive steps taken to improve the current and long-term quality of life for their residents.

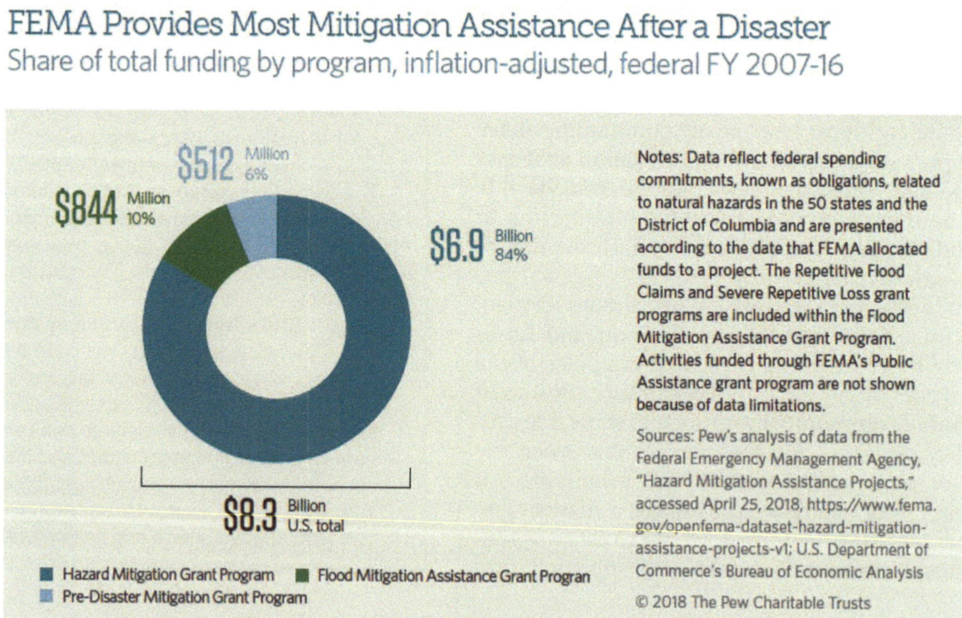

FIG. 1 FEMA Mitigation Expenditures, FY 2007–2016. (Credit: Courtesy, The Pew Charitable Trusts [2].)

The increasing threat to infrastructure systems also reveals the interconnectedness of community functions and the interdependencies that can add to a community's vulnerabilities. In Chapter 3, Brashear offers a management process to expose these interdependencies to infrastructure owners and managers and to develop a cooperative approach to funding and implementing strategies to address the vulnerability of these points of intersection.

SUSTAINABILITY AND RESILIENCE

Frequently, questions arise as to the relationship between sustainability and resilience. Both are focused on achieving higher levels of performance, both require thinking beyond business-as-usual, both address concerns beyond the individual building or project, and both have a long time horizon. But, how do they fit together?

The shocks and stresses associated with climate change provide an important lens to examine this relationship. In this case, sustainability and resilience are opposite sides of the same coin. Climate change mitigation in the form of reduced greenhouse gas emissions reduces the risk of climate change–induced events. Climate change adaptation reduces the impacts of climate change–induced events. Just concentrating on one of these efforts will push the other further and further out of reach.

Put another way, communities cannot be resilient without considering sustainability and vice versa. Without a focus on resources like water, air, and energy, a community can become an undesirable place to live or work. Ineffective management of these resources can trigger their own set of shocks or stresses. Energy policies contribute to the social resilience of a community and can help avoid significant burdens on vulnerable populations. Pollutants from energy generation can create or enhance health-related vulnerabilities such as asthma.

In the event of extreme heat or cold, energy infrastructure can be significantly stressed [7]. The polar vortex in 2014 caused increases in natural gas demand, which could not be met by many utility systems. A 2016 Southern California heat wave ended up leaving 5,300 households without power for several hours, as Los Angeles saw peak demand reach 50% higher than average [8]. Buildings constructed to be energy efficient maintain temperatures longer and require less energy to provide heating or cooling, resulting in less stress on the grid. This may allow the grid to remain functional during such an event, resulting in decreased overall impact to the entire community. A natural gas utility in Michigan experienced distribution challenges when a compressor station failed due to a fire. Residents were asked to lower their thermostats during freezing weather to allow continued service [9]. Without energy efficiency measures, the required use reductions would have been significantly higher, causing further reductions in service.

The concept of passive survivability has also emerged at the intersection of sustainability and resilience. Passive survivability is the ability for a building to remain habitable in the face of an event or crisis, resulting in the loss of energy, water, or sewage services. The need for passive survivability may surface during extreme heat or cold events when the grid is severely taxed or secondary to other hazard events. Temperature extremes can stress the grid, resulting in blackouts. Incorporating measures related to passive survivability can help support resilience on two ends—reducing energy demands through increased efficiency thus reducing grid strain, and keeping buildings occupiable for longer periods reducing shelter or other emergency services needs [10].

Nature-based solutions for addressing water-related vulnerabilities can also reap many cobenefits as discussed in Chapter 8.

Finally, more resilient structures mean less need for rebuilding after a disaster, reducing the need to landfill or recycle building material and lost property and new material for reconstruction.

Whether the words themselves become just buzzwords and are overused, the underlying concepts are vitally important. Ideally, by the time attention moves to another topic du-jour, the actions and knowledge developed during its heyday are embedded into professional practice and become a common offering in the marketplace. In many communities, green buildings have reached this tipping point. Hopefully, resilience is not far behind.

In Chapter 2, Hartke outlines how the sustainable buildings movement evolved and became a market-driven solution to address national- and community-level needs. Resilience in the built environment is currently in the early stages of its own evolution. Pathways forged by green building rating programs help lead the way for rating systems and tools such as the ANCR Community Resilience Benchmarks, the US Resiliency Council, RELi, and REDi.

The ongoing recognition by building owners, policy makers, and designers that the built environment either hinders or helps a community thrive in the face of everyday and occasional challenges will continue to drive improvements. This concept is covered in many of the chapters that appear elsewhere in this book.

REFERENCES

[1] Swiss Re Group, Preliminary Sigma Estimates for 2018: Global Insured Losses of USD 79 Billion Are Fourth Highest on Sigma Records, December 18, 2018. https://www.swissre.com/media/news-releases/nr_20181218_sigma_estimates_for_2018.html.

[2] Pew Charitable Trusts, Natural Disaster Mitigation Spending Not Comprehensively Tracked, September 2018.

[3] T. Lukasik, Alaska hails building codes after quake, Building Safety Journal (December 12, 2018).

[4] Mitigation Framework Leadership Group, National Mitigation Investment Strategy, August 2019.

[5] Department of Housing and Urban Development, Allocations, Common Application, Waivers, and Alternative Requirements for Community Development Block Grant Mitigation Grantees, 84 FR 45838, August 30, 2019.

[6] Federal Emergency Management Agency, 2018-2022 Strategic Plan, 2018. https://www.fema.gov/strategic-plan.

[7] US Department of Energy, U.S. Energy Sector Vulnerabilities to Climate Change and Extreme Weather, 2013.

[8] D. Ribeiro, T. Bailey, Indicators for Local Energy Resilience, American Council for an Energy Efficiency Economy, June 2017.

[9] J. Wisely, C. Hall, How Consumers Energy Customers Helped Avert a Michigan Gas Crisis, Detroit Free Press, January 31, 2019. https://www.freep.com/story/news/local/michigan/2019/01/31/consumers-lower-heat-emergency-alert/2732252002/.

[10] Alliance for National & Community Resilience and International Code Council, The Important Role of Energy Codes in Achieving Resilience, November 2019.

CHAPTER 1

Resilient Infrastructure: Understanding Interconnectedness and Long-Term Risk

ALICE C. HILL, JD • WILLIAM KAKENMASTER, BA

INTRODUCTION

An essential fact of today's infrastructure is its intense interconnectedness. The world's infrastructure operates like a gigantic chain linking together critical sectors including energy, communications, information technology, waste management, water, transportation, financial services, and public health. These interconnected systems aid the distribution of goods, services, and information from one sector to others that need them [1]. Factories and office buildings need electricity to keep the lights on. Urban transportation systems require roads to keep buses, trains, and planes working. Healthcare providers need functional facilities to deliver urgently needed care. Financial services depend on ready access to the internet. Moreover, failure in one infrastructure sector can often lead to failures in others. This interdependence increases infrastructure systems' vulnerability: one break in the chain can lead to far-reaching disruptions, cascading collapses, and unexpected consequences.

A break in this chain can occur as a result of something as simple as human error, as happened on September 8, 2011, when an electrical technician performing regular maintenance at an electrical substation in Yuma, Arizona forgot to replace a piece of equipment. The technician's mistake overloaded a major power line as well as several smaller lines and set in motion a series of infrastructure failures throughout the American Southwest and northern Mexico [2]. In just 11 minutes' time, with outdoor temperatures ratcheting past 100°F, life went dark for seven million people across California, Arizona, Baja California, and Sonora. In San Diego, the hardest-hit area of all, the massive blackout caused hospital emergency rooms to lose power, gas stations to stop pumping fuel, and traffic lights to stop working [3]. Without electricity, water treatment plants leaked millions of gallons of contaminated water, two nuclear reactors went offline, and all flights out of San Diego Airport were suspended. Eventually, the lights flickered back on and power was restored, but not before the technician's mistake cost over $100 million, mainly due to lost business [4].

Intentional disruption can also cause cascading failures in critical infrastructure. In 2011, with tensions burning white hot during the height of the Arab Spring, the Egyptian government curtailed virtually all internet access to its 80 million citizens. In a nanosecond, 90% of all data traffic in and out of the country vanished [5]. Egypt's presence in the virtual world disappeared instantly. The move was a deliberate attempt to disrupt antigovernment opposition and protest movements, but it also carried a host of cascading consequences. Online commerce and trade in digital products and services plummeted [6]. Hospitals and factories sputtered when online records and information became inaccessible. Education was disrupted as teachers and students lost access to online platforms. For 5 days, Egypt reverted to a premodern era, which cost the country roughly $90 million, according to the Organization for Economic Cooperation and Development (OECD) [7].

While direct human action can lead to infrastructure failures with severe consequences, natural disasters can as well. As of 2018, the two largest power failures in the world have come because of extreme weather events. In 2013, Typhoon Haiyan claimed the title of world's largest blackout, causing 6.1 billion hours of lost power [8]. Four years later, Hurricane Maria came in second with 3.4 billion hours lost, marking the largest power outage in US history. Because of Maria, hundreds of thousands of homes and businesses went without

Optimizing Community Infrastructure. https://doi.org/10.1016/B978-0-12-816240-8.00001-X

electricity for months, relying on costly generators and other temporary fixes until the grid finally came back online [9]. The consequences were stark: schools closed, businesses shuttered, and hospitals struggled to keep up with the tide of injuries and diseases the storm brought about. When Maria struck Puerto Rico, its electrical grid was already suffering. Roughly $9 billion in debt, the Puerto Rico Electric Power Authority filed for bankruptcy in July 2017 and had greatly reduced its maintenance, inspection, and repair efforts to cut costs [10].

However, even when the electric grid is in better condition, extreme weather can wreak havoc. In the northeastern United States in 2012, Hurricane Sandy affected over 200 energy assets across 10 states [11]. A punishing combination of wind, rain, and storm surge swept through and cut off power for over eight million people. Eventually, Sandy caused the third largest blackout in US history and posed significant challenges for multiple infrastructure sectors [8]. In New York, schools closed, hospitals were evacuated, planes and trains stopped running, and residents struggled to navigate flooded roads littered with debris. New York's comparative affluence could not prevent its more developed electrical grid from struggling, straining, and ultimately failing when presented with the shock of an extreme storm.

All three of these storms—Haiyan, Maria, and Sandy—have occurred since 2012. Climate change is pushing us toward a future of extremes and posing greater and greater risks to infrastructure. Existing infrastructure was not designed to withstand the effects of a changing climate, such as extreme heat, sea-level rise, extended drought, more wildfire, and more intense storms and precipitation. It was built to withstand the extremes of the past, not those of the future. Escalating conditions brought about by climate change dramatically increase the chances that any given infrastructure investment will fail and disrupt the operations of all others that depend on it. As new and greater risks emerge from the impacts of climate change, they add stress to an already tightly pulled chain connecting the world's infrastructure. This chain stands to grow tighter and tighter in the coming years, especially as the world's population grows by almost 2 billion people by 2040. Global Infrastructure Hub, a G20 initiative focused on developing the world's infrastructure, estimates that the world will need to invest $94 trillion in infrastructure by 2040 to keep up with economic and demographic changes, sustain economic growth, and mitigate the effects of climate change and disasters [12]. Strengthening this interdependent system's resilience can help to mitigate the risk of disruptions and failures due to climate change. The need to promote resilience to climate change and extreme weather has arguably never been greater or more urgent.

This chapter explores risks to infrastructure stemming from a changing climate and how interdependencies among infrastructure systems magnify those risks. It describes the impacts the globe is already experiencing, as well as the greater extremes climate change is expected to breed. It discusses how climate change can cause infrastructure sectors to suffer disruptions and failures, and how the failure of even just one critical sector can cause cascading failures in other, interdependent sectors. It ultimately offers a call to action, arguing that urgent and meaningful action today can help to prepare for future risks, and that building resilience to climate change makes infrastructure more resilient to other hazards.

THE SETTLED SCIENCE OF CLIMATE CHANGE

For over half a century, scientists at the Mauna Loa Observatory in Hawaii have studied in earnest our planet's climate. Teams first began their research at Mauna Loa in 1958. Ever since, they have worked to measure the atmospheric concentration of gases such as water vapor, carbon dioxide (CO_2), methane, and others through a variety of cutting-edge scientific methods. Over 11,000 feet above sea level and far from the mainland United States, the site makes for an ideal location to observe changes in the atmosphere. In 2013, scientists at Mauna Loa detected, for the first time, the highest levels of carbon dioxide gas in the atmosphere in at least the last 800,000 years, and some think since the Pliocene Epoch—a time many millions of years ago when the average temperature on Earth was 5 to 6°F warmer than today and seas were dozens of feet higher [13]. This new measurement of 400 parts per million (ppm) was not good news. More carbon in the atmosphere traps more heat energy, which contributes to warmer temperatures on Earth—a phenomenon, scientists have observed since the 1800s.

By 2019, CO_2 levels detected by scientists at Mauna Loa had climbed to 415 ppm. These measurements of atmospheric CO_2 are the highest levels recorded in all of human history [14a]. According to the Intergovernmental Panel on Climate Change (IPCC), established in 1988 by the World Meteorological Organization and the United Nations Environment Programme,

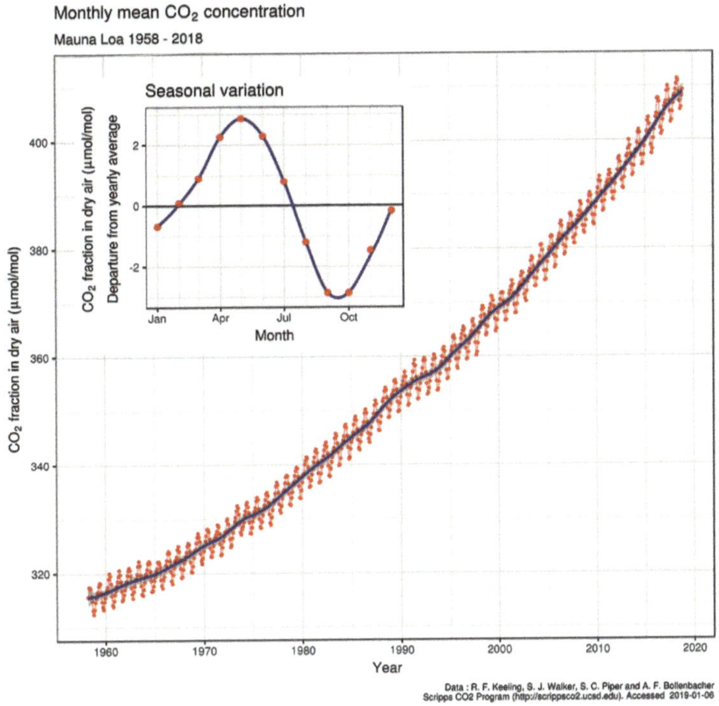

FIG. 1.1 Monthly mean carbon dioxide concentrations measured at Mauna Loa Observatory, 1958–2018. (Credit: Delorme – Own work. Data from Dr. Pieter Tans, NOAA/ESRL and Dr. Ralph Keeling, Scripps Institution of Oceanography., CC BY-SA 4.0, https://commons.wikimedia.org/w/index.php?curid=40636957.)

increasing concentrations of these gases are pushing the climate to new extremes, including a rise in global average temperatures [14b].[1] Since the 1880s, when modern temperature records began, the average global temperature has risen approximately 1.8°F. Each of the last three decades has been warmer than the previous one, and all of them have been warmer than any decade before 1850. Additionally, temperature data recorded by the National Oceanic and Atmospheric Administration (NOAA) show that, as of 2018, 6 of the 10 hottest years since record-keeping began have occurred since 2010 [15]. As the IPCC's Fifth Assessment Report plainly states: "Warming of the climate system is unequivocal" [14].

With warmer temperatures comes sea-level rise resulting from the expansion of ocean waters and the melting of glacial ice. Much like temperature rise, sea-level rise is occurring at a rate faster than it has for thousands of years because of climate change [14]. Warming ocean waters, combined with melting glacial ice, have caused the world's seas to rise by about seven and a half inches on average since 1901. Just since 1993, the average rate of sea-level rise has doubled from the 20th century average [14,16]. Some of the world's coastal regions now experience rates of sea-level rise three to four times higher than the global average [14,17]. Scientists warn that if current trends continue,

[1]In 2018, leading scientists at the National Center for Climate Restoration in Melbourne, Australia released a report claiming that the IPCC's approach to studying climate change and most climate research in general is based on overly "conservative projections and scholarly reticence" [14b]. Indeed, the IPCC has had to revise its projections of future climate scenarios as the data have outpaced its estimates time and time again. Just take projections of future sea-level rise, for instance. In 2001, the IPCC warned of just 0.08 inch of steady sea-level rise per year, and yet the average rate of sea-level rise had climbed to 30.13 inch by 2007. In 2007, the IPCC predicted anywhere between seven and 23 inches of sea-level rise by 2100, however later estimates suggested that the world would see at least 20 inches. In 2014, the IPCC projected just 21 inches even as organizations like the National Oceanic and Atmospheric Administration (NOAA) warned of over 8 feet. Because the IPCC operates according to consensus, it often errs on the "side of least drama" as some put it. So, while the IPCC remains the world authority on climate change, its conservative estimates ought to give one pause in considering possible future climate scenarios.

the world could have to cope with over 8 feet of average sea-level rise by century's end [18]. In some places, like Rhode Island, seas could rise close to 9 feet by century's end.

In addition to soaring temperatures and rising seas, climate change also produces more extreme weather that can lead to more punishing disasters. As temperatures increase, heat extremes will continue to break new records. In the 2000s, daily temperatures reached new record highs twice as often as new record lows [19,20]. The IPCC is "virtually certain" that there will be more extremely hot days because of rising temperatures [14]. One study conducted at NOAA's Geophysical Fluid Dynamics Laboratory in 2012 concluded that the average number of extremely hot days in the United States could increase anywhere from 20% to 100% by the end of this century if current warming trends continue [21]. Such extreme heat, along with drier conditions in some areas leads to deeper, more pervasive droughts, leeching water from streams, rivers, lakes, topsoil, and plant life. From 2012 to 2014, California suffered the worst drought in the last 1200 years, according to researchers who specialize in ancient climates [22]. According to other research, climate change worsened that drought by 8% to 27% [23].

Perhaps unsurprisingly, these hotter, drier conditions that come because of climate change contribute to greater wildfire risk. In the western United States, wildfires occurred roughly five times as frequently, burned over 12 times the total area, and lasted almost 10 times as long in the years between 2003 and 2012 compared to the years between 1973 and 1982 [24]. Globally, the average fire season has grown 18.7% longer since 1979 [25]. The IPCC has also found strong evidence linking climate change to more severe wildfires across the globe, including in Alaska, parts of the Mediterranean, and eastern Africa [26]. In California, which has long battled wildfires, 11 of the 20 most destructive wildfires since the state began keeping records in 1932 have occurred since 2007 [27]. During the 2017 fire season, 9,000 wildfires burned 1.2 million acres, an area roughly the size of Delaware, and, in 2018, fires scorched 1.9 million acres and torched over 18,000 structures [28–30]. Recent research in the field of attribution science has found that climate change is responsible for as much as half of the total acreage burned by forest fires since 1984 [31].

Climate change also brings greater extremes in precipitation and storms, which often means added risk of flooding. Because warmer air can absorb more moisture than cooler air, the rising temperatures brought about by climate change have coincided with an increase in extreme precipitation [32]. Between the 1900s and the 2010s, the continental United States has witnessed a more than 30% increase in the number of heavy downpours. In coastal areas, tropical storms and hurricanes that accumulate in this warmer atmosphere also travel across warmer ocean waters and are all the stronger for it. Houston, Texas, bore witness to the increasing extremes that climate change can bring when Hurricane Harvey dumped over 50 inches of rain on the city in just 1 week. Attribution studies have estimated that climate change increased precipitation levels from Harvey by as much as 40% [33–35]. Climate model predictions in 2017 estimated that extreme precipitation events in most regions of the world will grow in intensity by 3% to 15% for each additional 1.8°F that global temperatures rise [36].

FIG. 1.2 The La Tuna Fire, the largest wildfire in Los Angeles, California, USA in over 50 years, 2017. (Credit: Scott L, CC BY-SA 2.0, https://commons.wikimedia.org/wiki/File:La_Tuna_fire_and_cityscape_1.jpg.)

All of these impacts—new heat extremes, prolonged drought, more frequent wildfires, more extreme precipitation, and worsening storms—come with a growing price tag. NOAA keeps an annual tally of so-called billion-dollar weather and climate disasters occurring within the United States—those costing at least $1 billion in damages. The annual frequency of those disasters has risen from just a handful in the 1980s to 16 in 2017, the year that tied the record set in 2011. Moreover, for the first time ever, the cumulative losses from billion-dollar disasters in the United States crossed the $300 billion mark in 2017. According to researchers from the World Bank and OECD, worldwide losses from coastal flooding—just one of the impacts of climate change—in major cities could rise from approximately $6 billion in 2005 to between $60 billion and $1 trillion in 2050, depending on the adaptation efforts undertaken, if any [37]. Three of the five cities with the highest flood risk are located in the United States—Miami, New York, and New Orleans—and they make up 31% of the total assets at stake.

Additionally, the impacts of climate change could be far more severe in the future depending on the actions we take today. In 2018, the IPCC concluded that the impacts of climate change would be less drastic across the board if temperature rise were limited to 2.7°F compared to 3.6°F [38]. Heat extremes, drought, wildfire, sea-level rise, storm intensity, and extreme precipitation—all worsen as temperatures rise because of climate change. The debate is over, the science is settled. Climate change is already occurring and is having very real impacts across the globe.

AN ALREADY FRAGILE SYSTEM

Every 4 years, the American Society of Civil Engineers (ASCE) grades US infrastructure on its condition, funding, resilience, and a host of other indicators of infrastructure performance. In 2017, US infrastructure received a D+ overall for its sorry state, another D+ in 2013, and a D in 2009. In fact, the ASCE has not given the US infrastructure anything above a D since 1998, when it received a C-. Over two out of every five miles of road in the United States are congested, and one out of every five miles are in "poor condition" [39]. Almost 40% of the United States' more than 600,000 bridges are over 50 years old, and 1 in 10 are considered structurally deficient. Roughly 35% of tracks and rails in public transportation systems, along with 37% of stations, are "not in a 'state of good repair.'" As climate change strains an already fragile the US infrastructure system, it could carry disastrous results.

When Hurricane Sandy approached the United States in 2012, it stretched roughly 1,000 miles—over twice the width of Hurricane Katrina—and whipped its winds at speeds over 100 mph. Before Sandy made landfall on the night of October 29th, states along the eastern seaboard—including New York and New Jersey, where the storm hit first—had taken precautions. Homeowners boarded up their windows, businesses sandbagged their front doors, and people everywhere stocked up on groceries and everyday supplies. In New York City, Mayor Michael Bloomberg shut down the city's mass transit system and ordered 375,000 people to evacuate [40]. Governor Andrew Cuomo declared a state of emergency, calling on state employees to prepare public works for the coming storm [41]. New Yorkers everywhere braced for impact.

By the time it struck, NOAA had downgraded Sandy to a posttropical cyclone, but that meant little as the storm's capacity for destruction became apparent. Sandy arrived at high tide in lower Manhattan, carrying one of the tallest storm surges New York City had ever seen, clocking in at almost 14 feet [42]. With water pouring in, floods soon racked the city's infrastructure. In Manhattan's East Village, the surging floodwaters overwhelmed an electrical substation, whose flood walls only reached 12 feet [43]. The station exploded in a fireball that could be seen as far away as Brooklyn and the Lower East Side [44]. Instantly, the city that never sleeps plunged into darkness.

Over eight million people lost power in 15 states plus the District of Columbia 24 hours after Sandy reached the United States, often with devastating consequences for the region's critical infrastructure [12,45]. Without electricity, transportation ground to a halt. All seven subway tunnels underneath the East River flooded. New York's Metropolitan Transportation Authority (MTA) had to fall back on diesel-powered pump trains, private contractors, and the federal government to drive the water out because, when Sandy took the electrical grid offline, it also knocked out the subway's pump system [46–48]. Above ground, storm debris meant that tanker trucks struggled to get to gas stations and, even when they could, many stations lacked power [49]. As it turns out, New York learned, gas stations need power to pump fuel.

Flooding and the lack of power also caused wastewater treatment facilities to fail, fouling nearby waters. Approximately 11 billion gallons of wastewater—roughly one third of which was untreated—spilled

FIG. 1.3 Hurricane Sandy, observed from a NOAA satellite on October 28, 2012. (Credit: NASA Goddard Space Flight Center from Greenbelt, MD, USA, CC BY-SA 2.0, https://commons.wikimedia.org/wiki/File:NASA_Satellites_See_Sandy_Expand_as_Storm_Intensifies_(8131409082).jpg.)

into rivers, canals, and bays, mostly in New York and New Jersey [50]. Some of the spillage even bubbled up into city streets as sewers struggled to cope with the flow. Sandy racked up $2 billion worth of damage to wastewater treatment plants in New York alone, costing the state of New Jersey another $1 billion in damages and $1.7 billion to rebuild facilities more resiliently.

The lack of power hit the health sector particularly hard. Between the flooding and the power outage, hospitals across the city evacuated some 6,500 patients [51]. New York University's Langone Medical Center suffered large-scale power failures in the emergency room, the transplant unit, and other areas after flooding in the lower floors took out the hospital's primary and secondary backup generators [52,53]. Staff at NYU Langone evacuated 20 newborns in the neonatal intensive care unit, some of which required battery-powered respirators [54]. They also used hand-held flashlights to evacuate patients in the intensive care unit, located on the hospital's upper floors, down darkened stairways [55,56]. The power outage meant that auxiliary medical facilities lost power and could not operate—a quarter of dialysis patients in need of regular treatment missed sessions as out-patient treatment centers closed due to lack of power [57].

Although Sandy's floodwaters wreaked havoc on the electrical grid and the many infrastructure sectors dependent on it, they also posed a problem for individual infrastructure investments themselves. Before the storm's arrival, the National Hurricane Center warned New Yorkers to prepare for between 6 and 11 feet of storm surge [58]. What this estimate—and subsequent preparation efforts—failed to consider, however, was the more than 12 inches of sea-level rise New York City had experienced since 1900 due to climate change [59]. Moreover, this lack of flood preparation ran even deeper than the immediate preparations for Sandy—98% of all buildings destroyed by the storm were built before 1983, when the city adopted mandatory flood-proofing standards for buildings in the 100-year floodplain [60]. More recent projects suffered disastrous flooding, as well.

The South Ferry subway station, which opened in 2009, was not built to withstand a storm surge like the one Sandy brought, so when Sandy sent 15 million gallons of seawater rushing toward it, floodwaters quickly overcame the station's defenses [61]. New York has since rebuilt the South Ferry station more resiliently—including by installing 14-feet-high flood doors—but its "plywood-and-plastic-sheeting barricade" made no difference as water levels reached 80 feet in the station during the storm [62,63]. Sandy's unprecedented storm surge turned New York's subway system into a giant "fish tank" [64,65]. According to Joe Lhota, the MTA Chairman at the time, the devastation caused by Sandy exceeded anything the century-old subway system had ever faced [64,65].

Yet even where preparations had protected areas of New York, they could not necessarily guarantee that business would continue as planned. In advance of Sandy, one of the nation's premier investment banking firms, Goldman Sachs, stacked 25,000 sandbags around the base of its lower Manhattan headquarters [66]. It also installed generators, and when the surrounding buildings went dark, the tower shone brightly with virtually every floor lit up against the blackened New York City skyline. Then-chief operating officer of Goldman Sachs, Gary Cohn, assured the public that the building had survived the disaster [67]. There was just one problem: the streets below the skyscraper's lit windows were still flooded several feet deep, and its employees could not get to work. In a world of interconnected infrastructure, even islands of resilience like the Goldman Sachs headquarters could not perform as intended in the face of cascading infrastructure failures.

Climate change amplified the impact of Hurricane Sandy. Sea-level rise and warmer ocean temperatures fueled the storm's precipitation levels by as much as 35% [68]. Some researchers argue that warming temperatures and climbing sea levels, both resulting from climate change, added tens of billions of dollars to the storm's total damage. Stronger storms and higher floodwaters can disrupt a number of infrastructure investments and usher in a chain of events that causes a series of cascading failures. The interconnectedness that characterized much of the infrastructure affected by Sandy made for a delicate web of electricity, transportation, water, and public health systems each vulnerable to even one disruption in another sector. The ever-greater extremes fueled by climate change pose escalating risks to an already fragile infrastructure network where the failure of individual infrastructure investments—each built to withstand the extremes of yesterday, not the extremes of the future—jeopardizes not only the investment itself, but also all others that depend on it.

CLIMATE CHANGE RISKS TO INFRASTRUCTURE

When it comes to the choices we make with respect to infrastructure—where it is built, how it is built, and what it is connected to—climate change raises the stakes. Higher temperatures, more extreme heat, drought, and wildfire strain infrastructure by imposing harsher conditions than those in which infrastructure investments were designed to operate. If the choices we make fail to account for the future risk of climate change, then we lose the chance to reduce that risk and mitigate potential damage to infrastructure investments. We need to ensure that the built environment not only withstands the impacts of climate change already manifesting themselves, but also those materializing in the foreseeable future. The catastrophes we have experienced to date, including those described in this chapter, are but a prelude to the potential damage that climate change can inflict on our infrastructure. As the Third National Climate Assessment made clear, "Climate change once considered an issue for a distant future has moved firmly into the present" [69]. The impacts we have already experienced provide valuable illustrations of what is at stake.

Take the transportation sector, for example, where hotter temperatures can severely disrupt the operations and functioning of planes, trains, and automobiles. The United States has the largest rail network of any country in the world [70]. Railroad tracks crisscross the United States, allowing trains to carry billions of tons of freight and tens of millions of passengers each year [71]. However, when temperatures soar, the more than 200,000 miles of rail risk failing. High temperatures make steel tracks vulnerable to buckling, which can cause trains to derail—a problem known in the rail industry as a sun kink. Sun kinks have caused more than 2,100 train derailments in the United States since the mid-1970s [72].

With increasing temperatures, derailments have begun to occur more frequently. In 2012, then the hottest year on record in the continental United States, sun kinks caused so many derailments that the Federal Railroad Administration issued a special safety advisory warning operators to be on the look-out for them [73]. In 2017, California's Bay Area Rapid Transit slowed trains for fear of derailments caused by sun kinks [74]. Sun kinks also carry high costs as the amount of money needed to repair buckled sections of track, fix derailed cars, and replace damaged cargo mounts. Costs can reach upwards of $1 million per derailment in some cases, even before taking into account injuries, fatalities, and evacuations [72]. Researchers studying sun kinks predict that if we continue on our current path of emissions, delays from extreme heat and warming temperatures could cost the rail industry $45 to $60 billion cumulatively by 2100 [75].

Nor will climate change spare air transportation. The aviation industry and its customers can expect costly disruptions due to higher temperatures. In 2012, a plane stuck to the tarmac at Ronald Reagan National Airport in Washington, DC when the asphalt melted around its wheels [76]. In 2017, in Phoenix, Arizona, it got so hot that American Airlines canceled more than 40 flights in a single day [77]. In 2018, a 50-year-old runway at the Hannover Airport in Germany buckled as the thermometer bubbled to almost 100°F [78]. A 2017 study in the journal *Climatic Change* found that, based on an analysis of 19 major airports around the world, rising temperatures could mean that airlines will have to require weight reductions of 10% to 30% for flights departing at the hottest time of the day by 2050 [79]. The reductions would be small, but a reduction of even one half of 1% could mean that larger aircrafts like the Boeing 737, for example, would have to cut 722 pounds—or three passengers—from each flight [80].

In addition to the potential disruptions posed by rising temperatures, the Third National Climate Assessment, published in 2014, found that 13 of the United States' 47 largest airports had at least one runway within reach of a moderate to high storm surge [69]. Low-lying airports such as these are especially vulnerable to the coastal impacts of climate change which, by 2050, could cause a 1-in-500-year flood to become a 1-in-5-year flood in places like New York City [81]. In fact, New York's LaGuardia Airport is one of the most vulnerable airports in the United States to sea-level rise, coastal storms, and related flooding. During Hurricane Sandy, floodwaters over 10 feet high hit LaGuardia [51]. Had the storm struck the airport exactly at high tide, those floodwaters could have been at least 40% higher. With creeping sea-level rise from climate change, New York could have to deal with up to 2.5 feet more water washing over low-lying LaGuardia and other airports like it by the year 2050.

In 2012, a research team at Stanford University surveyed 160 members of the International Association of Ports and Harbors and the American Association of Port Authorities—the first worldwide study of port authorities that explicitly addressed climate change adaptation [82]. Ninety-three agencies representing seaports on every major continent responded. Most survey respondents ranked sea-level rise and heightened storm severity high on their list of concerns, and most ports who responded said that they were actively planning for a 1-in-100-year storm. However, only 6% of port authorities that responded said that they intend to build hurricane barriers in the next decade, fewer than 18%

had plans to embark on other protective structures on any time scale, and only 5% said that they were planning for anything stronger than a 1-in-100-year storm. Later research from the same team spelled out the possible implications of climate change for the world's ports: approximately 80% of the world's trade travels by sea, making seaports crucial in supporting the needs of ordinary people and a lynchpin in the global economy [83,84]. Were climate change to disrupt the functioning of ports, it would also disrupt the flow of goods and services around the globe.

Rising seas and stronger storms also threaten roads and bridges. In August 2011, Tropical Storm Irene led to widespread flooding in the US state of Vermont when as much as 11 inches of rain fell in just 2 days and forced 146 segments of the state highway system to close along with over 200 bridges [69]. Additionally, over 2,000 municipal roads and 200 miles of state-owned rail required some form of repair because of Irene. Climate scientists have found that Irene swept over oceanwaters anywhere from 1°F to 3.5°F warmer than average, which led to more extreme precipitation levels [85]. A report commissioned by the US Climate Change Science Program in 2008 found that over half of all major highways in the Gulf Coast region lie below 23 feet in elevation and could be vulnerable to flooding and damage from hurricane storm surges [86]. That same report found that the Gulf Coast is home to over 8,200 bridges serving those vulnerable interstates, meaning that they face many similar risks from climate change already bearing down on the region's highway system.

Similar to the transportation sector, water infrastructure faces a suite of risks from climate change. California had a particularly alarming preview of these impacts in 2017. Just as the state was pulling out of a severe drought, warmer-than-usual winter temperatures caused roughly 13 inches of rain to fall on the Feather River Basin between February 6 and 10 [87]. The increased water flows encountered rock-hard soil, parched from years of drought. The land simply could not absorb the high volumes of sudden runoff. The water found its way to the Oroville Dam—a dam built in 1968, well before the risks of climate change were fully understood and appreciated. The water level behind the dam swelled by 50 feet in a matter of days, prompting the dam's operators to evacuate nearby communities—approximately 200,000 people—as they opened both the primary and auxiliary spillways [88]. As water poured in, the main spillway collapsed and the auxiliary spillway began to erode. Debris from the erosion forced California's Department of Water Resources—the state agency responsible for the dam—to shut down its power plant and ramp up efforts to protect the auxiliary spillway with bags of rocks and sand in hopes of curbing the runoff. Fortunately, the dam held. However, the United States currently has over 87,000 dams, the overwhelming majority of which are more than 50 years old—well past their designated service life [89]. Virtually no US dam was designed to handle extreme precipitation or extreme drought. Like the Oroville Dam, those dams could collapse under the growing stress of climate change.

Yet, even if the dams hold, the drinking waters within them may be at risk. Wildfires, which can burn at over 2,000°F, often leave a scarred, barren landscape in their wake. Absent plant and animal life, and with soil baked into a hard, dense mass, wildland areas can become prone to severe flooding in the aftermath of a fire. Heavy rains can erode silt and sediment from the denuded landscape and deposit them into freshwater reservoirs and dams. Scientists with the US Geological Survey concluded in 2017 that sediment from erosion after wildfires could more than double by 2050 for roughly one third of all western watersheds in the United States [90]. More aggressive wildfires certainly pose a danger in and of themselves, but they can also pollute dams and reservoirs and disrupt the supply of clean, potable water.

Wastewater treatment, another aspect of water infrastructure, is already threatened by climate change. Urban planners typically locate wastewater treatment facilities at the lowest elevation possible. However, for many coastal cities, the wastewater plants sit just above

FIG. 1.4 Damaged Oroville Dam Spillways and debris field, 2017. (Credit: Dale Kolke / California Department of Water Resources, Wikimedia Commons, https://commons.wikimedia.org/wiki/File:Oroville_Dam_spillway_debris_in_Feather_River_27_February_2017.jpg.)

the water's edge. As sea levels rise, that infrastructure can come under increasing stress. A study commissioned by California's Bay Area Association of Clean Water Agencies in 2018 squarely addressed the risks to three dozen wastewater treatment plants in coastal areas of the state [91]. It found that 15 facilities were then vulnerable to regular flooding due to sea-level rise and that, if seas rise by 6 feet, all of the remaining plants would face similar vulnerabilities by century's end. A separate study from the same year came to similar conclusions, finding that over 30 million people in the United States could lose access to wastewater services given 6 feet of sea-level rise [92]. In 2018, Hurricane Florence dumped almost 36 inches of rain in North Carolina, leading to widespread flooding that caused wastewater treatment plants in towns such as Benson, Wilmington, and Georgetown to fail and discharge millions of gallons of contaminated water into nearby rivers and streams [93–95].

The electrical grid is also highly vulnerable to climate impacts. According to a 2011 report from the California Energy Commission, growing wildfire risk in California over the next three decades could increase the probability of power lines catching fire by as much as 40% under the most extreme warming scenario [96]. Power lines have already sparked 6 of the 18 most destructive California wildfires with known causes, while 2 more were under investigation during the time of writing [27]. However, as conditions worsen, they open up the state's infrastructure investments to a vicious cycle of growing wildfire risk and infrastructure disruption. Power lines, high-voltage transformers, power plants, and substations all lie in the path of rampaging wildfires growing in size and duration.

Electricity generating facilities, including nuclear power plants, are similarly at risk from climate change. Warming temperatures affect the ability of nuclear plants to cool their reactors. If ambient temperatures warm the waters used to cool plants, they may have to shut down. In 2012, a nuclear power facility in Connecticut shut down due to a problem its designers in the 1960s never imagined [97]. The water temperature in the Long Island Sound was too high to cool the reactors. According to the station's safety protocols, cooling water must never exceed 75°F and, until an unusually warm Sunday in August, it never had. However, when temperatures in the Long Island Sound rose to nearly 77°F, its operator, Dominion Power, was forced to shut down operations. Additionally, the likelihood of nuclear plants flooding increases along with sea-level rise. Stanford University researchers in 2013 analyzed the flood protections of 89 nuclear power stations located along the world's coastlines [98]. They flagged four as particularly vulnerable to flooding as seas rise beyond the height planned for in the design of their flood protections. Of the four plants the researchers identified, climate change has already disrupted the operations of one.

Whether in dry, arid regions, or along the coast, climate change threatens the world's infrastructure investments with more extreme heat, drought, and wildfires, as well as stronger storms, sea-level rise, and coastal flooding. Climate change packs a punch that our current infrastructure simply cannot take. Infrastructure interdependencies, moreover, can amplify disruptions caused by climate change, leading to cascading failures and creating a ripple effect. However, the potential for climate change to damage and destroy the world's interdependent infrastructure is not inevitable.

BUILDING RESILIENT INFRASTRUCTURE

The relationship between climate change and infrastructure is a double-edged sword. Climate change often poses a grave risk to infrastructure, threatening to disrupt its functioning and potentially even destroying infrastructure investments altogether. However, infrastructure can also serve to protect vulnerable populations and ordinary people from the increasingly extreme impacts of climate change when those impacts are actively, consciously considered. Building resilience in infrastructure investments and interdependent systems can increase protection against other types of hazards as well. Fortunately, there are numerous positive examples that can guide efforts to adapt infrastructure to a changing climate.

Today, the Dutch are world leaders in finding innovative ways to handle water, but it was not always that way. In 1953, in the middle of a January night, floodwaters overwhelmed scores of dikes and dams designed to protect the northwest coast of the Netherlands. To this day, people in the Netherlands commemorate the 1,800-plus lives lost in the *Watersnoodramp*, or "water emergency disaster," of 1953. In the wake of the storm, the Dutch have invested deeply in flood control. In a country where just about one quarter of the land lies below sea level, Dutch officials have passed standards requiring all new flood infrastructure be capable of withstanding water levels so severe they only have a probability of occurring once every 10,000 years [99]. They also created the Maeslantkering, an enormous floodgate outside The Hague. The Maeslantkering boasts 70-foot-tall storm surge barriers at the end of two mechanical arms each roughly the size of the Eiffel Tower that open and close depending on flood

FIG. 1.5 Thames flood barrier, London, UK. (Credit: shirokazan, CC BY-SA 2.0, https://www.flickr.com/photos/shirokazan/7766800984.)

conditions [100]. The Maeslantkering protects the port of Rotterdam, which serves tens of thousands of ships from around the world each year and provides more than 90,000 jobs to Dutch residents.

Dutch officials have also worked to find ways to accommodate floodwaters rather than struggle endlessly against them, creating outdoor parks that double as backup reservoirs and floating neighborhoods that rise with the seas. Green spaces in the Netherlands provide additional benefits in terms of curbing runoff and helping rainwater to infiltrate the ground. As the Dutch have prepared for sea-level rise and coastal flooding, so too have their efforts made them more prepared for extreme precipitation. The mayor of Rotterdam, Ahmed Aboutaleb put the Dutch approach to flooding this way in 2017: "We have no choice. We must learn to live with water ... That's just common sense."

Also affected by the 1953 North Sea Flood, London has built soaring infrastructure to cope with its flood risk. The Thames Barrier, completed in the 1980s, boasts 10 steel floodgates, each of which rises over five stories to protect more than 30,000 acres of Central London. Originally, English officials planned the Thames Barrier based on the annual rate of sea-level rise at the time, which in the 1970s was less than one tenth of one inch per year [101]. Since then, the annual rate of sea-level rise near London has doubled, and so has the number of instances the Thames Barrier has had to close [101,102]. However, the barrier was built to withstand an absolute maximum of around 0.3 inch of sea-level rise per year, as well as a 1-in-1,000-year storm surge [101]. In 2012, with these revised measurements of sea-level rise, the UK government set out a plan for the Thames Barrier through the year 2100. Their plan provides for almost 3 feet of sea-level rise by century's end [102]. Still, the British government has said that it expects the Thames Barrier to function properly through 2070 [101]. Whether or not the Thames Barrier can cope with the pace of climate change, future sea-level rise has been a major driver of planning and policy in London.

However, of course, flooding does not just affect the Netherlands and England. Ever since catastrophic floods struck the Malaysian capital of Kuala Lumpur in 1971, the government began developing ways to increase the city's ability to cope with flooding from heavy rains and the rivers that flow through the city. Over time, the government deployed a wide range of conventional flood management techniques, such as widening and deepening the channels through which rivers flowed. However, these proved increasingly inadequate the more urban and developed Kuala Lumpur became. Eventually, the city developed infrastructure to tackle the problems of intense flooding, namely the Stormwater Management and Road Tunnel, or SMART [103].

SMART stretches underground for seven miles and contains three separate sections: two roadways and a stormwater tunnel. Under normal conditions—or when rainfall is minimal enough that safe driving conditions can be maintained—SMART operates as a regular traffic tunnel, while the stormwater system beneath stays shut. In the case of moderate storms, the stormwater tunnel opens, diverting would-be floodwaters underneath the trafficways. During an impending flood, SMART shuts off access to traffic and all sections of the tunnel fill to the brim with diverted floodwaters—even the roadways. SMART also serves to reduce roadway congestion, which helps to reduce the likelihood of traffic-related disruptions such as accidents and delays. Malaysia has consciously planned for escalating risk and turned Kuala Lumpur's weakness—its vulnerability to flooding—into a strength. Malaysia's planning for climate change with SMART, moreover, carries added protection for motorists and supports the resilience of its roads.

Similarly, the Texas Medical Center (TMC) has taken aggressive action to mitigate its flood risk. When Tropical Storm Allison slammed into Texas in 2001, it was the deadliest and costliest tropical storm in US history [104]. Its heavy rains pummeled TMC, the world's largest medical complex, and caused over $2 billion in damage to the institution alone. The over 5-foot-high floodwaters inundated underground parking garages and tunnels, as well as critical infrastructure below ground [105]. In the Baylor College of Medicine, located on TMC's campus, flooding destroyed 25 years of research data stored on servers in lower levels. To avoid future catastrophes,

TMC took drastic steps to mitigate its risk. It widened culverts, installed 50 watertight flood doors, built four stormwater tanks that hold three and a half billion gallons of water each, elevated electrical equipment, lab animals, and research experiments to higher floors, and so on [106]. When Hurricane Harvey struck 16 years later, surpassing the level of rain Tropical Storm Allison dumped on the area, every hospital except one on TMC's campus was fully operational [107]. These resilience measures also help to guard against other threats. Elevating equipment and experiments, for example, can lessen the chances of their being damaged and destroyed by nonflood-related disasters or tampering by unauthorized personnel. In the case of TMC, action taken to mitigate the risk of future flooding paid off, saving time, money, and lives.

In addition, plagued by urban flooding, the Chinese government initiated in 2015 what it calls its Sponge City Initiative. Currently, only about 20% to 30% of rainwater infiltrates the ground in China's urban areas, disrupting the natural water cycle and making surface flooding more likely [108]. The basic premise of the sponge city program is simple: work with nature to reduce the risk of urban flooding. China plans to transform 30 cities across the country into makeshift sponges to absorb or reuse rainwater and reduce the risk of flooding. Plans are tailored to the different needs of each city, and sponge cities like Nanhui, a city in Shanghai entirely planned around this concept, are applying resilience principles to their stormwater management and flood control [108]. Since the Sponge City Initiative began, Shanghai has invested $119 million in resilience projects such as permeable surfaces, raised walkways, and rooftop gardens, as well as wetland preservation and natural ecosystem restoration to help the city to soak up floodwaters during severe storms [109].

By 2020, China hopes its sponge cities will soak up at least 70% of all rainwater that lands in them. Moreover, restoring ecosystems and increasing the amount of green space across China's urban landscape can combat nonclimate change-related hazards, such as erosion and land subsidence, and help to absorb CO_2 gas in heavily polluted areas. A comprehensive survey of China's sponge cities found that obstacles remain, including planning models not specific enough to cities' local needs, scarce eco-friendly resources, and "ambitious goals without sound research basis" [110]. Researchers remain hopeful about the success of China's Sponge City Initiative but only with better coordination, research, and funding.

In Washington, DC, risks to water treatment facilities have prompted city officials to begin planning for climate change. Across the nation's capital, most wastewater drains to the Blue Plains Advanced Wastewater Treatment Plant which, like most wastewater treatment plants, was built at the lowest point in the city [111]. This conventional siting of Blue Plains helps to save energy costs by having water flow downhill, rather than pumping it uphill. However, it also makes the facility particularly vulnerable to sea-level rise. Sea levels around the mid-Atlantic rose by over 10 inches between around the 1930s and the 2010s, and nuisance flooding has increased by 373% in Washington, DC since the 1950s [112]. To understand what this would mean for Washington's infrastructure, the city constructed a series of different models of future flood conditions given predicted sea-level rise and decided to move proactively on climate change [111]. DC Water ultimately approved plans to build a $13.2 million seawall around the facility [113]. At 17 feet high, the wall will protect Blue Plains—and the millions of people it supports—from a 1-in-500-year flood [113].

Two years after Hurricane Sandy, Consolidated Edison, New York's largest electric utility and the company whose electrical substation blow-out plunged New York City into darkness, agreed to undertake an immediate, comprehensive risk assessment of the threat that climate change poses to its power supply [114]. It further agreed to integrate climate science into every aspect of its long-term planning, construction, and budgeting. These agreements were made as part of a settlement between the New York Public Service Commission (NYPSC) and Con Edison. In return, NYPSC permitted Con Edison to spend $1 billion for storm hardening, a win-win for climate resilience [115]. In the settlement order, the NYPSC offered a warning to other New York utilities as well. "We expect the utilities to consult the most current data to evaluate the climate impacts anticipated in their regions over the next years and decades," the settlement order reads, "and to integrate these considerations into their system planning and construction forecasts and budgets" [116].

Similar examples of infrastructure resilience exist throughout the world. An airport in New Zealand has extended one of its runways to accommodate the effect of higher temperatures on takeoffs and landings. Philadelphia is building rain gardens, green roofs, wetlands, and other green infrastructure investments to prevent stormwater from overwhelming its sewer system. Florida is elevating roads to cope with sea-level rise. Hospitals in California are installing air filters to ensure they have a fresh supply of clean air in the event of a wildfire. The Los Angeles—San Diego—San Luis Obispo rail corridor in California, the second busiest intercity rail corridor in the United States, plans to build bridges and other infrastructure so that they can be raised as sea levels climb. These examples can guide future efforts to prepare for climate change and encourage us to think of resilience as a way to safeguard against a panoply of risks, some of which may not stem from climate change at all. Although we must rise to meet the monumental challenge of climate change, many places around the world have already begun to embark on a path toward resilience and can help to show us the way.

CONCLUSION

In an ideal world, infrastructure owners and operators, as well as the communities that depend upon them, would actively plan for the future that climate change threatens. They would incorporate climate projections into their operational, maintenance, and construction plans. They would model how a failure in one sector could cause cascading failures elsewhere and seek to address risks and vulnerabilities accordingly. They would scrutinize the implications of climate change for individual infrastructure investments by conducting thorough risk assessments, designing and constructing facilities stronger, building redundancies into the system, and investing deeply in resilience. Experts would harness the best available science to forecast a range of possible futures and realize the need to prepare for the most extreme scenarios. All of this would aid in reducing vulnerabilities and doubling down on defenses. Perhaps most importantly, it would also help to recognize climate change for what it is—a major strain on our interdependent infrastructure. It is time that the world moves closer to that ideal. We need to treat the threats posed by interdependency and climate change with the urgency they deserve. An aggressive strategy to strengthen infrastructure resilience today can help to foster a future safer from the extreme consequences of climate change.

REFERENCES

[1] S.M. Rinaldi, J.P. Peerenboom, T.K. Kelly, Identifying, understanding, and analyzing critical infrastructure interdependencies, IEEE Control Systems Magazine 21 (6) (2001) 11–25.
[2] FERC, NAERC, Arizona-southern California Outages on September 8, 2011: Causes and Recommendations, Federal Energy Regulatory Commission, Washington, 2012.
[3] A. Burgin, K. Davis, R. Hawkins, K. Kucher, J. Lebron Kuhney, L. Leung, H. Persinger, P. Repard, G. Robbins, S. Shroder, N. Scharn, Unprecedented outage left millions in the dark, San Diego Union-Tribune (September 8, 2011). Available at: http://www.sandiegouniontribune.com/news/sdut-widespread-power-outages-across-san-diego-county-2011sep08-htmlstory.html.
[4] NUSIPR, Economic Impact of September 9th Power Outage: Conservatively Estimated at $97 to $118 Million, National University System Institute for Policy Research, San Diego, CA, 2011.
[5] M. Richtel, Egypt cuts off most internet and cell service, New York Times (January 29, 2011). Available at: https://www.nytimes.com/2011/01/29/technology/internet/29cutoff.html.
[6] D.M. West, Internet Shutdowns Cost Countries $2.4 Billion Last Year, Brookings Institution, Washington, 2016.
[7] T. Reynolds, A. Mickoleit, The Economic Impact of Shutting Down Internet and Mobile Phone Services in Egypt, Organisation for Economic Co-operation and Development, Paris, 2011. Available at: http://www.oecd.org/countries/egypt/theeconomicimpactofshuttingdowninternetandmobilephoneservicesinegypt.htm.
[8] T. Houser, P. Marsters, The World's Second Largest Blackout, Rhodium Group, New York, 2018. Available at: https://rhg.com/research/puerto-rico-hurricane-maria-worlds-second-largest-blackout/.
[9] A.R. Hernández, W. Leaming, A. Murphy, Sin luz: life without power, Washington Post (December 14, 2017). Available at: https://www.washingtonpost.com/graphics/2017/national/puerto-rico-life-without-power/?utm_term=.ee9da3679173.
[10] M. Weissenstein, D. Coto, Turmoil slows rebuilding of Puerto Rico's power grid, AP News (July 19, 2018). Available at: https://www.apnews.com/11893d2477e04e7f9b9c8befbea295ad/Turmoil-slows-rebuilding-of-Puerto-Rico's-power-grid.
[11] US Department of Energy, Comparing the Impacts of Northeast Hurricanes on Energy Infrastructure, US Department of Energy, Washington, 2013. Available at: https://www.energy.gov/sites/prod/files/2013/04/f0/Northeast%20Storm%20Comparison_FINAL_041513b.pdf.
[12] Global Infrastructure Hub, Global Infrastructure Outlook, Global Infrastructure Hub, Sydney, 2017.
[13] P. Tans, R. Keeling, Trends in Atmospheric Carbon Dioxide, 2013. Available at: https://www.esrl.noaa.gov/gmd/ccgg/trends/data.html.
[14] IPCC, in: Core Writing Team, R.K. Pachauri, L.A. Meyer (Eds.), Climate Change 2014: Synthesis Report. Contribution of Working Groups I, II and III to the Fifth Assessment Report of the Intergovernmental Panel on Climate Change, IPCC, Geneva, Switzerland, 2014.
[14a] D. Rice, Earth's carbon dioxide levels are highest they've been in millions of years, USA Today, 2019. [online] Available at: https://www.usatoday.com/story/news/nation/2019/06/04/carbon-dioxide-levels-hawaiis-mauna-loa-soar-record-peak/1337163001/ [Accessed 6 Jul. 2019].
[14b] D. Spratt, I. Dunlop, What lies beneath: the scientific understatement of climate risks, Breakthrough National Centre for Climate Restoration, Melbourne, 2018.
[15] NOAA National Centers for Environmental Information, State of the Climate: Global Climate Report for Annual 2017, National Oceanic and Atmospheric Administration, Silver Spring, MD, 2018.

Available at: https://tidesandcurrents.noaa.gov/waterlevels.html?id=8518750&units=standard&bdate=20121028&edate=20121031&timezone=GMT&datum=MLLW&interval=6&action=data.

[16] R.S. Nerem, B.D. Beckley, J.T. Fasullo, B.D. Hamlington, D. Masters, G.T. Mitchum, Climate-change–driven accelerated sea-level rise detected in the altimeter era, Proceedings of the National Academy of Sciences of the United States of America 115 (9) (2018) 2022–2025.

[17] A.H. Sallenger Jr., K.S. Doran, P.A. Howd, Hotspot of accelerated sea-level rise on the Atlantic coast of North America, Nature Climate Change 2 (2012) 884–888.

[18] W.V. Sweet, R.E. Kopp, C.P. Weaver, J. Obeysekera, R.M. Horton, E.R. Thieler, C. Zervas, Global and Regional Sea Level Rise Scenarios for the United States, NOAA Technical Report NOS CO-OPS 083, NOAA/NOS Center for Operational Oceanographic Products and Services, Silver Spring, 2017. Available at: https://pubs.giss.nasa.gov/abs/sw01000b.html.

[19] B. Kirtman, S.B. Power, J.A. Adedoyin, G.J. Boer, R. Bojariu, I. Camilloni, F.J. Doblas-Reyes, A.M. Fiore, M. Kimoto, G.A. Meehl, M. Prather, A. Sarr, C. Schär, R. Sutton, G.J. van Oldenborgh, G. Vecchi, H.J. Wang, Near-term climate change: projections and predictability, in: T.F. Stocker, D. Qin, G.-K. Plattner, M. Tignor, S.K. Allen, J. Boschung, A. Nauels, Y. Xia, V. Bex, P.M. Midgley (Eds.), Climate Change 2013: The Physical Science Basis. Contribution of Working Group I to the Fifth Assessment Report of the Intergovernmental Panel on Climate Change, Cambridge University Press, Cambridge, United Kingdom and New York, NY, USA, 2013.

[20] G.A. Meehl, C. Tebaldi, G. Walton, D. Easterling, L. McDaniel, Relative increase of record high maximum temperatures compared to record low minimum temperatures in the U.S, Geophysical Research Letters 36 (23) (2009) L23701.

[21] N.C. Lau, M.J. Nath, A model study of heat waves over North America: meteorological aspects and projections for the twenty-first century, Journal of Climate 25 (2012) 4761–4784.

[22] D. Griffin, K. Anchukaitis, How unusual is the 2012–2014 California drought? Geophysical Research Letters 41 (24) (2014) 9017–9023.

[23] A. Park Williams, R. Seager, J.T. Abatzoglou, B.I. Cook, J.E. Smerdon, E.R. Cook, Contribution of anthropogenic warming to California drought during 2012–2014, Geophysical Research Letters 42 (16) (2015) 6819–6828.

[24] A.L.R. Westerling, Increasing western US forest wildfire activity: sensitivity to changes in the timing of spring, Phil. Trans. R. Soc. B 371 (2016), 20150178.

[25] W.M. Jolly, M.A. Cochrane, P.H. Freeborn, Z.A. Holden, T.J. Brown, G.J. Williamson, D.M.J.S. Bowman, Climate-induced variations in global wildfire danger from 1979 to 2013, Nature Communications 6 (2015), 7537.

[26] W. Cramer, G.W. Yohe, M. Auffhammer, C. Huggel, U. Molau, M.A.F. da Silva Dias, A. Solow, D.A. Stone, L. Tibig, Detection and attribution of observed impacts, in: C.B. Field, V.R. Barros, D.J. Dokken, K.J. Mach, M.D. Mastrandrea, T.E. Bilir, M. Chatterjee, K.L. Ebi, Y.O. Estrada, R.C. Genova, B. Girma, E.S. Kissel, A.N. Levy, S. MacCracken, P.R. Mastrandrea, L.L. White (Eds.), Climate Change 2014: Impacts, Adaptation, and Vulnerability. Part A: Global and Sectoral Aspects. Contribution of Working Group II to the Fifth Assessment Report of the Intergovernmental Panel on Climate Change, Cambridge University Press, Cambridge, United Kingdom and New York, NY, USA, 2014, pp. 979–1037.

[27] California Department of Forestry and Fire Protection, Top 20 Most Destructive California Wildfires, California Department of Forestry and Fire Protection, Sacramento, 2018.

[28] California Department of Forestry and Fire Protection, 2017 Fire Statistics, California Department of Forestry and Fire Protection, Sacramento, 2018.

[29] California Department of Forestry and Fire Protection, 2018 Fire Statistics, California Department of Forestry and Fire Protection, Sacramento, 2018b.

[30] BBC, California Wildfires: Number of Missing Falls to 25, BBC News, December 2, 2018. Available at: https://www.bbc.co.uk/news/world-us-canada-46421177.

[31] J.T. Abatzoglou, A. Park Williams, Impact of anthropogenic climate change on wildfire across western US forests, Proceedings of the National Academy of Sciences of the United States of America 113 (42) (2016) 11770–11775.

[32] J. Walsh, D. Wuebbles, K. Hayhoe, J. Kossin, K. Kunkel, G. Stephens, P. Thorne, R. Vose, M. Wehner, J. Willis, D. Anderson, S. Doney, R. Feely, P. Hennon, V. Kharin, T. Knutson, F. Landerer, T. Lenton, J. Kennedy, R. Somerville, Ch. 2: our changing climate, in: J.M. Melillo, T.(T.C.) Richmond, G.W. Yohe (Eds.), Climate Change Impacts in the United States: The Third National Climate Assessment, US Global Change Research Program, 2014.

[33] M.D. Risser, M.F. Wehner, Attributable human-induced changes in the likelihood and magnitude of the observed extreme precipitation during Hurricane Harvey, Geophysical Research Letters 44 (24) (2017) 12457–12464.

[34] G.J. van Oldenborgh, K. van der Wiel, A. Sebastian, R. Singh, J. Arrighi, F. Otto, K. Haustein, S. Li, G. Vecchi, H. Cullen, Attribution of extreme rainfall from hurricane Harvey, August 2017, Environmental Research Letters 12 (2017), 124009.

[35] S.-Y.S. Wang, L. Zhao, J.-H. Yoon, P. Klotzbach, R.R. Gillies, Quantitative attribution of climate effects on Hurricane Harvey's extreme rainfall in Texas, Environmental Research Letters 13 (2018) 054014.

[36] S. Pfahl, P.A. O'Gorman, E.M. Fischer, Understanding the regional pattern of projected future changes in extreme precipitation, Nature Climate Change 7 (2017) 423–427.

[37] S. Hallegatte, C. Green, R. Nicholls, J. Corfee-Morlot, Future flood losses in major coastal cities, Nature Climate Change 3 (2013) 802–806.

[38] IPCC, Global warming of 1.5°C. An IPCC Special Report on the impacts of global warming of 1.5°C above pre-industrial levels and related global greenhouse gas emission pathways, in: V. Masson-Delmotte, P. Zhai, H.O. Pörtner, D. Roberts, J. Skea, P.R. Shukla, A. Pirani, W. Moufouma-Okia, C. Péan, R. Pidcock, S. Connors, J.B.R. Matthews, Y. Chen, X. Zhou, M.I. Gomis, E. Lonnoy, T. Maycock, M. Tignor, T. Waterfield (Eds.), The Context of Strengthening the Global Response to the Threat of Climate Change, Sustainable Development, and Efforts to Eradicate Poverty, World Meteorological Organization, Geneva, Switzerland, 2018.

[39] ASCE, 2017 Infrastructure Report Card: A Comprehensive Assessment of America's Infrastructure, American Society of Civil Engineers, Reston, VA, 2017. Available at: https://www.infrastructurereportcard.org/.

[40] City of New York, Mayor Bloomberg Issues Order for Mandatory Evacuation of Low-Lying Areas as Hurricane Sandy Approaches, 2012. Available at: https://www1.nyc.gov/office-of-the-mayor/news/377-12/mayor-bloomberg-issues-order-mandatory-evacuation-low-lying-areas-hurricane-sandy.

[41] Governor of New York, Governor Cuomo Declares State of Emergency in New York in Preparation for Potential Impact of Hurricane Sandy, 2012. Available at: https://www.governor.ny.gov/news/governor-cuomo-declares-state-emergency-new-york-preparation-potential-impact-hurricane-sandy.

[42] NOAA, Water Levels: 8518750 the Battery, NY, National Oceanic and Atmospheric Administration, Silver Spring, MD, 2018. Available at: https://data.worldbank.org/indicator/IS.RRS.TOTL.KM.

[43] S. Gandel, How Con Ed turned New York City's lights back on, Fortune (November 12, 2012). Available at: http://fortune.com/2012/11/12/how-con-ed-turned-new-york-citys-lights-back-on/.

[44] A. Plitt, The Night the Lights Went Out in Manhattan, Curbed, (October 29, 2017). Available at: https://ny.curbed.com/2017/10/29/16560706/hurricane-sandy-anniversary-manhattan-power-outage-photos.

[45] D. Sandalow, Hurricane Sandy and our energy infrastructure, in: Speech, Columbia University Energy Symposium, 30 November, 2012. Available at: https://www.energy.gov/articles/hurricane-sandy-and-our-energy-infrastructure.

[46] M. Flegenheimer, Flooded tunnels may keep city's subway network closed for several days, New York Times (October 30, 2012). Available at: https://www.nytimes.com/2012/10/31/nyregion/subways-may-be-shut-for-several-days-after-hurricane-sandy.html.

[47] Kaufman, S., Qing, C., Levenson, N., and Hanson, M. Transportation during and after Hurricane Sandy. New York: Rudin Center for Transportation, NYU Wagner Graduate School of Public Service. Available at: https://wagner.nyu.edu/files/rudincenter/sandytransportation.pdf [Accessed 7 Sep. 2018].

[48] R. Kolker, How did the MTA restore subway service in time for Monday's rush hour? New York Magazine (November 5, 2012). Available at: http://nymag.com/daily/intelligencer/2012/11/how-did-the-mta-restore-subway-service.html.

[49] K. Zerinke, Gasoline runs short, adding woes to storm recovery, New York Times (November 1, 2012). Available at: https://www.nytimes.com/2012/11/02/nyregion/gasoline-shortages-disrupting-recovery-from-hurricane.html.

[50] A. Kenward, D. Yawitz, U. Raja, Sewage Overflows from Hurricane Sandy, Princeton: Climate Central, 2013. Available at: http://www.climatecentral.org/news/11-billion-gallons-of-sewage-overflow-from-hurricane-sandy-15924.

[51] City of New York, A Stronger, More Resilient New York, City of New York, New York, 2013. Available at: https://www1.nyc.gov/site/sirr/report/report.page.

[52] J. Goodman, C. Moynihan, Patients evacuated from city medical center after power failure, New York Times (October 30, 2012). Available at: https://www.nytimes.com/2012/10/30/nyregion/patients-evacuated-from-nyu-langone-after-power-failure.html.

[53] CBS/AP, What caused generators to fail at NYC hospitals? CBS News (November 2, 2012). Available at: https://www.cbsnews.com/news/what-caused-generators-to-fail-at-nyc-hospitals/.

[54] R. Mulford, M. Dorian, A. Sechrist, Nurses who saved NICU babies remember harrowing, triumphant hurricane night, ABC News (November 3, 2012). Available at: https://abcnews.go.com/Health/nicu-nurses-saved-babies-remember-harrowing-triumphant-hurricane/story?id=176
32993.

[55] L. Italie, M. Marchione, Sandy prompts harrowing NYC hospital evacuation, The San Diego Tribune (October 30, 2012). Available at: http://www.sandiegouniontribune.com/news/sdut-sandy-prompts-harrowing-nyc-hospital-evacuation-2012oct30-story.html.

[56] E. Peralta, At New York University medical center, a dramatic, critical evacuation [Blog], The Two-Way (October 30, 2012). Available at: https://www.npr.org/sections/thetwo-way/2012/10/30/163961325/at-new-york-university-medical-center-a-dramatic-critical-evacuation.

[57] N. Murakami, H.B. Siktel, D. Lucido, J.F. Winchester, N.B. Harbord, Disaster preparedness and awareness of patients on hemodialysis after Hurricane Sandy, Clinical Journal of the American Society of Nephrology 10 (8) (2015) 1389–1396.

[58] National Hurricane Center, Hurricane Sandy Advisory Number 29, National Oceanic and Atmospheric Administration, Silver Spring, MD, 2012. Available at: https://www.nhc.noaa.gov/archive/2012/al18/al182012.public.029.shtml.

[59] New York State Sea Level Rise Task Force, Report to the Legislature, New York State Department of Environmental Conservation, New York, 2011. Available at: https://www.dec.ny.gov/energy/45202.html.

[60] NYC Planning, Hurricane Sandy: Initial Lessons for Buildings, Department of City Planning, City of New York, New York, NY, 2012. Available at: https://www1.nyc.gov/assets/planning/download/pdf/plans-studies/climate-resilience/presentation_sandy.pdf.

[61] MTA, Fix&Fortify: Rebuilt South Ferry Station Reopens after Superstorm Sandy Destruction, New York Metropolitan Transportation Authority, New York, NY, 2017. Available at: http://www.mta.info/press-release/nyc-transit/fixfortify-rebuilt-south-ferry-station-reopens-after-superstorm-sandy.

[62] D. Rivoli, MTA reinforces South Ferry station with steel doors to shield from Sandy-like storms, New York Daily News (October 27, 2017). Available at: http://www.nydailynews.com/new-york/south-ferry-new-steel-doors-shield-sandy-like-storms-article-1.3594216.

[63] R. Sullivan, Could New York City subways survive another hurricane? New York Times (October 23, 2013). Available at: https://www.nytimes.com/2013/10/27/magazine/could-new-york-city-subways-survive-another-hurricane.html.

[64] E. Deprez, M. Braun, NYC Subway-Station-Turned-Fish-Tank Poses $600 Million Dilemma, Bloomberg, (December 6, 2012). Available at: https://www.bloomberg.com/news/articles/2012-12-06/nyc-subway-station-turned-fish-tank-poses-600-million-dilemma.

[65] E. Honan, M. Geller, E. Beech, P. Simao, Sandy Leaves Unprecedented Challenges for New York City Subways, Reuters (October 30, 2012). Available at: https://www.reuters.com/article/us-storm-sandy-subway/sandy-leaves-unprecedented-challenges-for-new-york-city-subways-idUSBRE89T0SU20121030.

[66] F. Salmon, How Goldman Sachs Protects Itself from a Hundred-Year Storm, Reuters (October 29, 2012). Available at: https://www.reuters.com/article/idUS258330703720121029.

[67] K. Roose, Goldman Sachs was not washed away in Hurricane Sandy [updated], New York Magazine (October 30, 2012). Available at: http://nymag.com/daily/intelligencer/2012/10/goldman-sachs-survived-hurricane-sandy.html.

[68] K. Trenberth, J. Fasullo, T. Shepherd, Attribution of climate extreme events, Nature Climate Change 5 (2015) 725–730.

[69] J.M. Melillo, T.(T.C.) Richmond, G.W. Yohe (Eds.), Climate Change Impacts in the United States: The Third National Climate Assessment, US Global Change Research Program, 2014.

[70] World Bank, World Development Indicators: Rail Lines (Total Route-Km), The World Bank Group, Washington, DC, 2018. Available at: https://data.worldbank.org/indicator/IS.RRS.TOTL.KM.

[71] Bureau of Transportation Statistics, Seasonally Adjusted Transportation Data, US Department of Transportation, Washington, DC, 2018. Available at: https://www.bts.gov/topics/rail.

[72] Federal Railroad Administration, 3 – Train Accidents, US Department of Transportation, Washington, DC, 2018. Available at: https://safetydata.fra.dot.gov/officeofsafety/default.aspx.

[73] Federal Register, Federal Railroad Administration, Federal Register vol. 77 no. 136, 41881-41882, US Department of Transportation, Washington, DC, 2012. Available at: https://www.gpo.gov/fdsys/pkg/FR-2012-07-16/pdf/2012-17343.pdf.

[74] KPIX 5, BART, Caltrain Slowed Down during Heat Wave over Track Concerns, CBS San Francisco (September 1, 2017). Available at: https://sanfrancisco.cbslocal.com/2017/09/01/bart-trains-heat-wave-track-concerns/.

[75] P. Chinowsky, J. Helman, S. Gulati, J. Neumann, J. Martinich, Impacts of climate change on operation of the US rail network, Transport Policy 75 (2019) 183–191.

[76] C. Simpson, Your plane was delayed because the tarmac melted, The Atlantic (July 2012). Available at: https://www.theatlantic.com/national/archive/2012/07/your-plane-was-delayed-because-tarmac-melted/326183/.

[77] A.B. Wang, It's so hot in Phoenix that airplanes can't fly, Washington Post (June 20, 2017). Available at: https://www.washingtonpost.com/news/capital-weather-gang/wp/2017/06/20/its-so-hot-in-phoenix-that-airplanes-cant-fly/.

[78] AP News, Extreme heat damages runway, forces German airport to close, AP News (July 24, 2018). Available at: https://apnews.com/ae71d118e2314cba808c3cac0fdd8f15.

[79] E. Coffel, T. Thompson, R. Horton, The impacts of rising temperatures on aircraft takeoff performance, Climatic Change 144 (2) (2017) 381–388.

[80] M. Ives, In a warming world, keeping the planes running, New York Times (September 30, 2017). Available at: https://www.nytimes.com/2017/09/30/business/airports-climate-change-global-warming.html.

[81] A. Garner, M. Mann, K. Emanuel, R. Kopp, N. Lin, R. Alley, B. Horton, R. DeConto, J. Donnelly, D. Pollard, Impact of climate change on New York City's coastal flood hazard: increasing flood heights from the preindustrial to 2300 CE, Proceedings of the National Academy of Sciences of the United States of America 114 (45) (2017) 11861–11866.

[82] A.H. Becker, S. Inoue, M. Fischer, B. Schwegler, Climate change impacts on international seaports: knowledge, perceptions, and planning efforts among port administrators, Climatic Change 110 (1–2) (2012) 5–29.

[83] A.H. Becker, M. Acciaro, R. Asariotis, E. Cabrera, L. Cretegny, P. Crist, M. Esteban, A. Mather, S. Messner, S. Naruse, A.K.Y. Ng, S. Rahmstorf, M. Savonis, D.-W. Song, V. Stenek, A.F. Velegrakis, A note on climate change adaptation for seaports: a challenge for global

ports, a challenge for global society, Climatic Change 120 (4) (2013) 683–695.

[84] A.H. Becker, A.K.Y. Ng, D. McEvoy, J. Mullett, Implications of climate change for shipping: ports and supply chains, WIREs Climate Change 9 (2) (2018) e508.

[85] C. Wilson, W. Abdullah, R.S. Reddy, W. Walters, Ocean-atmospheric interactions, heavy precipitation, and hurricane predictive index (HPI) associated with land-falling Hurricane Irene over the eastern coast of the United States, in: 2012 IEEE International Geoscience and Remote Sensing Symposium, Munich, 2012, pp. 1944–1947.

[86] R.S. Kafalenos, K.J. Leonard, D.M. Beagan, V.R. Burkett, B.D. Keim, A. Meyers, D.T. Hunt, R.C. Hyman, M.K. Maynard, B. Fritsche, R.H. Henk, E.J. Seymour, L.E. Olson, J.R. Potter, M.J. Savonis, What are the implications of climate change and variability for Gulf Coast transportation? in: M.J. Savonis, V.R. Burkett, J.R. Potter (Eds.), Impacts of Climate Change and Variability on Transportation Systems and Infrastructure: Gulf Coast Study, Phase I. A Report by the U.S. Climate Change Science Program and the Subcommittee on Global Change Research US Department of Transportation, Washington, DC, 2008.

[87] California Data Exchange Center, Daily 24-hour Rainfall Totals for Selected Automatic-Reporting Rain Gages, California Department of Water Resources, Sacramento, 2018. Available at: https://cdec.water.ca.gov/cgi-progs/prevprecip/DLYPCP.

[88] D.A. Graham, How did the Oroville Dam crisis get so dire? The Atlantic (February 2017). Available at: https://www.theatlantic.com/national/archive/2017/02/how-did-the-oroville-dam-get-so-bad/516429/.

[89] J. Rowland, Aging Dams and Clogged Rivers: An Infrastructure Plan for America's Waterways, Center for American Progress, Washington, 2016.

[90] J.B. Sankey, J. Kreitler, T.J. Hawbaker, J.L. McVay, M.E. Miller, E.R. Mueller, N.M. Vaillant, S.E. Lowe, T.T. Sankey, Climate, wildfire, and erosion ensemble foretells more sediment in western USA watersheds, Geophysical Research Letters 44 (17) (2017) 8884–8892.

[91] L. Mesbeh, Bay Area Clean Water Agencies, Nutrient Optimization and Upgrade Project: Sea Level Rise Assessment, Bay Area Clean Water Agencies, Oakland, 2018. Available at: https://bacwa.org/wp-content/uploads/2018/02/February-16-2018-BACWA-Executive-Board-Meeting-Packet.pdf.

[92] M.A. Hummel, M.S. Berry, M.T. Stacey, Sea level rise impacts on wastewater treatment systems along the U.S. coasts, Earth's Future 6 (4) (2018) 622–633.

[93] Associated Press, The latest: waste treatment plant disabled due to flooding, Washington Post (September 26, 2018). Available at: https://www.washingtonpost.com/national/energy-environment/the-latest-waste-treatment-plant-disabled-due-to-flooding/2018/09/26/1a020a44-c1ee-11e8-9451-e878f96be19b_story.html?noredirect=on&utm_term=.9cffed5b9f87.

[94] J. Murawski, 5.25 million gallons of waste water spills into Cape Fear River after power outage, The News & Observer (September 15, 2018). Available at: https://www.newsobserver.com/news/local/article218476205.html.

[95] J. Murawski, More hog lagoons overflowing; 300,000 gallons of raw sewage spills in Johnston County, The News & Observer (September 19, 2018). Available at: https://www.newsobserver.com/news/local/article218672475.html.

[96] J. Sathaye, L. Dale, P. Larsen, G. Fitts, K. Koy, S. Lewis, A. Lucena, Estimating Risk to California Energy Infrastructure from Projected Climate Change, California Energy Commission publication no. CEC-500-2011-XXX, California Energy Commission, Berkeley, 2011.

[97] M.L. Wald, Heat shuts down a coastal reactor, New York Times (August 13, 2012). Available at: https://green.blogs.nytimes.com/2012/08/13/heat-shuts-down-a-coastal-reactor/.

[98] P.Y. Lipscy, K.E. Kushida, T. Incerti, The Fukushima disaster and Japan's nuclear plant vulnerability in comparative perspective, Environmental Science & Technology 47 (12) (2013) 6082–6088.

[99] A. Higgins, Lessons for U.S. from a flood-prone land, New York Times (November 15, 2012). Available at: https://www.nytimes.com/2012/11/15/world/europe/netherlands-sets-model-of-flood-prevention.html.

[100] M. Kimmelman, The Dutch have solutions to rising seas. The world is watching, New York Times (June 15, 2017). Available at: https://www.nytimes.com/interactive/2017/06/15/world/europe/climate-change-rotterdam.html.

[101] S. Connor, Sea levels rising too fast for Thames Barrier, The Independent (March 22, 2008). Available at: https://www.independent.co.uk/environment/climate-change/sea-levels-rising-too-fast-for-thames-barrier-799303.html.

[102] UK Environment Agency, TE2100 Plan: Managing Flood Risk through London and the Thames Estuary, UK Environment Agency, London, 2012. Available at: https://www.gov.uk/government/publications/thames-estuary-2100-te2100.

[103] A. Darby, A dual-purpose tunnel, Ingenia (March 30, 2007). Available at: https://www.ingenia.org.uk/Ingenia/Articles/c6bb95ed-3a11-42a1-87d5-baf1508919aa.

[104] P.E. Sirbaugh, R.N. Bradley, C.G. Macias, E.E. Endom, The Houston flood of 2001: the Texas Medical Center and lessons learned, Clinical Pediatric Emergency Medicine 3 (4) (2002) 275–283.

[105] RMS, Tropical Storm Allison, June 2001: RMS event report, Risk Management Solutions, Inc, Newark, CA, 2001.

[106] Z. Fang, G. Dolan, A. Sebastian, P.B. Bedient, Case study of flood mitigation and hazard management at the Texas Medical Center in the wake of Tropical Storm Allison in 2001, Natural Hazards Review 15 (3) (2014), 05014001.

[107] R.A. Phillips, R.L. Schwartz, W.F. McKeon, M.L. Boom, Lessons in leadership: how the world's largest medical center braced for Hurricane Harvey [Blog], NEJM Catalyst (2017). Available at: https://catalyst.nejm.org/lessons-leadership-texas-medical-center-hurricane-harvey/.

[108] H. Roxburgh, China's 'sponge cities' are turning streets green to combat flooding, The Guardian (December 28, 2017). Available at: https://www.theguardian.com/world/2017/dec/28/chinas-sponge-cities-are-turning-streets-green-to-combat-flooding.

[109] L. Garfield, China Is Building 30 'sponge Cities' that Aim to Soak up Floodwater and Prevent Disaster, Business Insider (November 2017). Available at: http://uk.businessinsider.com/china-is-building-sponge-cities-that-absorb-water-2017-11.

[110] H. Li, L. Ding, M. Ren, L. Changzhi, H. Wang, Sponge City construction in China: a survey of the challenges and opportunities, Water 9 (9) (2017) 594–611.

[111] J. Freeman, Protecting Critical Infrastructure in the Nation's Capital [Blog], US Climate Resilience Toolkit, 2017. Available at: https://toolkit.climate.gov/case-studies/protecting-critical-infrastructure-nation%E2%80%99s-capital.

[112] W.V. Sweet, J. Park, J. Marra, C. Zervas, S. Gill, Sea Level Rise and Nuisance Flood Frequency Changes Around the United States, NOAA Technical Report NOS CO-OPS 073, NOAA/NOS Center for Operational Oceanographic Products and Services, Silver Spring, 2014. Available at: https://tidesandcurrents.noaa.gov/publications/NOAA_Technical_Report_NOS_COOPS_073.pdf.

[113] E. Harball, Washington, D.C. Water utility adapts to global warming, Scientific American (July 13, 2014). Available at: https://www.scientificamerican.com/article/washington-dc-water-utility-adapts-to-global-warming/.

[114] E. Strell, Public service commission approves Con Ed rate case and climate change adaptation settlement [Blog], Climate Law Blog (February 21, 2014). Available at: http://blogs.law.columbia.edu/climatechange/2014/02/21/public-service-commission-approves-con-ed-rate-case-and-climate-change-adaptation-settlement/#_ftn1.

[115] Governor of New York, Governor Cuomo Announces Con Edison Rate Settlement, 2013. Available at: https://www.governor.ny.gov/news/governor-cuomo-announces-con-edison-rate-settlement.

[116] New York Public Service Commission, in: Proceeding on Motion of the Commission as to the Rates, Charges, Rules and Regulations of Consolidated Edison Company of New York, Inc. For Electric Service, New York State Department of Public Service, New York, 2014.

CHAPTER 2

Sustainable and Resilient Buildings: Essential Together

JASON HARTKE, PHD

INTRODUCTION: THE RESILIENCY AGENDA

The tragedy and sorrow associated with the recent disaster from tropical Cyclone Idai and the now all-too-familiar pattern of disasters that have occurred over the past several years all around the world dramatically underscores the importance of fashioning a new resiliency agenda. A central challenge of the 21st century is to advance a new resilience paradigm and develop strategies that can help society not only bounce back from potentially disastrous events but also bounce forward, recovering faster, greener, and stronger.

The concept of resilience is especially suitable in a world that is increasingly more interconnected, more urbanized, more complex, and yet more vulnerable than ever. As a new, developing paradigm, resilience requires foresight and broad societal understanding and support. Just as the protean dimensions around the concept of sustainability were explored and examined in depth in the early 1980s, resiliency will continue to go through a similar evolution, gaining conceptual and technical clarity and scope, generating new agendas and policy perspectives, and mobilizing a new generation of leadership. This agenda is the focus of this book, but it is also the work of the generations to come. It is a fundamental task of civilization.

The risk or vulnerability arises from adverse climate change impacts, earthquakes, hurricanes, extreme weather events, and security threats. It arises in a manifold of ways that can, often predictably, stress communities, cities, and entire countries to the breaking point. The organizing principle is how to be ready, not surprised, to continuously look over the horizon to see what plans are on the table, what preparations need to be made, what assets are in place to handle the foreseeable and unforeseeable crises, and then when these tragedies do occur, ensuring that the right plans are in place and that the right resources and assistance are ready to deploy to help these communities recover.

Therefore, national and subnational leaders and countless other stakeholders will need to be able to understand and address the complex range of issues that arise in any full event-cycle analysis. There is a growing need for strategies and technical solutions to prevent and mitigate disasters to the extent possible, consider the plans and preparations for the inevitable events that will come, and develop the appropriate tools and resources to rebound smarter, greener, and better. Thus, although oversimplified, some have used the more colloquial term of "future proofing" to describe the path forward. However, there is no question that the future of a sustainable world now rests on fully embedding preparedness and resilience across every aspect of human life. As to the focus of this chapter, no sector is more visible, more pressing, and more urgent than the built environment. Buildings—where people live, work, learn, and play—are also where people spend nearly 90% of their time [1].

It would be difficult to understate the enormous task of delivering resilience to homes, schools, hospitals, office buildings, stores, plants, warehouses, and other facilities. Today, the United States has 118 million homes [2] and 5.6 million commercial buildings [3]. Variability may be the most notable aspect across both sectors—residential and commercial—on characteristics such as location, design, structure, features, products, and equipment. Clearly, the prescription for a "resilient" building is not one-size-fits-all, and solutions, therefore, will need to account for the diffuse, complex, and variable nature of the building sector. The critical question of this chapter is how buildings can be more resilient while also continuing the current trajectory of decreasing environmental impact. The

Optimizing Community Infrastructure. https://doi.org/10.1016/B978-0-12-816240-8.00002-1

challenge is one of integration, advancing sustainability *and* resilience in buildings.

Resilient and Sustainable: A Great Convergence

> *Mitigation alone won't work, because the climate is already changing, we're already experiencing impacts .... A mitigation only strategy would be insanity.*
>
> JOHN HOLDREN, FORMER SCIENCE ADVISOR TO PRESIDENT OBAMA [4].

This chapter explores the rise of resilience as a key strategy to protect and ready the built environment in the face of increasing risk and vulnerability from the effects of climate change, extreme weather, natural disasters, and other shocks and disruptions. In a normative sense, the building community is fortunate that the story is mostly one of convergence. The chronology is such that sustainable building practices took root first, driven by the pursuit of mitigating environmental impacts. As the urgency and call for more resilient buildings became louder and louder, thought leaders worked hard to try to ensure that sustainability and resilience were considered together and that solutions and strategies were appropriately integrated. Assessing today whether or not full convergence will occur is more difficult. Although sustainability and resilience are complementary and should be forged together, more and more is demanded of buildings—safe, accessible, high-performing, climate-friendly, resilient, healthy, smart, connected, cybersecure, and more.

Green Building as an Early Catalyst for Change

Over the past 25 years, the built environment has been a central focus for creating change and driving sustainability. The reason is clear—buildings have an enormous impact on the environment—from climate change to water to materials. Buildings, according to the US Department of Energy, are the single largest energy-consuming sector in the US economy, accounting for more than 70% of electricity use and nearly 40% of total energy demand. Thus, in the context of combating climate change, mitigation tools and strategies targeting the built environment began to proliferate. Today, the use of such systems, such as ENERGY STAR and the Leadership in Energy and Environmental Design (LEED) Green Building Rating System and other high-performance building strategies, are now ubiquitous in the market and continue to build an impressive, proven track record of spurring significant energy savings and other sustainable outcomes in buildings. Again, the primary focus and driver was mitigating the environmental impact.

Responding to "Changes in Climate"

About 15 years ago, as the quote above suggests, that focus had begun to shift. There is no question this shift tracked how leaders in the building community—architects, engineers, environmentalists, and others—responded to unprecedented disasters, from Hurricane Katrina to Hurricane Sandy to the 2007 two-mile-wide tornado that destroyed 95% of the buildings of the small town of Greensburg, Kansas. In addition, key studies now show how the early effects of climate change were much more pervasive than previously thought. In fact, these studies argued that effects were already here and significantly affecting everything from weather patterns to agriculture to human health [5]. With extreme weather on the rise, technical experts and advocates alike began to rewrite the sustainability agenda to include both mitigation and adaptation—that achieving the former was impossible without also succeeding on the latter; in other words, noting that a sustainable future is inherently dependent on the ability to also secure a resilient one.

In that same speech by John Holdren referenced at the beginning of this section, Holdren went on to say: "Changes in climate are already harming human well-being. This is not just a problem for our children and grandchildren, it's a problem for us now. We're seeing more and bigger floods in regions prone to flooding, we're seeing more and bigger droughts in regions prone to those. We're seeing worse wildfires; more powerful storms; worse outbreaks of forest pests like the pine bark beetle and the spruce budworm; more coral bleaching events; increased coastal erosion; damage to structures and roads from thawing permafrost in the far north, and a lot more" [4]. The takeaway was clear: not only do the agendas of these two efforts overlap, but how to affect their outcomes and develop solutions should be taken up together.

The devastation caused by these disasters and their societal costs have been strong drivers for finding ways to improve resilience and reduce costs—a tough task in the face of more frequent, more costly events. Since 1980, according to the National Oceanic and Atmospheric Administration (NOAA), the United States has endured 241 weather or climate disasters causing damage of $1 billion or more, totaling more than $1.6 trillion in damages [6]. In 2018, 14 of the 241 "billion-dollar disasters" occurred, ranging from wildfires to cyclones to droughts. The historical context of recent disaster trends is even more alarming:

> *The most recent years of 2018, 2017, and 2016 have all been historic in the number of billion-dollar disasters that have impacted the United States—totaling 45 separate events. This is a 3-year average of 15 disaster events/year, the highest on record, and well above the annual inflation-adjusted average of 6.2 events per year (1980–2018). In addition, this 3-year average even exceeds the annual inflation-adjusted average of 12.6 events per year over the past 5 years (2014–18) [6].*

No longer could building experts and practitioners ignore vulnerabilities facing the built environment. Research in the area accelerated as did new ideas and solutions designed to mitigate hazard risk.

THE SEEDS OF RESILIENCE

One of the earliest shifts in this direction came in the aftermath of one of the most destructive natural disasters in American history, Hurricane Katrina. Louisiana, Mississippi, and Florida suffered through the costliest and third most deadly storm in US history, killing nearly 1000 people, damaging more than a million housing units in the Gulf Coast region, and causing $150 billion in damages [7]. Questions about storm risk and vulnerability swirled as did sharper questions about the ability to effectively deploy assistance and recovery support.

At the time, the green building movement was experiencing a major growth period thanks in large part to the innovative work of the US Green Building Council (USGBC) and the success of its LEED Green Building Rating System. Green building market penetration was seeing tremendous growth from year to year, and excitement for sustainable, green building was palpable. Hurricane Katrina had hit just months before thousands of building practitioners were set to gather in Atlanta for the USGBC's annual Greenbuild conference. In an effort to support the recovery and address underlying issues of how to become more resilient to natural disasters, the USGBC and some of its chapters organized a series of planning charrettes to foster and cultivate open and honest discussion. Close to 200 people participated, including a large subset from the Gulf Coast. Following the charrette process, the USGBC published several reports, the first called "The New Orleans Principles: Celebrating the Rich History of New Orleans Through Commitment to a Sustainable Future" [8].

In many respects, the report, with its emphasis on protection and readiness, was an early precursor to the development of the concept of resilient building. In this early report, new terms now more fully embedded in resiliency thinking were developed and pioneered. One such term was "passive survivability," credited to building expert Alex Wilson. In the report, passive survivability was one of the principles, described this way: "Homes, schools, public buildings, and neighborhoods should be designed and built or rebuilt to serve as livable refuges in the event of crisis or breakdown of energy, water, and sewer systems." The report provided the first examination of design strategies that could help achieve passive survivability, including backup emergency water systems, distributed infrastructure, solar electric systems, solar water heating, and burying or protecting electric and gas lines. "These buildings should be designed to maintain survivable thermal conditions without air conditioning or supplemental heat through the use of cooling-load-avoidance strategies, natural ventilation, highly efficient building envelopes, and passive solar design," said the report [8].

Across academia, researchers also set their sights on further examining and exploring risk and vulnerability in buildings. As an example, in 2010, Haigh and Amartunga provided a more in-depth review of the literature and the important role of the built environment to increasing community resilience [9]. Although they found that relevant research is often scattered across multiple disciplines, they show that more and more attention has been focused on important research questions about building resilience. Experts were now making appropriate linkages between resilience and sustainability. Although progress was being made, the path forward was not altogether clear. Difficult and complex technical questions certainly remained.

Bridging Sustainability and Resilience: Key Conceptual Linkages

In the middle of 2012, another Obama Administration official took note of the need to converge on thinking about the issue and its solutions. Craig Fugate, the Administrator of the Federal Emergency Management Agency (FEMA), while acknowledging the progress of green building, argued that the building community also needed to begin to more thoughtfully consider and integrate resilience in the design, construction, and retrofit of buildings. He argued for convergence, not creating something brand new. He said that sustainable building practices, in and of themselves, were a key part of the way forward. "Being green is one part of being resilient," Fugate said [10]. In other words, the technical solutions outlined in green building provided part of the solutions, and as this aspect of the solution was further along, it also represented a key starting point.

Leading environmental stakeholders in the built environment were listening and quietly making significant progress toward this end. In 2011, the US Green Building Council, in partnership with the University of Michigan, released a first-of-its-kind report showing the myriad ways that green building strategies support and advance a more resilient built environment. The in-depth study, *Green Building and Climate Resilience*, sought to identify and assess the linkages between specific green building strategies and desired resilient building outcomes [11].

The seminal research achieved two important objectives. First, the study helped demonstrate that sustainable building practices were foundational toward achieving resilience indicators and objectives. Second, the study revealed that buildings were currently being designed and engineered based on past weather and indicator data, forgoing forward-thinking and ignoring predictive data based on how climate is changing future indictors. At an event in 2012, one of the study's authors, Dr. Chris Pyke, explained the importance of the findings this way:

> *Every building is designed for a specific range of conditions, such as peak temperature, storm surge, and average precipitation. Climate change has the potential to undermine some of these assumptions and potentially increase risks to people and property. Fortunately, there are practical steps we can take to understand and prepare for the consequences of changing environmental conditions and reduce potential impacts. This can help green buildings meet and exceed expectations for comfort and performance long into the future [12].*

Building design, then, must look forward, not backward, accounting for what increasingly are predictable changes in local weather conditions and patterns. For example, the size of heating and cooling systems, which are calibrated for past climate conditions, could pose new vulnerabilities in the face of future changes in climate. Pavement surfaces, if designed based only on yesterday's indicators, could create new risks in the face of heavier, more intense precipitation events of tomorrow. Storm water management systems could be overwhelmed by more frequent and more powerful storms. However, our building systems, as the report explains, have the power to improve resilience if the design community accounted for changing temperatures, stronger storms, more intense precipitation events, more flooding, and more frequent and lengthier droughts.

For the first time, the report signaled a simple, but fundamental, point—that building design and construction needed to be done differently, with more foresight, preparing buildings for tomorrow. In other words, the buildings of today need to be built, retrofitted, and operated for the environment and the risks of today as well as those on the horizon. In addition, the study showed how to adopt a whole-building design approach to better and more cost-effectively integrate strategies that mitigate the impact of disasters and climate change.

From a technical aspect, and equally important, the report confirmed synergies with specific energy conservation measures and other green building strategies and mapped their utility in building resilience. To do so, the report first outlined the many vulnerabilities of buildings, such as warmer temperatures, extreme heat events, sea level rise, expanding pest ranges, increased frequency, and intensity of storms. In the "envelope category," for example, the report says, "The no-regrets envelope strategies focus on improving the thermal properties of the building shell to respond to temperature-driven impacts. The resilient strategies in this category help the shell respond to other climate impacts such as termite damage (pests) or wind-driven rain (storms)." [11] For example, beyond code insulation of walls and roofs were classified as "no regrets," whereas strategies such as design for increased wind, oversized roof drainage, and pitched roof were classified as "resilient." The report conducted this mapping for green building strategies across all relevant categories, including envelope; siting and landscaping; heating, cooling, and lighting; water and waste; equipment; and process and operations [11].

Ultimately, the report, as well as similar studies and efforts, paved new ground by essentially baselining green building strategies and their relationship to building resilience (i.e., "no regrets" or "resilient" classification). What came next is not all that surprising. Organizations and building experts worked to develop strategies, indicators, and measures that would go beyond green building and what it prescribes.

FROM RESEARCH TO IMPLEMENTATION

The idea and study of resilience has now permeated research far and wide, spanning numerous approaches across the qualitative and quantitative spectrum. Zooming out on the larger picture of resilience research, it is easy to see just how diverse and vast it is, with deep tracks across multiple disciplines from sociology to natural science. Although unable to adequately summarize the full literature, this chapter touches briefly on research applicable to building resilience and therefore studies focusing on physical vulnerability. Sven Fuchs and his coauthors put it this way, "... natural scientists often view vulnerability in terms of the likelihood of occurrence of specific process scenarios and associated impacts on the built environment" [13]. Getting to resilience, in other words, has everything to do with better understanding vulnerability. The research in this area has grown considerably over the years. Again, Fuchs does a good job of describing the course of the research, juxtaposing the approach in the social sciences with that of the natural sciences:

> *From the perspective of social sciences, vulnerability is understood as a predisposition and potential of communities or individuals to be harmed; consequently, vulnerability does not change if the hazard is more intense or not—it is in contrast the exposure that might change and that influences the degree of being at risk. Initially, the focus in vulnerability research was on the internal side of vulnerability (exposure of communities or elements at risk exposed) and on the external side (coping capacity of people or systems) and to societal structures [14]. Subsequently, it was increasingly regarded as more important to focus on the capacity of people to cope with and adapt to hazardous events and processes [15]. Complementary, natural scientists define vulnerability as the expected degree of loss resulting from the impact of a hazardous phenomenon of a given magnitude and frequency on elements at risk exposed. Natural science approaches therefore mostly focus on the susceptibility of physical elements at risk to natural processes [16,17] to provide information necessary for operational risk analyses [18] and mitigation [19,20]. Through the rise of research on the consequences of climate change, only recently an additional categorization gained relevance that includes the components of exposure, sensitivity, and adaptive capacity [21].*

Today, a rich base of research exists focused on examining risk and vulnerability by and across various natural hazards as well as other risk factors or disruptive events. For the purposes of this chapter, it is important to note again the depth and diversity of this research, which not only highlights the shear complexity of vulnerability but also underscores how difficult it is to effectively address these vulnerabilities and therefore build resilience.

The Emergence of Resilient Building Systems and Ratings

Although it is one thing to define resilience and advance the research that helps explain and understand its importance for buildings and infrastructure, it is altogether different to operationalize it for building professionals and put it to practice. Operationalizing resilience in buildings, therefore, soon became a shared objective and exigent pursuit. Multiple organizational actors stepped up and worked to develop policy prescriptions as well as specialized tools and systems that would allow the building community to distinguish building assets as "resilient." For many of these efforts, the path forward often imitated the same market leadership model pioneered by the USGBC and exemplified in LEED.

One of the early groups to begin to bring resilience into the context of a larger, multidisciplinary approach was the National Institute of Building Sciences (NIBS). Through its Whole Building Design Guide, which focuses on several areas of "design objectives," NIBS built out new sections to address resilient design and construction methods based on risk and vulnerability science and research [22]. The Design Guide section on resilience leans on early intellectual and technical work from the National Infrastructure Advisory Council (NIAC), which identified four "R's" to denote key characteristics of resilience: robustness, resourcefulness, rapid recovery, and redundancy [23].

NIAC sought to define infrastructure resilience, doing so this way: "Infrastructure resilience has the ability to reduce the magnitude and/or duration of disruptive events. The effectiveness of a resilient infrastructure or enterprise depends on its ability to anticipate, absorb, adapt to, and/or rapidly recover from a potential disruptive event" [23]. Most of these efforts and interested stakeholders noted that the first, foundational step forward was starting with strong codes and standards.

> *Traditionally, building codes have regulated life safety issues. New building codes and standards should extend beyond life-safety aspects to include resilient design concepts in a performance-based approach as well as continuity of operations. They should rely on common and widely adopted methods of measurement, provide a flexible framework to address different facility types, address types of structures (from residential to large commercial and industrial structures), and recognize the differing levels of performance that are required. Uniform adoption of resiliency objectives by jurisdictions requires including resiliency requirements in the current model building codes, educating regulators and their constituents, and incentivizing the application, inspection, and regulation of resiliency approaches. This process begins with the development of criteria, codes, and standards that address resiliency objectives and the supporting tools and validation for their use [24].*

However, it was not long before certain actors looked to beyond-code solutions that targeted specific hazards or other risks more common to the built environment.

How to Fortify

In the mid-2000s, one group that had long identified the vulnerability of homes and buildings, the Insurance Institute for Business and Home Safety (IBHS) was working to make resilient building easier. The group, which works closely with the insurance industry, had become an established leader supporting and developing the needed research, science, and understanding in building safety, which very naturally extended into natural hazard mitigation. IBHS not only recognized the enormous impact severe weather has on buildings, but it worked to develop a market-based, voluntary solution. IBHS, determined to play an active role in accelerating more resilient structures, followed the successful green building model and worked to develop a resilience-oriented rating system for homes.

In 2010, IBHS launched the FORTIFIED Home Hurricane Program for retrofitting existing residential structures, a rating system analogous to LEED [25]. The program developed in such a way that its systems focused on building type—new or existing—and then targeted and certified to specific vulnerabilities, including hurricane, high wind and hail, and high wind. The hurricane system focuses on securing the chimney, sealing the roof deck, gable end bracing, installing impact-rated windows, and other envelope products. The system also includes guidance for accounting for additional wind loads for solar photovoltaic systems. Today, there are more than 10,000 homes that have certified to FORTIFIED [26], and multiple states have passed policies providing incentives to reduce the cost of insurance utilizing or referencing the FORTIFIED system, including in Alabama, Mississippi, and North Carolina [27].

Other Systems and Approaches

About 5 years ago, the USGBC, in a continuation of its resilience-oriented work, organized a group of resilient building experts to develop LEED pilot credits on resilient design. In late 2015, three new resilient credits were approved—assessment and planning for resilience; design for enhanced resilience; and passive survivability and functionality during emergencies. "In a nutshell, these three credits are designed to ensure that a design team is aware of vulnerabilities and addresses the most significant risks in the project design, including functionality of the building in the event of long-term interruptions in power or heating fuel," said Alex Wilson, who helped lead development of the credits [28].

About 8 years ago, a group of organizations came together to create a more holistic resilience rating system applicable to buildings, neighborhoods, homes, and infrastructure. These groups, which included Perkins and Will, C3 Living Design, Capital Markets Partnership, and the University of Minnesota, came together to form the RELi Collaborative, which then developed and built the RELi resilient rating system. RELi focuses on resilient design criteria across eight categories: panoramic approach, hazard preparedness, hazard adaptation, community vitality, productivity/health + diversity, energy/water + food, materials + artifacts, and applied creativity. It was also designed to work alongside other sustainability guidelines and systems such as LEED. The system debuted in 2014. Three years later, the Green Business Certification Inc. (GBCI), the same organization that oversees and certifies LEED buildings, acquired RELi and became the system's certifying body [29].

Around the same period, Arup, a global engineering and consultancy company, developed another similarly resilience-focused system. In development of the system, Arup had a particular focus, resilience to seismic events, creating the Resilience-based Earthquake Design Initiative or REDi rating system. The tool was created to give architects, engineers, and building project teams the framework and prescriptive criteria that addresses areas such as resilience planning, seismic hazards, enhanced structural design, structural analysis, and general assessment guidelines. The focus is on measures that help lessen damage and accelerate recovery. The approach is described as, "Resilience-based earthquake design is a holistic process which identifies and mitigates earthquake-induced risks to enable swift recovery in the aftermath of a major earthquake—this exceeds code-intended performance objectives and typical performance-based design objectives" [30].

A New Political Urgency to Solutions

In late 2012, when the East Coast was hit by Hurricane Sandy, which caused more than $70 billion in damages, government leaders at all levels called for a new urgency to mature the concept of resilience and move faster to deploy solutions. Fully bridging sustainability and resilience had become a new imperative. Much of the conversation took place in the policy arena where leaders looked to policy that would accelerate solutions and steer resources to fill gaps. Mayor Michael Bloomberg was one of the first to act, announcing the first-of-its-kind comprehensive resilience plan, *A Stronger, More Resilient New York*. The plan included more than 250 recommendations to improve resilience across sectors, from infrastructure to buildings [31].

The Obama administration spearheaded several resilience-focused initiatives, including the creation in 2013 of a Task Force on Climate Preparedness and Resilience, the Partnership for Resilience and Preparedness, and the Climate Data Initiative, an open-data initiative for climate resilience. At the Department of Housing and Urban Development (HUD), the administration created a National Disaster Resilience Competition that awarded nearly $1 billion toward funding for disaster recovery and long-term community resilience, awarding eight states and five localities.

Philanthropy, too, was making resilience a key pillar in its giving strategy. The Rockefeller Foundation and its president at the time Judith Rodin, who did much of the earliest intellectual work to call attention to the issue, made significant investments to pioneer new thinking, new methodologies, and new solutions to improve resilience. In 2013, the foundation created the 100 Resilient Cities to support cities and deliver the resources to become more resilient. That same year, a group of nonprofits working to build awareness about how local leaders can take action to champion a more resilient community joined with 44 mayors to launch the Resilient Communities for America campaign—an effort that grew to more than 200 mayors. A race had begun to move from awareness to plans to action, and leaders at all levels of government had jumped in.

CONCLUSION

In addition to changing weather and more frequent and intense violent weather, the country is also facing a silent onslaught that further complicates its current vulnerabilities and has the potential to jeopardize future security—underinvesting in infrastructure. The country's infrastructure has often suffered disrepair and hovered near failing. The state of roads, bridges, and buildings—the very bones of the national economy—are in dire need of attention.

The American Society of Civil Engineers (ASCE) has been ringing the alarm bell for years. In its latest *report card* assessment, the overall grade was a "D+," signifying "poor" condition, one rung up from "F," which signifies failing. Of 16 sectors, only four received a grade higher than a "D." [32] Schools, for examples, received a "D+," which is a slight improvement from previous reports. "The nation continues to underinvest in school facilities, leaving an estimated $38 billion annual gap," says the report, continuing, "As a result, 24% of public school buildings were rated as being in fair or poor condition" [33]. Overall, the ASCE report signals an infrastructure figurative diagnosis of severe osteoporosis. In addition to several policy prescriptions, the ASCE calls for bridging a $2 trillion investment gap over the next 10 years, adding calcium to the bones, making the country stronger and more prepared, and unleashing opportunities to lead and innovate.

As this chapter outlines, the tools to transform the built environment—to be both more sustainable and more resilient—have been developed and are available. The investments we make—and the solutions we adopt—will be telling. Clearly, the future of buildings will require both sustainability and resilience. The changing nature of climate risk—and its impacts—is something communities around the country are beginning to understand all too well. If readiness and resilience is the answer, we have more work to do to achieve (1) a better understanding of vulnerability, (2) a fuller accounting of increasing risks, climate, and otherwise, and (3) charting a path to adopt a truly integrated solutions framework.

Back in 2012, FEMA Administrator Fugate noted how some synergies across sustainability and resilience at that point were happening serendipitously: "Through a [hazard] mitigation practice, if you're looking at one hazard—looking at roof attachments, how we built our roofs, how we built the envelope—the unintended consequence was in many cases we were actually improving energy efficiency." So, the big question is, as Fugate later noted, "How do we get the synergy of looking at multiple things?" [12]. Greater integration will allow us to do this effectively and at least cost. Maximizing and optimizing for technical synergies across sustainable and resilience outcomes will save money, save lives, and make buildings stronger to the inevitable threats ahead. Although significant and meaningful progress to date has been made to bridge sustainability and resilience in buildings, whether we will realize a full and integrated convergence across these solution sets is still in question.

Epilogue

Just as we now have a set of resilience tools that target buildings, experts have also been looking into advancing whole-community resilience. One of the first groups to undertake the development of an integrated framework for improving community resilience was the Community and Regional Resilience Institute (CARRI) led by Warren Edwards. A similar effort is now housed at the International Code Council, called the Alliance for National & Community Resilience (ANCR). Today, building on what CARRI started,

ANCR is working to create community resilience benchmarks across key sectors such as buildings, transportation, housing, water, energy, waste management, education, communications, public safety, and business. The goal is to measure resilience and "consolidate existing assessments, certifications, and research to serve as a useful, centralized tool to help communities, businesses, governments and people make decisions to become more resilient, based on consistent and comparable information—on where to live, where to invest, what to prioritize, and how to measure progress" [34].

Undoubtedly, the larger effort to forge greater resilience will not stop at the building scale, and so whether the focus is a building or a community, the critical task ahead is to marshal and integrate our collective and multidisciplinary know-how to increase resilience while simultaneously achieving sustainability goals—the two are essential together.

REFERENCES

[1] N.E. Klepeis, W. Nelson, W. Ott, J. Robinson, A. Tsang, P. Switzer, J. Behar, S. Hern, W. Engelmann, The National Human Activity Pattern Survey: A Resource for Assessing Exposure to Environmental Pollutants, Lawrence Berkeley National Laboratory, 2001. https://indoor.lbl.gov/sites/all/files/lbnl-47713.pdf.

[2] U.S. Energy Information Administration, Residential Buildings Energy Consumption Survey, 2015.

[3] U.S. Energy Information Administration, Commercial Buildings Energy Consumption Survey, 2012.

[4] Climate Science & Policy Watch, Text of Remarks by Obama Science Advisor John Holdren to the National Climate Adaptation Summit, 2010. http://www.climatesciencewatch.org/2010/05/28/text-of-remarks-by-obama-science-adviser-john-holdren-to-the-national-climate-adaptation-summit/.

[5] C. Rosenzweig, G. Casassa, D.J. Karoly, A. Imeson, C. Liu, A. Menzel, S. Rawlins, T.L. Root, B. Seguin, P. Tryjanowski, Assessment of observed changes and responses in natural and managed systems, in: M.L. Parry, O.F. Canziani, J.P. Palutikof, P.J. van der Linden, C.E. Hanson (Eds.), Climate Change 2007: Impacts, Adaptation and Vulnerability, Contribution of Working Group II to the Fourth Assessment Report of the Intergovernmental Panel on Climate Change, Cambridge University Press, Cambridge, UK, 2007.

[6] A.B. Smith, 2018's Billion Dollar Disasters in Context, National Oceanic and Atmospheric Administration, 2019. https://www.climate.gov/news-features/blogs/beyond-data/2018s-billion-dollar-disasters-context.

[7] A. Plyer, Facts for Features: Katrina Impact, The Data Center, 2016. https://www.datacenterresearch.org/data-resources/katrina/facts-for-impact/.

[8] U.S. Green Building Council, The New Orleans Principles: Celebrating the Rich History of New Orleans through Commitment to a Sustainable Future, 2005. https://www.resilientdesign.org/wp-content/uploads/2013/02/NewOrleans_Principles_LowRes1.pdf.

[9] R. Haigh, D. Amaratunga, An integrative review of the built environment discipline's role in the development of society's resilience to disasters, International Journal of Disaster Resilience in the Built Environment 1 (1) (2010) 11–24. https://doi.org/10.1108/17595901011026454.

[10] M. Comstock, FEMA Administrator Fugate Applauds Green Building as a Cornerstone of the Resiliency Agenda, USGBC blog. https://www.usgbc.org/articles/fema-administrator-fugate-applauds-green-building-cornerstone-resiliency-agenda.

[11] University of Michigan, U.S. Green Building Council, Green Building and Climate Resilience: Understanding Impacts and Preparing for Changing Conditions, 2011.

[12] J. Hartke, Resiliency & Sustainability: A Great Convergence and Synergies in Solutions, 2012. https://www.usgbc.org/articles/resiliency-sustainability-great-convergence-and-synergies-solutions.

[13] S. Fuchs, J. Berkmann, T. Glade, Vulnerability assessment in natural hazard and risk analysis: current approaches and future challenges, Natural Hazards (December 2012).

[14] R. Chambers, Vulnerability, coping and policy, IDS Bulletin 20 (2) (1989) 1–7.

[15] M. Anderson, Vulnerability to disaster and sustainable development: a general framework for assessing vulnerability, in: M. Munasinghe, C. Clarke (Eds.), Disaster Prevention for Sustainable Development: Economic and Policy Issues. The International Bank for Reconstruction and Development, The World Bank, Washington, 1995.

[16] D. Varnes, Landslide Hazard Zonation: A Review of Principles and Practice, UNESCO, Paris, 1984.

[17] M. Papathoma-Köhle, M. Kappes, M. Keiler, T. Glade, Physical vulnerability assessment for alpine hazards: state of the art and future needs, Natural Hazards 58 (2) (2011) 645–680.

[18] W. Carter, The disaster management cycle, in: W. Carter (Ed.), Disaster Management: A Disaster Manager's Handbook, Asian Development Bank, Manila, 1991, pp. 51–59.

[19] M. Holub, Fuchs, Mitigating mountain hazards in Austria—legislation, risk transfer, and awareness building, Natural Hazards and Earth System Sciences 9 (2) (2009) 523–537.

[20] M. Holub, J. Suda, S. Fuchs, Mountain hazards: reducing vulnerability by adapted building design, Environmental Earth Science 66 (7) (2012) 1853–1870.

[21] B. Turner II, R. Kasperson, P. Matson, J. McCarthy, R. Corell, L. Christensen, N. Eckley, J. Kasperson, A. Luers, M. Martello, C. Polsky, A. Pulsipher, A. Schiller, A framework for vulnerability analysis in sustainability science, Proceedings of the National Academy

of Sciences of the United States of America 100 (14) (2003) 8074–8079.
[22] Whole Building Design Guide, Building Resilience, National Institute of Building Sciences, 2018. http://www.wbdg.org/resources/building-resiliency.
[23] National Infrastructure Advisory Council (NIAC), Critical Infrastructure Resilience Final Report and Recommendations, NIAC, Washington, DC, 2009.
[24] U.S. Department of Homeland Security, Designing for a Resilient America: A Stakeholder Summit on High Performance Resilient Buildings and Related Infrastructure, 2010. https://www.dhs.gov/xlibrary/assets/designing-for-a-resilent-america-11302010-12012010.pdf.
[25] F. Malik, R. Brown, W. York, IBHS FORTIFIED Home Hurricane; Bronze, Silver and Gold: An Incremental, Holistic Approach to Reducing Residential Property Losses in Hurricane-Prone Areas, Insurance Institute for Business & Home Safety, 2010. http://disastersafety.org/wp-content/uploads/ATC-FORTIFIED_IBHS.pdf.
[26] J. Sharp, Alabama Leads Nation in Building the Strongest Hurricane-Resilient Homes, AL.com, 2019. https://www.al.com/hurricane/2019/02/alabama-leads-nation-in-building-stronger-hurricane-resilient-homes.html.
[27] Insurance Institute for Business & Home Safety, Regulatory Framework for FORTIFIED Insurance Incentives. http://disastersafety.org/wp-content/uploads/FORTIFIED-Home-Incentives_IBHS.pdf.
[28] A. Wilson, LEED Pilot Credits on Resilient Design Adopted!, Resilient Design Institute, 2015. http://www.resilientdesign.org/leed-pilot-credits-on-resilient-design-adopted/.
[29] Green Business Certification Inc, GBCI and RELi Resilience Standard Work Together, 2017. http://www.gbci.org/reli.
[30] I. Almufti, M. Willfrod, REDi Rating System: Resilience-Based Earthquake Design Initiative for the Next Generation of Buildings, Version 1.0, 2013, https://www.arup.com/-/media/arup/files/publications/r/redi_final-version_october-2013-arup-website.pdf.
[31] New York City, A Stronger, More Resilient New York, 2013. https://www1.nyc.gov/site/sirr/report/report.page.
[32] American Society of Civil Engineers, 2017 Infrastructure Report Card, 2017. https://www.infrastructurereportcard.org/americas-grades/.
[33] American Society of Civil Engineers, "2017 Infrastructure Report Card: Schools.", 2017. https://www.infrastructurereportcard.org/wp-content/uploads/2017/01/Schools-Final.pdf.
[34] Alliance for National & Community Resilience. http://www.resilientalliance.org.

Resilience Solutions

Creating problems is easy. We do it all the time. Finding solutions, ones that last and produce good results, requires guts and care.

HENRY ROLLINS, AMERICAN MUSICIAN

Persistence and resilience only come from having been given the chance to work through difficult problems.

GEVER TULLEY, AMERICAN WRITER

Introduction

Utilities (water and energy in particular) are so essential to human life that they are often called lifelines. The service utilities provide are so vital and engrained in daily life that most people take them for granted—until these essential services are disrupted. Multiple examples illustrate how disruptions in lifelines impact individuals and the economy. The 2003 black out in the northeastern US stranded commuters and brought the region's economy to a halt. Potable water and centralized wastewater treatment have led to significant advancements in public health.

The resilience of a community is predicated on the resilience of its infrastructure. Although not covered in-depth here, other lifelines such as transportation and communication networks are equally essential. The collapse of the I-35W Mississippi River Bridge in Minneapolis, Minnesota in 2007 highlights the impact of transportation networks on communities. The collapse killed 13 people and injured 145 more. The bridge was the third busiest in the state, carrying 140,000 vehicles a day. The Minnesota Department of Transportation estimated that the road-user costs due to the collapsed bridge were $400,000 a day. The loss to the Minnesota economy was about $17 million in 2017 and $43 million in 2008. These estimates do not include commuter time value [1]. Because of the importance of the bridge to the state, construction on a replacement began just 6 weeks after the collapse and opened 1 year later.

Despite this catastrophic wake up call, the current state of transportation in the United States remains poor. The American Society of Civil Engineers (ASCE) Infrastructure Report Card gives bridges a C+, roads a D, aviation a D, transit a D−, and rail a B (the highest score across all categories examined) [2]. To solve this challenge, ASCE recommends an infusion of funding—on the order of $2 trillion over 10 years across all infrastructure with surface transportation infrastructure making up more than half of that gap. Although such funding will address today's needs, a more comprehensive strategy that captures needs across the entire life cycle of the infrastructure is necessary. Public and private capital should be leveraged for both initial investment and long-term maintenance of projects. Part III of this book delves into the opportunities to finance resilience projects.

ASCE further recommends a focus on community resilience programs that establish communication systems and recovery plans that reduce the impacts that severe events have on local economies, quality of life, and the environment. Improved land-use planning (covered further in Chapter 9) should consider the function of existing and new infrastructure and balance the social, economic, and environmental needs of the local community. All such strategies should focus on improving the triple bottom line with economic, social, and environmental benefits [2].

Communication systems have become a fixture of daily life. The internet has become an essential tool for commerce. Smart phones and social media have unlocked new ways to communicate and share information in near real time. This has opened up huge opportunities to get messages to residents before, during, and after hazard events. Following the 2010 earthquake in Haiti, Twitter became a source for both information on the earthquake and a mechanism to mobilize recovery support [3]. Location-targeted messages can be sent to cellphone users on hazards or other important activities in their area from flash flood warnings to amber and silver alerts.

AT&T, one of the largest communication providers, has recognized the importance of investing in the resilience of their systems. Since 2016, natural disasters have cost the company $847 million, with about two-thirds of that coming in 2017 alone [4]. To reduce these costs into the future and to address the risks posed by climate-related events, AT&T is working with Argonne National Lab to develop a tool that will track flooding, hurricanes, and windstorms in the southeast. The tool will eventually be expanded to include other parts of the country and additional risks including wildfire and droughts. The project will allow projections up to 30 years through high-resolution climate models, local data, and additional information as it becomes available. The results will help to influence the installation of mitigation measures including elevating equipment or other strategies [4].

The interoperability of communications networks during disaster is essential to support emergency management. Unfortunately, the events of September 11 proved this point [5]. The U.S. Department of Homeland Security has developed an "Interoperability Continuum" to support effective emergency response. The Continuum identifies five critical success elements that must be addressed to achieve a sophisticated interoperability solution: governance, standard operating procedures, technology, training and exercises, and usage of interoperable communications [6]. The Continuum is illustrated in Fig. 1.

Existing and emerging communication technologies can be used to support disaster prediction and recovery. The Internet of Things (IoT), wireless sensors, and 5G wireless communication coupled with data analytics could enhance emergency management [7].

In the context of disaster relief, Shittu, Parker, and Mock have examined strategies for improved communication between relief organizations and the impacted community [8]. In the context of information process theory and organization design, they propose three strategies:

1. Improving the interoperability and information flow across organizational boundaries,
2. Increasing synergies between organizations on adapting new technology such as social media for the coordination of structured and unstructured data for use in decision-making, and
3. Increasing the flexibility of relief organizations to use external resources from areas not affected by disasters on an opportunistic basis.

In addition to the physical communication network, scholars have begun looking into the social fabric of communities and how communication among community members influences a community's resilience [9,10]. What message is being delivered is just as important as the mechanism deployed to deliver it. In addition, this book does not go in-depth into the social aspects of community resilience, but research on disaster communication is emerging.

RISK, INTERDEPENDENCIES AND EXTERNALITIES

In Chapter 3, Brashear proposes a risk management process that supports a holistic approach across lifelines. This process addresses many of the challenges that have hindered such an approach previously—inconsistencies in how various lifelines identify and manage risk, the lack of willingness to share potentially

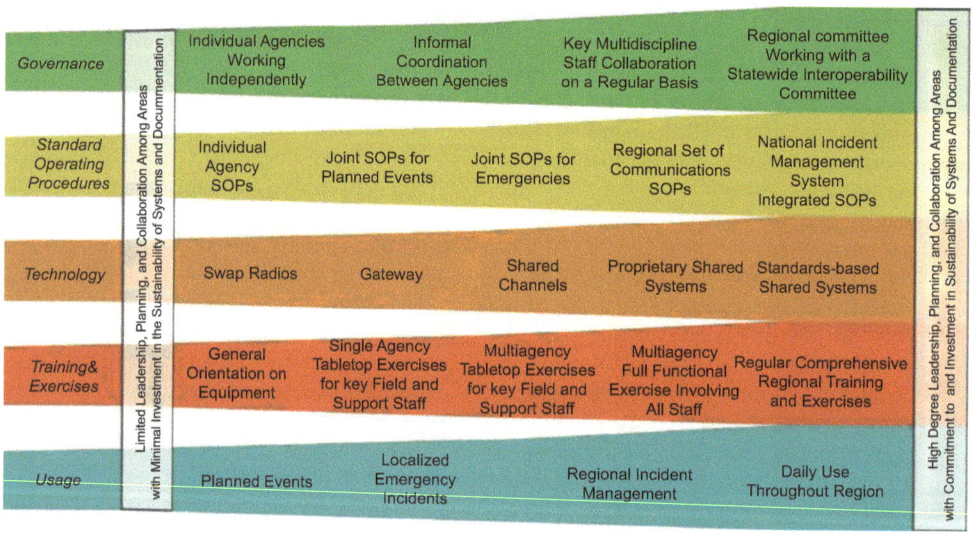

FIG. 1 The interoperability continuum. (Source: Reproduced from, U.S. Department of Homeland Security, Interoperability Continuum: A tool for improving emergency response communications and interoperability, n.d. https://www.dhs.gov/sites/default/files/publications/interoperability_continuum_brochure_2.pdf.)

sensitive information, and a disregard for the community-level benefits. Once applied, this process would lead to increased resilience within the individual lifelines and at a community level.

Alignment of risk management approaches across a community's lifelines provides a valuable side effect—the ability to execute mitigation measures that serve the community but may not approach the threshold for investment by any one lifeline. Multiple stakeholders can "buy-in" to mitigation measures consistent with the level of benefit they receive. This allows a level of coordination rarely seen before: a result where the whole is truly more than the sum of its parts.

The National Institute of Building Sciences (NIBS) through its Multihazard Mitigation Council and Council on Finance, Insurance, and Real Estate have developed a similarly powerful concept—a holistic approach to incentives that captures benefits that accrue to parties involved in a specific transaction or project. This concept of Incentivization looks at the risk exposure across the project owner, insurer, and financier and the community as a whole and the potential to bundle incentives whereby each stakeholder receives a benefit and contributes to the achievement of the overall benefit at an equitable rate [11,12]. Fig. 2 illustrates the costs and benefits associated with a mitigation investment and the entities that receive those benefits and could/should bear some of the costs. Incentivization is covered in greater depth in the introduction to Part VI.

ELECTRIC POWER INFRASTRUCTURE

The electricity sector is undergoing the biggest transition since the establishment of the electric grid itself. Moreover, its not just one focused on addressing a singular challenge—although the end result will be a more efficient and resilient system. A combination of economics, shifting consumer expectations, changing technology, and policy changes are driving this evolution. The grid is no longer just a one-way system providing power. The smart grid is enabling two-way communication (and power flow) between utilities and customers, unlocking a whole new set of opportunities.

On-site renewable energy generation can contribute to increased reliability and reduced greenhouse gas emissions when planned and implemented effectively. Thus far, building owners have focused primarily on the sustainability benefits of renewable energy and not the resilience ones. Few owners have implemented strategies to allow buildings to use their renewable energy generation capacity at times when the grid is unavailable. Islanding capabilities or interconnection to a microgrid is just beginning to be implemented for critical assets. On-site storage capacity is also just beginning to come online.

When more widely deployed, energy storage and microgrids or islanding can contribute to the everyday resilience of the grid and not just in response to a hazard event. Utilities can dispatch these systems as they would any other generation source to meet peak demand or trigger storage systems when excess power is being fed

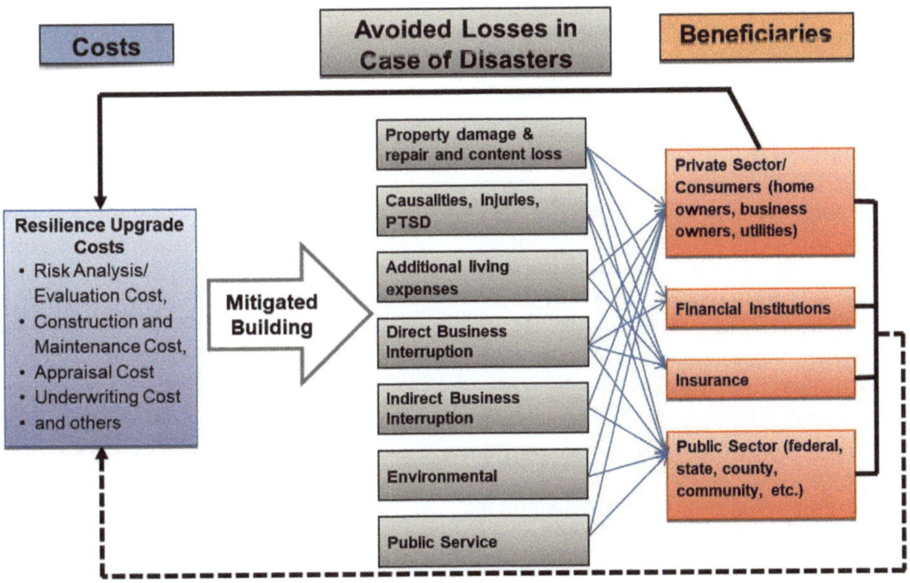

FIG. 2 Costs, benefits, and beneficiaries of mitigation measures. (Source: Courtesy, National Institute of Building Sciences.)

into the grid. Electric vehicles (EVs) add another dimension to the electric system—what if EV batteries can also be deployed as two-way energy devices? They could store energy when prices are cheap or the grid has excess capacity and discharge into the grid when electricity demand is high.

Today, the grid is largely a "just in time service" with limited capacity to store excess energy when it is not needed. As renewable energy generation increasingly connects to the grid, grid operators sometimes struggle with finding customers to use that energy at the time when it is generated. As a leader in renewable energy deployment, California has already seen this new paradigm shift. On occasion, the California Independent System Operator (ISO) has had to send power to neighboring states—sometimes paying the receiving utility to take it [13]. Increased deployment of energy storage will help to avoid the need for such approaches in the future.

The interconnection of the grid with other infrastructure systems is also becoming increasingly obvious and requires greater focus on these intersections. Vehicles are moving increasingly to electricity, buildings are becoming energy generators as they pursue zero-energy approaches, and technologies such as sensors and controls are interacting with the grid in new ways through the Internet of Things (IoT). The whole electric system is undergoing a recalibration to address these challenges and deliver the services necessary in today's world. These issues are not insurmountable, but do require a concerted, cross-disciplinary effort.

The buildings to grid intersection is becoming increasingly critical. To support increased energy efficiency and mitigate the release of greenhouse gases, leaders in the building community are moving increasingly to the widespread deployment of zero-energy buildings (sometimes also called zero net energy or net zero-energy buildings).

The Department of Energy defines a zero-energy building as, "An energy-efficient building where on a source energy basis, the actual annual delivered energy is less than or equal to the on-site renewable exported energy [14]." The definition clearly anticipates the push and pull of generation to and from the grid. As indicated earlier, when implemented with resilience measures in mind, this approach can serve the dual purpose of enhancing energy efficiency (and thus reducing greenhouse gas emissions) and serving as back-up power in the case of a grid-level outage.

Clearly, the role of buildings on the electricity grid is changing. A new set of solutions and strategies are necessary to support these changes and assure that the grid continues to meet our needs. A project called the GridOptimal Buildings Initiative is underway to develop metrics to help quantify and measure how individual buildings contribute to achievement of a safe, reliable, and resilient grid [15]. These types of metrics will ultimately support efforts by architects, engineers, and building owners to understand what steps they must take to be a good "grid-citizen."

In Chapter 4, Dagle focuses on the details of the electricity grid that contribute to its resilience. These features alone highlight just how complex electricity delivery is and the multiple nodes where it can fail. He identifies the emerging risks due to increased interconnectedness including cyber risk and cyber resilience and some of the nontechnical resilience solutions including mutual aid agreements. Many people outside the energy disciplines likely are unaware of these complexities. This chapter perfectly captures the intent of this book—build cross-disciplinary knowledge to allow greater collaboration to achieve resilience.

WATER SYSTEMS

Water systems are particularly vulnerable to both natural and man-made hazards—both chronic and acute. Following 9/11, concern grew about the potential introduction of pathogens or other harmful substances into the water supply. As Brashear mentions, this triggered a focus on risk management in the water sector. Water systems are also particularly vulnerable to flooding risk as they are often located near water sources or discharge points. More chronic risks include drought and the availability of water sources—this may become even more pronounced in some places due to climate change.

The NIBS Mitigation Saves Study examined several strategies to improve water system resilience. Four water and wastewater system mitigation projects provided a range of benefit cost ratios (BCR) between a $1.30 and $31 saved for each $1 invested [16].

- Elevating electrical equipment above the 500-year floodplain in Portsmouth, Virginia provided a benefit of $112 million from the initial $11.6 million invested, or a $10 to $1 BCR. Benefits primarily came from preventing business interruption.
- A Columbus Junction, Iowa water treatment plant was relocated at a cost of $4.6 million to move from within the 100-year floodplain to just outside the 500-year floodplain. The benefit was calculated at $5.9 million, mostly associated with business interruption, providing a BCR of $1.30 to $1.
- Iowa City, Iowa elected to reroute wastewater from its north treatment plant, which was located within the 100-year floodplain to the south plant where processes were elevated outside the 500-year

floodplain. The south plant underwent an expansion to accommodate the additional wastewater and mitigate some preexisting flood risk. The project produced $195 million in benefit at a cost of $54 million, for an overall BCR of approximately $4 saved per $1 invested. Property loss and loss of recreation on the neighboring Iowa River were the most significant benefits of the rerouting.

- Greenville, North Carolina undertook projects at both its water and wastewater treatment plants to reduce flood risk. The water treatment plant constructed a flood-protection berm and pumping station and the Northside Wastewater Treatment Plant raised a flood protection wall and a retaining wall. The project produced $212 million in benefit at a cost of $6.8 million, for an overall BCR of $31 saved per $1 invested. These relatively low-cost projects provided significant benefits in the form of business interruption and expenses for displaced residents.

In addition to examining actual projects attempting to mitigate flood risk, NIBS looked at a potential strategy for improving the water grid to enhance resilience to seismic risks. This involved addressing possible breakage and the need for fire-suppression following a seismic event. Trunk lines could be replaced with more resilient pipes (earthquake-resistant ductile iron pipe for instance) to reduce these risks. BCRs were produced for four western cities subject to seismic risk: San Francisco and Los Angeles, CA; Seattle, WA; and Portland, OR. The BCRs showed that such an approach could be cost effective in the former three cities with the BCR in Portland below $1 to $1. San Francisco saw an $8.30 benefit for every $1 invested.

The interdependence of community infrastructures and the importance of water systems to communities are increasingly obvious. As rebuilding gets underway in Paradise, California following the 2018 Camp Fire, the devastated water system is slowing the recovery. The system was irreparably damaged as meters and plastic pipes heated and melted, releasing benzene and other volatile organic compounds into the water and allowing in bacteria. The Paradise Irrigation District says it will be 2½ years before the water system is restored. Until then, the District will provide businesses and homeowners with water tanks [17].

As the NIBS study illustrates, the benefits and costs of mitigation vary considerably based on specific risks and types of strategies deployed. In Chapter 5, Hooker and his colleagues from the Onondaga County Water Authority provide a first hand look at how a water system identified its vulnerabilities and took steps to enhance its resilience.

REFERENCES

[1] Minnesota Department of Transportation and Department of Employment and Economic Development, Economic Impacts of the I-35W Bridge Collapse. http://www.dot.state.mn.us/i35wbridge/rebuild/pdfs/economic-impacts-from-deed.pdf.

[2] American Society of Civil Engineers, 2017 Infrastructure Report Card. http://www.infrastructurereportcard.org.

[3] Pew Research Center, Journalism & Media, Social Media Aid the Haiti Relief Effort. January 21, 2010. https://www.journalism.org/2010/01/21/social-media-aid-haiti-relief-effort/.

[4] E. Newberger, AT&T, Hit by Higher Natural Disaster Costs, Unveils 30-year Climate Change Model, CNBC, March 27, 2019. https://www.cnbc.com/2019/03/27/att-hit-by-higher-natural-disaster-costs-unveils-30-year-climate-change-model.html.

[5] J. Dwyer, K. Flynn, F. Fessenden, Fatal confusion: a troubled emergency response; 9/11 exposed deadly flaws in rescue plan, New York Times (July 7, 2002). https://www.nytimes.com/2002/07/07/nyregion/fatal-confusion-troubled-emergency-response-9-11-exposed-deadly-flaws-rescue.html.

[6] U.S. Department of Homeland Security, Interoperability Continuum: A tool for improving emergency response communications and interoperability, n.d. https://www.dhs.gov/sites/default/files/publications/interoperability_continuum_brochure_2.pdf.

[7] A. Adeel, M. Gogate, S. Farooq, C. Ieracitano, K. Dashtipour, H. Larijani, A. Hussain, A survey on the role of wireless sensor networks and IoT in disaster management, in: T. Durrani, W. Wang, S. Forbes (Eds.), Geological Disaster Monitoring Based on Sensor Networks, Springer Natural Hazards, 2019.

[8] E. Shittu, G. Parker, N. Mock, Improving communication resilience for effective disaster relief operations, in: Environment Systems and Decisions, Springer, June 2018.

[9] J.B. Houston, M.L. Spialek, J. Cox, M. Hardy, J. First, The centrality of communication and media in fostering community resilience, American Behavioral Scientist 59 (2015) 270–283, https://doi.org/10.1177/0002764214548563.

[10] J.B. Houston, Community resilience and communication: dynamic interconnections between and among individuals, families, and organizations, Journal of Applied Communication Research 46 (1) (2018) 19–22, https://doi.org/10.1080/00909882.2018.1426704.

[11] Multihazard Mitigation Council and Council on Finance, Insurance and Real Estate, Developing Pre-disaster Resilience Based on Public and Private Sector Incentivization, National Institute of Building Sciences, October 2015. https://www.nibs.org/resource/resmgr/MMC/MMC_ResilienceIncentivesWP.pdf.

[12] Multihazard Mitigation Council and Council on Finance, Insurance and Real Estate, An Addendum to the White Paper for Developing Pre-disaster Resilience Based on Public and Private Sector Incentivization, National Institute

of Building Sciences, September 2016. https://www.nibs.org/resource/resmgr/MMC/MMC_IncentivizationWB_Add.pdf.

[13] I. Penn, California invested heavily in solar power. Now there's so much that other states are sometimes paid to take it, Los Angeles Times (June 22, 2017). https://www.latimes.com/projects/la-fi-electricity-solar/.

[14] U.S. Department of Energy, A Common Definition for Zero Energy Buildings, Prepared by the National Institute of Building Sciences, September 2015, https://www.energy.gov/sites/prod/files/2015/09/f26/bto_common_definition_zero_energy_buildings_093015.pdf.

[15] New Buildings Institute, U.S. Green Building Council, The GridOptimal Buildings Initiative. https://newbuildings.org/resource/GridOptimal.

[16] Multihazard Mitigation Council, Natural Hazard Mitigation Saves: 2018 Interim Report. K. Principal Investigator Porter, C. co-Principal Investigators Scawthorn, C. Huyck, Investigators: R. Eguchi, Z. Hu, A. Reeder, P. Schneider, Director, MMC, National Institute of Building Sciences, Washington, D.C. https://www.nibs.org/resource/resmgr/mmc/NIBS_MSv2-2018_Interim-Repor.pdf.

[17] P.R. Russel, California towns rebuild after wildfires with resilience in mind, Engineering News-Record (April 10, 2019). https://www.enr.com/articles/46681-california-towns-rebuild-after-wildfires-with-resilience-in-mind.

CHAPTER 3

Managing Risk to Critical Infrastructures, Their Interdependencies, and the Region They Serve: A Risk Management Process

JERRY P. BRASHEAR, MBA, PHD

THE PRESENT SITUATION

The Challenge

Significant portions of the human, material, and financial losses from natural and man-made disasters occur because such events disrupt the delivery of vital services from lifeline critical infrastructures (CIs). These lifelines include energy, water/wastewater, transportation, and communications.[1] Without the basic lifelines, communities can neither recover from disasters nor long survive.

Many CIs have been marked by long-term underinvestment in maintenance, rehabilitation, replacement, and expansion, even as populations and demand for their services increase. This underinvestment has stretched existing infrastructures to meet higher demands by operating closer to and even above their design maximum loads and has kept aging facilities in service well beyond their original design lives. Such practices make them more vulnerable. One reason for systematic underinvestment is that the value of uninterrupted CI services is usually systematically and significantly undervalued in estimating the benefits of CI investments relative to their costs, including investments in risk/resilience improvements (more on this is in the Local CI and Regional Decision Context and Constraints, Regional Risk and Resilience and Conclusions and Implications sections).

Each lifeline CI is interdependent with the others, so the direct loss of one can be the failure that initiates a cascade of other infrastructure failures in a "chain reaction" that spreads losses widely throughout a region and beyond. In fact, the risk of disruption of the supply of lifeline services may be the single largest risk faced by individual CIs and the communities they serve. No stable, sustainable recovery from natural or man-made disasters is possible until these interdependent lifeline CIs are able to meet the minimum demands of other CIs and their collective customer base loads.

Secure, resilient CIs are necessary, but not sufficient, for secure, resilient modern communities. Sound risk/resilience management processes are necessary, but not sufficient, for CIs to be secure and resilient. Competent, effective risk management can guide efficient, systematic development of resilience and security of both CIs and communities. Effective risk/resilience management processes, then, may serve as a leverage point for a strategy to elevate the level of security and resilience of metropolitan-scale regions. Doing so across all major regions could constitute a sound, concrete plan for achieving national-level, yet locally tailored, physical security and resilience.

Relatively few CIs, however, routinely conduct sound risk/resilience management processes for their own assets and virtually none conducts risk/resilience management of the risks the CIs pose to their

[1] In addition to the four lifelines, the other CI sectors identified by the US Department of Homeland Security are chemical; commercial facilities; critical manufacturing; dams; defense industrial base; emergency services; financial services; food and agriculture; government facilities; healthcare and public health; and nuclear reactors, materials, and waste [15]. The risk/resilience management process in this chapter has been developed for the lifelines because of lifelines' centrality to resilience, but they apply to all CIs and many other entities whose mission is to provide goods and services through a physical process.

Optimizing Community Infrastructure. https://doi.org/10.1016/B978-0-12-816240-8.00003-3

communities. The challenge, then, is to develop and deploy management processes that identify and quantify risks to CIs and regions, design and value risk-reduction and resilience-enhancement options, implement the best options, and evaluate their performance in reducing risks once implemented. The approach must capture the risks posed by interdependencies among CIs and enable regional communities to participate in choosing among options and to contribute to the financing of cost-effective CI and regional security and resilience.

This chapter explores how CIs and the communities they serve can understand their risks due to natural hazards, human attacks, and dependencies and interdependencies and leverage that understanding to select cost-effective portfolios of risk/resilience improvement options to increase resilience for the individual CIs and the community as a whole. Data sharing; solid, common methodologies supporting comparability across CIs; analyses and investment decision-making; and the engagement of community or regional level champions are necessary to achieve community resilience in the most cost-effective manner.

Critical Infrastructures and Communities at Risk from Interdependencies and Resource Constraints

Risk is a function of the consequences of an unwanted incident and the likelihood of its occurrence (i.e., likelihood-weighted consequences, or expected values). The consequences of concern are typically financial losses to the utility, costs of recovery and restoration, human casualties, and loss of economic activity of the community. *Resilience* is the ability to withstand major negative events and continue meeting demand for service or to return service quickly if it is impossible to avoid temporary outage. The degree of resilience can be measured as the risk of outage. The lower the risk, the more resilient, to a limit of zero, where no outage is possible (i.e., perfect resilience). Most, but not all, options to reduce risk also enhance resilience. To make this chapter more readable, it uses the language of risk and risk reduction, and implicitly includes resilience and its enhancement.

A complete risk/resilience management process systematically completes five necessary and discrete phases:

1. Defines which specific assets are critical to carry out the CI's mission and the specific threats to them that could disrupt that mission;
2. Calculates current baseline risk of specific threats to specific assets;
3. Devises and estimates the value of risk-reduction options;
4. Invests in the options that meet value criteria; and
5. Evaluates the actual performance of the funded options in the field, continuing, redirecting or terminating programs based on their level of success.

Risk management at the level of specific threats to specific assets or subsystems stimulates definition of very specific operational/engineering options to reduce risk and enhance resilience. These options can be valued by conventional management criteria. Yet, very little true CI risk management has been undertaken beyond the first two phases. Such risk management is often done to comply with high-level federal mandates requiring risk analysis, or as a formality to fulfill a condition for eligibility for federal grants.

No risk management process is complete for an individual CI unless it fully accounts for its dependencies and interdependencies (D&Is) on other CIs and key suppliers. Interruption of services provided by others may render the CI inoperable. Moreover, any regional risk management approach must include D&Is because the potential for cascading CI failures due to D&Is is central to regional risk and options for managing it. Only when the full set of potential CI outages is defined can their impact on regional well-being be estimated or managed. Because CIs, especially the lifelines, are essential to regional well-being, calculation of the true risk and the benefits of risk reduction to the community must be considered.

The implication is that some level of cooperation and information sharing is necessary for any of the CIs to complete even their own risk management processes. Most CIs are highly reluctant to share information about their vulnerabilities or possible unreliability of service under stressed conditions. Fear of losing public confidence, divulging information that could be used for malevolent or competitive purposes, and potential legal liabilities all limit the willingness of CI managers to enter into agreements to exchange the information required to calculate and manage D&Is.

Recent developments and new insights in CI and regional risk management, however, suggest solutions to these limitations. The development of a pragmatic, bottom-up, fine-grain engineering/operational risk management process and tools to execute it, the emergence of highly secure information sharing processes, pragmatic approaches to D&I analysis and management, and greater attention to the social and economic risks born by the community are at hand or on the horizon. This chapter describes an integrated risk management process (RMP) that has been developed and field tested at the CI level over the past decade and a half and how it has been extended to address D&Is and regional community risks from an engineering/operational basis. This chapter is confined to physical and cyber/physical risks (e.g., physical results of tampering with process control systems). Work has begun on integrating it with the cybersecurity work of the National Institute of Standards and Technology.

Local CI and Regional Decision Context and Constraints

Rare is the infrastructure that uses risk/resilience management on a regular basis in its routine planning and budgeting processes—where it can create the greatest value. The author has yet to identify one that has completed all five phases. Rarer still is the CI that systematically addresses risks arising from its D&Is on other CIs and key suppliers, even though these are widely acknowledged to pose some of the greatest risks CIs face. Furthermore, operational-level CI risk management conducted from the perspective of regional community well-being is virtually nonexistent. This is despite the fact that the potential risks of losses of CI services to the community are typically orders of magnitude greater than the risks of losses to the CIs themselves.

These are some of the more important conclusions drawn from more than 100 interviews with decision-makers and analytical staff in lifeline CIs, local agencies, and regional public–private coalitions. There are exceptions that are encouraging, including the following:

- *Emergency managers and responders* in the public sector are required to meet a number of Federal Emergency Management Agency (FEMA) guidelines, including applying a very crude "risk analysis" that is primarily geared to "preparedness" in the form of planning emergency responses, stockpiling emergency response materials and building capabilities.
- *Water utilities* serving more than 3300 people were required by the Bioterrorism Act of 2002 to conduct a vulnerability assessment [1]. Several tools emerged to assist in meeting the requirement and a significant number of water sector personnel were introduced to risk analysis, if not risk management, at that time. The America's Water Infrastructure Act of 2018 [2] renews this mandate, requiring all community water systems serving more than 3300 persons to conduct risk/resilience analysis—covering both physical and cyber risks—on a staggered schedule of deadlines beginning March 31, 2020, with emergency response plans, based on the risk/resilience analyses, to follow within 6 months. The sole recognized developer of voluntary consensus standards for the water industry, the American Water Works Association, has developed a risk analysis and planning process that meets the Act's requirements and is preparing a 2019 updated version.
- *Electricity transmission owners* have recently become much more interested in the physical security of their assets. For example, the North American Electric Reliability Corporation (NERC) is a regulatory authority with responsibility for reliability of the electricity supply. NERC issued NERC-CIP 014: Critical Infrastructure Protection which was mandated for use by 2014 [3]. A former NERC official involved in writing the standard explained that the independent assessments were included "to inject into the compliance process…deeper understanding of physical security and a fresh set of eyes" [4]. NERC-CIP 014 advances six requirements similar to the five phases of risk/resilience management outlined earlier, but with a strong emphasis on independent verification and assessment:
  1. Initial identification and risk assessment of critical assets;
  2. Independent, third-party verification review of the initial risk assessment;
  3. Coordination between operator and owner of critical facilities about control centers;
  4. Threat and vulnerability assessment;
  5. Development and implementation of a physical security plan;
  6. Independent, third-party assessment of steps 4 and 5 [3].
- *Highways and bridges* are required by the Moving Ahead for Progress in the 21st Century Act (MAP-21) to conduct risk-based asset management plans to be submitted to the Secretary of Transportation for the entire National Highway System [5].

The most frequently observed pattern triggering attention to risk was a seriously negative incident, usually with a significant incident-caused outage, casualties or massive damage, that demonstrated a need for remedial investment. Few CIs currently use formal risk analysis. Of those who do, fewer were satisfied with their current processes. A number of these potential users reported conducting a process simply to comply with requirements from higher authorities, usually to qualify for grants (that do not depend on the results of the risk assessment), rather than actually basing decisions on the results.

Most of those who do some form of risk analysis limit it to estimating risk (often with questionable methodology). That is, they perform the first two phases of the five needed. Without conducting a net benefit or benefit/cost analysis (i.e., without knowing that the options actually efficiently reduce risk) they may be selecting the wrong projects to pursue. This is likely to lead to the misallocation of limited resources and foregoing available benefits. Almost all of the potential users, however, have business processes that could readily contribute to and use the results of risk analysis including asset management; continuity planning; strategic, capital, and operational planning; operating and capital budgeting; and performance appraisal.

A few very forward-thinking jurisdictions and utilities have adopted more sophisticated risk management into their routine management processes. Many of the more advanced systems use risk methods that are unique, proprietary, methodologically questionable, or narrowly threat-specific, and cannot readily be transferred or integrated. Outside of these, lifelines and local jurisdictions have actually performed very little risk analysis, and no resilience analysis beyond business continuity/continuity of government planning. They generally see resilience as synonymous with reliability or as equivalent to risk reduction. The availability of personnel with time for risk management and/or funds for hiring consultants to undertake substantive analysis is sharply limited. Requirements from an external authoritative source (e.g., higher government, industry standards, regulatory agency) can ease arranging the allocation of the time and limited funds to risk analysis because it removes the need to justify the work.

One reason for this limited use is the widely held belief among local agencies and many lifeline operators that if disaster strikes, the federal or state governments will step in to pay for recovery and restoration. They doubt the value of investing in prevention, protection, or pre-event mitigation. One respondent went so far as to say, "Investing 100-cent dollars of local taxpayer or ratepayer money *before* a future event that most likely won't happen (at least on my watch) seems irrational compared to paying 25-cent dollars of local taxes [the typical local share, with 75% from the federal government] *after* the event has become a certainty, if and when it ever does." Clearly, the business case must be made for using risk analysis and investing in risk mitigation at all—or require high-quality risk management as an eligibility criterion for federal and state assistance.

At the same time, state and local officials and infrastructure operators increasingly recognize the need to better understand the impact of major hazards on interdependent CIs. They have expressed interest in using a simple, low-cost, transparent, and manageable process to prioritize and justify actions and investments in security and resilience. Ideally, their own staffs could routinely carry out such a process, perhaps with a minimal level of training and the availability of technical assistance as needed. Increasingly, their focus is on pre-event prevention, protection, and mitigation (including resilience), as well as post-disaster collaborative response and recovery and restoration of critical assets and systems. Those organizations that are interested appreciate expert advisors for both process and substantive suggestions on risk-assessment options, but cost and time remain serious constraints.

A near universal issue, especially in the private sector, is the fear of legal liability and negligence suits associated with conducting risk analyses and then experiencing casualties or damages due to a known risk that was determined to be too low a priority to justify investment. Another issue is the costs associated with identifying risks that require substantial investment to mitigate, but generate little or no incremental revenue, while reducing potentially catastrophic losses if the threat becomes a reality.

Most infrastructure managers were sensitive to the essential role their service plays in the well-being of their community. Several public-sector owners spontaneously raised the issue of balancing risk reduction for their own systems with maintaining or restoring service rapidly to their customers. Several indicated that it is crucial to address the economic impacts on the community as well as the utility as part of the risk analysis, "especially when there's not enough return on investment to make the business case using only direct impacts to the utility," as one local utility official said. However, none of the respondents had a way to estimate the magnitude of impacts on the community.

This contributes to a second near universal issue: the systematic understating of risks and benefits of available options to mitigate risk. This arises from the CI's exclusive focus on internal, direct losses from an incident, consistent with standard management practices. One of the major direct impacts is loss of revenue due to service outages, which of course is valued at the prices they would have received had there been no incident. Prices for electricity and water/wastewater services, however, are set by regulation and based on recovering operating costs, debt service, and limited dividends (in investor-owned utilities)—not on the actual *value* of the commodity. Recently, there has been an increasing recognition that cost-based prices significantly understate the true value of the service [6].

The internal, direct focus misses all the "spin-offs" and "ripple effects"—the externalities in economists' terms. Different from most industries, these externalities can be orders of magnitude greater than the direct, internal consequences, leading to systematic underestimation of risks and benefits. The true value is the sum of all the values added by customers and customers' customers and on down the line—virtually the entire regional community—enabled by the CIs' services. Undervaluing the benefits, especially of risk mitigation investments that lack revenue, makes these investments seem less attractive than they should be to be funded.

Virtually all the respondents were keen to better understand their risks and possibilities of outages of the CIs and suppliers they rely on. All were acutely aware of their dependencies and interdependencies, especially between electricity and water, and many have taken steps to reduce this vulnerability with back-up electricity generators and interconnections with other water systems. Few have done actual risk assessments of the threat of utility or supplier outages, so the scale of the back-ups may or may not correspond to the respective threats.

For the most part, resilience is equated to reliability of service delivery, continuity of business, continuity of operations planning or continuity of government, and addressed by continuity plans and exercises. Across the nation, numerous utilities and service providers are incorporating resilience into their continuity planning and are beginning to join with other organizations and associations focusing on community and regional resilience. A CI's rapid return to service affects their employees and their families, which in turn can affect the employees' attendance and effectiveness at work.

Several CI owners suggested linking any new risk management methods directly with on-going internal processes such as asset management, continuity planning, and/or economic and community development. Integrating approaches increases the likelihood that the methods would be sustained over time, potentially leads to savings in the costs of the analysis efforts and facilitates synergies in the investments. The water, electricity, highways and other sectors have taken up asset management mainly to address the risks of seriously aging assets, but some have also extended it to include other hazards, including financial and cyber ones. Many mentioned the need to find a way to measure security (risk) and resilience (fragility or expected outage) in ways that can be reported to and understood by top management, rate-setting boards, local governments, customers, the general public and state and national agencies, especially those that provide grants.

Most CI operators had not thought about whether risk and resilience tools should be comparable across sectors. Those who had considered comparability expressed the view that comparability would have many advantages, especially in information sharing, conducting interdependencies analyses, and in better educating elected officials and their budget staffs, rate-setting bodies, and the general public. For large investments in long-term security and resilience, selling these enhancements is necessary for the investments to be made.

Many respondents expressed significant concern about the locally pressing manifestations of climate change. Additionally, along the east, west, and Gulf Coasts, the concern is coastal storm surge, tsunamis, and sea-level rise. In the Midwest and South, the issues are severe ice storms and snow in winter, major flooding with spring snow melt, and tornadoes and derechos in summer. Much of the West has been experiencing extreme drought, rampant wildfires and occasional earthquakes. Virtually all of them are seeking solutions, but formal risk analysis and option valuation are seldom seen as parts of the solution.

In most areas, the relevant agencies, for example, emergency management, public health, public works, and the respective utilities (whether publicly or investor owned) are "siloed" from one another, with little or no interaction, so interdependencies are virtually never analyzed beyond the decision of whether to acquire back-up electricity generators. CI managers are acutely aware of the issue, but lack the tools and data to address it.

All this suggests that there is a small but potentially growing demand for risk management if it can be simple, acceptable to peers and risk experts, include D&Is and all hazards explicitly, capture the interests of the broader community, and supports quantitative analysis not only of current risk and resilience but the value of risk reduction and resilience-enhancement options in terms that can be related to dollar costs for use in making the business case to invest.

GOAL, OBJECTIVES, AND DESIGN REQUIREMENTS FOR AN INTEGRATED CI-REGIONAL RMP

Goal and Objectives

The overall goal of a risk and resilience management process for CIs and regional communities is to achieve the greatest amount of risk reduction and resilience enhancement possible for the available amount of resources. Said another way, the goal is to reduce the loss of lives and impacts on health, livelihoods, property, and community function due to unwanted man-made and natural events as much as possible within reasonable, available budgets. A competent, integrated risk management process for the system of CIs in a region is one that:

- Manages "all-hazard" risk to each CI and is accepted into ongoing management and decision processes so as to achieve the greatest direct risk reduction possible at any given level of available budget. This includes calculating the CI's baseline risk, valuing and choosing risk-reduction options and evaluating the actual performance, with appropriate adjustments for effectiveness and efficiency.
- Captures the most significant dependencies and interdependencies at least among the lifeline CIs and their major suppliers and supports D&I risk management at the CI level.
- Manages the regional risks CI outages present to the community's well-being, including major D&Is, to achieve the greatest regional community risk reduction possible at any given level of available budget. This entails and supports management of regional risks at baseline, option valuation, and performance evaluation.

Design Considerations

Accordingly, the RMP should operate at three integrated levels (Fig. 3.1) as covered in greater depth in the Risk Management Process Description section:

- *The individual CI*, the prime mover that manages all-hazards risks defined as specific threats to specific assets or subsystems at the level of operational/engineering assessment, as these risks directly affect the CI;
- *Dependencies and interdependencies*, which require collaboration with other CIs and suppliers to calculate each CI's risk of service outages due to direct threats or outages by key sources of critical materials; and
- *The regional community*, which examines the CI- and D&I-level results as they affect the well-being of the population of the region served by the CIs.

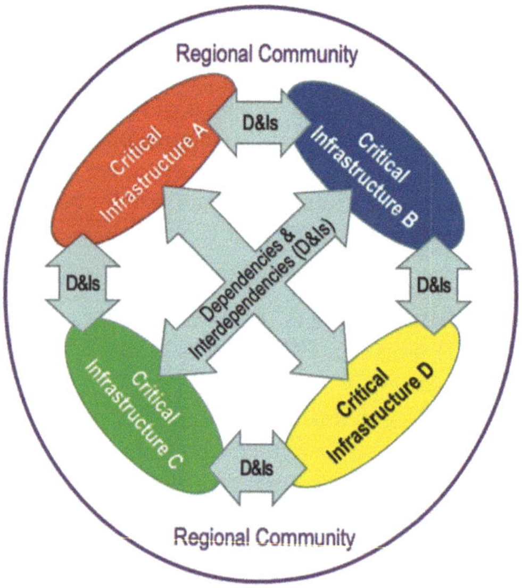

FIG. 3.1 Analysis proceeds from each CI's threat–asset risks, then to their interdependency risks, and then to the regional community's risks.

Each level should comply with generally accepted risk management standards, be transparent and repeatable, and include all five phases of the risk management process. The three levels are necessarily integrated, as each provides information to the process that other levels require, and each requires information that only others can provide. Most CI risk approaches address only the first of these—ignoring D&I risks and risks to the regional community well-being—so they are necessarily incomplete and can mislead decision-makers into significant misallocation of risk-reduction resources.

Of course, an RMP with results that are not used is a failure. Adoption of the RMP requires that it be cost-effective to employ; its results are credible, understood, and accepted by both the personnel who conduct the assessments and the decision-makers who commit resources to reduce identified risk; its process is transparent, logical, repeatable, and within the bounds of contemporary risk management principles.

To assure cost-effectiveness, the RMP relies on any models available to the analysts. Of primary import are the "mental models" carried in the heads of the operators, engineers, and line managers who run the CI on a daily basis and deal professionally with maintenance and periodic rehabilitation and the models they currently have in place. Their experience, often acquired over decades of service, and models that they routinely use generally support very subtle and flexible anticipation of the behavior of their systems under diverse and sometimes extreme conditions [7]. The RMP assumes that they can make as quick and accurate—or quicker and more accurate—estimates of key risk elements than most built-for-purpose formal simulation models of the CI that could be built at reasonable cost. Most CIs have some form of system simulation model(s) that they use for other, but related, purposes that can be used to buttress the judgment of the professionals (e.g., the hydraulic models of water systems).

Early in the development of the RMP, it became apparent that a trade-off would be necessary to meet the expectations of contemporary risk management principles and the pragmatic requirement that the process be used for actual decisions. Contemporary risk management principles require capturing the inherent uncertainties in the estimated elements of risk and including them expressly in risk, resilience, and benefit calculations. A draft RMP methodology based on these principles was distributed for review by CI professionals representing the target users. The reviewers rejected the approach entirely because relatively few are familiar with probability distributions combined through some type of simulation.[2] This aspect of RMP in particular caused extensive and absolute rejection of even trying basic risk analysis methods. If the process is rejected, it cannot contribute to improvement.

Decisions affecting the risks CIs face are made at least annually and often more frequently, *whether or not* an RMP is used. In the absence of an RMP, CIs tend to ignore risks or to focus investments on threats with the most horrific consequences, regardless of their likelihood of occurrence, significantly distorting the resulting resource allocation. Because of the design team's belief that even a simplified RMP would produce better decisions and more risk reduction than no risk analysis at all, the decision was made to "keep it simple" by initially using simple point estimates of the needed terms, linear functions for the risk and resilience, and a simple product form for risk and resilience. This ignores, for the present, their uncertainties except for sensitivity analysis. This made the approach acceptable to the potential users and its results understandable to decision-makers.

To capture the uncertainties in the estimates, sensitivity analysis is suggested. Such analysis defines how far off the estimates would need to be to change the resulting decisions and whether that value is plausible. A later, more advanced version of the RMP will explicitly capture uncertain quantities as probability distributions, modeling correlations among the variables as dependent uncertainties, and replace the product function with a simulation. The output of these improvements would display risk as a probability curve. Almost always, once the basic, single point/linear approach is mastered, organizations seek greater sophistication without prodding. Operators and engineers become frustrated with trying to make single point estimates of terms that are inherently uncertain, so they begin to ask to use ranges, then likelihood distributions, contingent correlations and finally, some way to combine them into the equivalent of the simple product function. In these cases, the more sophisticated approach is seen as a solution to *their* frustrations and *their* problem and the more advanced method is adopted readily. Even if this approach is not adopted, the simplified approach will still provide a more effective resource allocation than no risk analysis at all.

A final consideration is the "granularity" of the analysis. The RMP adopts a "grass roots," "bottom-up," engineering/operational approach. The CI-level analysis is based on simple scenarios of a specific threat, defined in detail, and a specific asset, defined by name and function in the context of the overall CI system. The threats would be a common set across all participating CIs in a region to support analysis of D&Is and to support D&I analyses and to allow aggregation of risks and benefits. The assets could be single pieces of machinery, a function, a group of personnel, or a geographically contiguous subsystem of the overall CI system. These features make the analysis pragmatic and readily understandable to both analysts and decision-makers. They also drive sound, concrete risk reduction options, based on the detailed understanding of the operators, engineers, and line managers who run the system.

The basic three-level system has had only one limited testing at the regional, multi-CI level that was interrupted for unrelated reasons, but proceeded far enough to determine the basic feasibility, acceptability, and effectiveness of the approach using a common risk analysis process and a common threat set across diverse lifeline CIs in a major metropolitan region [8].

Risk and Resilience Definitions

For use with CI and regional security and resilience, *risk* is the loss of something of value weighted by the likelihood of an event leading to its loss. Other definitions of risk, such as financial market volatility, colors reflecting rankings on a matrix (a "heat chart") of likelihood and consequences, and a variety of "qualitative" risk methods are not used in the current RMP and are not encouraged. More formally, risk here is defined as a function of the likelihood of an unwanted event and the magnitude of its consequences:

$$Risk = f(\textit{Likelihood of occurrence}, \textit{Magnitude of consequences})$$

This basic relationship has been the foundation for gaming, insurance, and corporate and personal risk

[2]Such a simulation, called a Monte Carlo simulation, is a technique for estimating a quantity where key attributes are uncertain. It randomly samples from probability distributions of the uncertain variables, makes whatever calculations are necessary, and then repeats the process a large number of times (100, 1000 or more) to produce a combined calculation expressed as a distribution that captures the uncertainties in the underlying variables. The mean of this distribution is its expected value.

management since the Renaissance [9]. In risk analysis pertaining to security and resilience where some of the threats are man-made, it is useful and common to divide the likelihood of occurrence into two parts—the likelihood the event will occur at all (called "threat likelihood") and the conditional likelihood that, *given the event has occurred*, the likelihood that the estimated consequences will result (called "vulnerability"). Vulnerability of a malevolent attack can also be interpreted as the likelihood of success of the attacker. So,

$$Risk = f(Threat\ Likelihood,\ Vulnerability, Consequences)$$

In the simplified process using point estimates of these terms, the risk function becomes the product:

$$Risk(R) = Threat\ Likelihood(T)\ \times\ Vulnerability(V)\ \times\ Consequences(C)$$

This definition is often summarized as "R = TVC" or "R = T * V *C." The consequences of concern in the RMP, and the respective risks, are (1) human casualties (fatalities and serious injuries), (2) direct financial losses to the CI, (3) service outages, an indicator of the absence of resilience, and (4) economic losses to the regional economy. These discrete risks can be combined when it helps decision-makers to make the comparisons they need to allocate their resources into a combined direct loss to the CI and a combined loss of regional well-being, as discussed in the Regional Risk and Resilience section.

Resilience is the ability to withstand an undesirable event without an interruption of service or, if that is not possible, to restore service rapidly, both reflected in a minimal risk of service outage. A fully resilient asset or system would, of course, have an outage risk of zero across all threats considered.

Unless aggregated, risk is specific to a particular consequence, threat and asset, for example, the risk to human life from a wildfire, or the risk of financial loss from a 7.0 earthquake on a 36-inch water main running from point *x* to point *y*. In addition, risks can be *combined* to support effective individual decision-making based on (1) the overall *combined direct risks to a CI* and (2) the broader *combined risks to the regional community's well-being*. Combined risks include all the consequences by converting human casualties to dollar quantities.

For CI-level analysis, casualties are valued by "human capital" (lost productivity, broadly comparable in magnitude to direct liability awards) [10,11]. For regional analysis, casualties are valued by the "statistical value of a life" based on willingness to pay, as estimated by the the most relevant federal agency [12,13].

Because risks by this definition are statistically "expected values"—that is, probability-weighted consequences, not "expectations" in common use—they may be carefully added together to summarize risk at any useful level of aggregation (e.g., by hazard type, by organization, by group of organizations, or by the regional whole).

RISK MANAGEMENT PROCESS DESCRIPTION

Overview of the RMP

As outlined previously, the RMP is a three-level process. Each level leads to the decisions required to reduce risk and enhance resilience—CIs' direct risks, risks due to D&Is among CIs, and the risks faced by the regional community.

The *first level* begins with each individual CI estimating current baseline risk to the CI, the valuation of risk-mitigation options, choices, implementation and execution of options, and evaluation of actual performance. It focuses on the direct risks and benefits to the CI. The process is "all hazards" in considering man-made as well as natural threats and hazards.

It starts with a standardized set of threats to establish a common basis for information sharing, interdependencies analysis and regional risk analysis. Table 3.1 shows the range of standard or "reference" threats used in one risk management standard.

The parties to dependency analysis must reference a common set of threats. Users may add or delete specific threats to assure the most important ones are included while keeping the total number of threats manageable. The types of response options considered include prevention, protection, consequence mitigation, emergency response, and recovery/restoration. This level of CI risk management is designed to integrate with a CI's existing risk management, asset management, business continuity, and emergency response planning for continuing attention to risk reduction.

The *second level* of the RMP addresses dependency/interdependency/proximity risks. This analysis builds on the individual CI's RMP results to manage D&Is *among* CIs, their suppliers, and their high-priority customers. This level is made up of the members of supply chains, with special emphasis on lifeline CIs. Dependencies risk—the risk that necessary inputs will be cut off due to an external event—can be the greatest risk facing a CI. For example, loss of electricity due to a major storm or earthquake may cause a water system to fail—impacting the whole community—if prudent precautions have not been taken. Interdependencies represent mutual reliance (e.g., the electricity distribution system that

TABLE 3.1
Reference Threats Used in ANSI/AWWA J100-10

DIRECTED REFERENCE THREATS BY THREAT CATEGORY								
Aircraft	Assault Team	Maritime	Vehicle Borne Bomb	Contamination of Product	Directed / Sabotage	Theft or Diversion	Cyber Insider	Cyber Outsider
(A1) Helicopter	(AT1) 1 Assailant	(M1) Small Boat	(V1) Car	C(B) Biotoxin	S(PI) Physical—Insider	T(PI) Physical—Insider	(CI1) Insider	(CO1) Cyber Outsider Attackers
(A2) Small Plane	(AT2) 2–4 Assailants	(M2) Fast Boat	(V2) Van	C(C) Chemical	S(PU) Physical—Outsider	T(PU) Physical—Outsider	Trusted Insider / Accidental (CI2)	(CO2) Criminal Group
(A3) Regional Jet	(AT3) 5–8 Assailants	(M3) Barge	(V3) Midsize Truck	C(C) Explosive	AS – Active Shooter	T(CI) Cyber Insider		(CO3) Terrorist
(A4) Large Jet	(AT4) 9–16 Assailants	(M4) Deep Draft Ship	(V4) Large Truck (18 Wheeler)	C(P) Pathogen		T(CU) Cyber Outsider		(CO4) Foreign Intelligence Service
				C(R) Radionuclide				

RANDOM (NATURAL DISASTER) REFERENCE THREATS BY HAZARD CATEGORY						DEPENDENCY & PROXIMITY THREATS	
Earthquake	Floods	Hurricanes	Ice Storms	Tornadoes	Wildfires	Dependency	Proximity
EQ1 - PGA 0.0 – 0.2	F1 = – 100 Year Flood	H1 - Category 1	I0 - Minimal Damage	T0 - Fujita 0	W1 - FRG1	D(T) - Transportation	D(P) - Proximity
EQ2 - 0.2 – 0.4	F2 – 500 Year Flood	H2 - Category 2	I1 - Isolated outages	T1 - Fujita 1	W2 - FRG2	D(C) - Key Customers	
EQ3 - PGA 0 – .4 – .08		H3 - Category 3	I2 - Scattered outages	T2 - Fujita 2	W3 - FRG3	D(E) - Key Employees	
EQ4 - PGA 0.8 – 1.1		H4 - Category 4	I3 - Numerous Outages	T3 - Fujita 3	W4 - FRG4	D(S) - Key Suppliers	
EQ 5 - PGA > 1.1		H5 - Category 5	I4 - Prolonged Outages	T4 - Fujita 4	W5 - FRG5	D(U) - Utilities	
			I5 - Widespread Outages	T5 - Fujita 5			

powers the water system requires a continuous supply of water to cool the computers and high-rise buildings that control power distribution). CIs must collaborate with one another and with their suppliers and key customers to manage dependency risks. Proximity risks are due to co-location with other risky facilities or activities (e.g., railroad tracks used to carry toxic gases that pass through residential and commercial areas).

The *third level* is the *regional public's* security and resilience. This level of analysis and decision-making is necessitated by dependencies among CIs and the risk of regional cascading failures. It is also necessitated by the common situation where response options have high *public* benefits but smaller direct CI benefits. Such options would likely be rejected by the CIs due to low or negative direct cash returns or to budget constraints, causing significant public benefits to be foregone, often without consideration. These benefits are "external" to the CI's conventional decision scope, so they are what economist call "externalities" or "market failures" to achieve the highest total benefit affordable at any given level of cost. The regional analysis estimates current regional baseline risk, benefits of options to the regional public (both CI options and options based on other considerations), and regional performance evaluation. Based on the CI's estimates of their own service outages, the regional economy is modeled to estimate the lost gross regional product and the impacts on the regional industries most affected by the threat. These results assist CIs, local and state governments, public–private coalitions and fully private coalitions to participate in choosing options and providing incentives for CIs and others to optimize regional public benefits even while optimizing each CI's direct benefits. In brief, each level optimizes its own investments based on its *respective* benefits, costs, and sources of funding.

Having these three levels of risk analysis allows constrained optimization of the allocation of resources to achieve the greatest benefit within the available resources, serving both the fiscal integrity of the CIs (and other participating organizations) and capturing the most benefits to the public possible at any given level of available resources.

Basic Method Selection

To identify a management process engineering approach to risk and resilience management, the author evaluated all available public domain, nonproprietary, and/or readily available critical infrastructure risk analysis/management tools, processes, and methods [14]. Federal agencies with responsibility for critical infrastructure protection under the 2013 National Infrastructure Protection Plan [15] nominated the methods they had sponsored for evaluation as risk management tools. The primary selection criteria for the candidate methods were the following:

- Support for all five phases of risk management decision-making or readily made to do so.
- Generally comply with contemporary risk management standards, including the following:
  - Use of ratio scales of measurement (numerical scales having equal intervals between numbers and natural zero) *instead of* ordinal scales (ratings, rankings, rankings named by numbers, e.g., 1 to 5 or 1 to 10 scales) in estimating risk and resilience. Ordinal scales cannot be used to support the calculations needed for quantitative optimization, as required by the overall goal [16]. Ordinal scales may be used for ranking of threats, assets, etc., to focus on the most important assets and threats.
  - Be defensible, repeatable, and transparent.
  - Be amenable to upgrading to full uncertainty estimation and Monte Carlo simulation.
- Include both man-made and natural hazards.
- Support D&I information sharing and analyses.
- Guide "optimal" risk mitigation and resilience-enhancement investment decisions under budgetary and other constraints. "Optimal" is in quotation marks because it is always a constrained optimum due to limited information, prior commitments, limited resources, etc.
- Be implemented in accessible, open-source software or have a reasonable expectation such software will be developed.

Of a total of 21 methods evaluated, 10 were eliminated immediately because they were not risk analysis methods—three estimated useful elements of risk but were not complete risk methods; and seven were detailed surveys producing index scores for benchmarking but did not calculate risk or risk-reduction benefits. Of the remaining 11, five used ordinal scales, which, as noted, cannot support quantitative optimization as the objectives require [16]. Of the six remaining tools, five assumed the threat likelihood to be 1.0, or certain to happen, calling this "conditional risk."[3] Because threat likelihoods can vary by many orders of magnitude, especially in "all-hazards" analysis including both natural and man-made events, these

[3] Instead of $R = TVC$, they assume risk is the product of vulnerability and consequences given that threat likelihood is certain (or unknowable, so assumed away), that is, $R = VC$ and $T = 1$, risk is the product of vulnerability and consequences, given the threat likelihood is 1.0.

tools also cannot support resource allocation. Eliminating these left only American National Standards Institute/American Water Works Association Standard J100-10 Risk and Resilience Management of Water and Wastewater Systems (J100-10) [17] as viable for the objectives of risk/resilience management.

AWWA J100-10 is a recognized American National Standard [17]; the product of more than 15 years of development, testing and adaptation from Risk Analysis and Management for Critical Asset Protection, version 3 (RAMCAP Plus, see text box) [18]; adapted expressly to the water sector [19]; and designated under the SAFETY Act to reduce user liability to certain threats [20]. It has been applied extensively in the water and wastewater sector for actual risk/resilience management since its 2010 release. It is currently being updated for release as ANSI/AWWA J100-19 in 2019. The newer version has been used for this chapter and is referred to as simply J100.

As noted earlier, the America's Water Infrastructure Act of 2018 (42 U.S.C. 300i-2, Section 1433) [2] mandates risk and resilience assessments and emergency response plans by all water systems serving more than 3300 people. Its deadline for water systems serving more than 100,000 people is March 31, 2020. The Act directs the Environmental Protection Agency Administrator to recognize technical standards developed by voluntary consensus standard organizations. AWWA J100 is recognized by the American National Standards Institute (ANSI) as such a standard, the only one expressly for water and wastewater utilities. J100-19 will be released in time to support the Act.

RAMCAP Plus, A Generic RMP

Shortly after the terrorist attacks of 9/11, at White House request, senior managers of industry and infrastructure convened to set a national strategy for managing infrastructure risks, with emphasis on terrorism risk. The result was to develop a generic RMP that could, with little adaptation, be used by *all* critical infrastructures and many other organizations. An iterative development process of design, testing, and refinement produced Risk Analysis and Management for Critical Asset Protection (RAMCAP; its third major revision was called RAMCAP Plus). The basic approach of the current RMP was set as part of that strategy, including a seven-step process, use of threat–asset pairs as the basic focus of analysis, the formulation of $R = T \times V \times C$, common threat sets, estimation of a baseline risk and risk-reduction benefits of risk mitigation options, comparability of results for aggregations of risks and comparisons within and across infrastructures. RAMCAP or RAMCAP Plus was successfully field tested at multiple sites each for nuclear power plants and nuclear waste systems; chemical manufacturing; oil refineries; liquified natural gas terminals; college campuses; water/wastewater systems; electricity distribution; police emergency response; fire suppression; emergency medical services; emergency dispatch; and highways, roads, and bridges. It has also supported regionally coordinated executive decision-making. RAMCAP is no longer being developed, but has continued in the form of ANSI/AWWA J100 and in several proprietary methods and tools.

CI Level RMP in Brief: Five Critical Decisions

J100 is a seven-step process used iteratively to carry out the five phases of risk/resilience management. J100-10 is implemented by open-source commercial software developed by AEM Corporation called Program to Assist in Risk and Resilience Examination [21], and by software developed by the Environmental Protection Agency, the Vulnerability Self-Assessment Tool [22]. Although their developers intend to update their tools once J100-19 is released, the current versions will support the following analysis with a few simple adaptations.

Risk/resilience management—at either the CI level or the regional level—in its simplest form is to make five critical decisions—corresponding to the five phases introduced earlier—in a transparent, logical, and data-driven manner (Fig. 3.2; note: the ovals labeled "D&I Analysis" and "D&I" are discussed in the Dependencies and Interdependencies (D&Is) section):

1. **Scoping:** *What is the scope of the risk management process?* Which assets and which threats are to be included in the analysis? This decision requires a review of the utility's mission and functions to define which assets and subsystems are most critical to the mission and concentrating on those. Which threats to consider? After reviewing an initial, standardized set of man-made and natural threats and hazards—adding and deleting as necessary to assure the most important ones are included—define which pose the greatest threat to the mission-critical assets. The highest ranking "threat–asset pairs" (or "TAPs) are basic units that are analyzed through the whole process.
2. **Baseline Risk Analysis:** *What are the most significant current risks, and which are the most important to reduce?* Each term in the risk equation is systematically estimated for each TAP and combined to calculate overall risk and resilience levels if nothing further were done. These risks are sorted as to those the organization will accept, those it will transfer (e.g., by buying insurance), and those it will consider for direct risk-reduction options. The

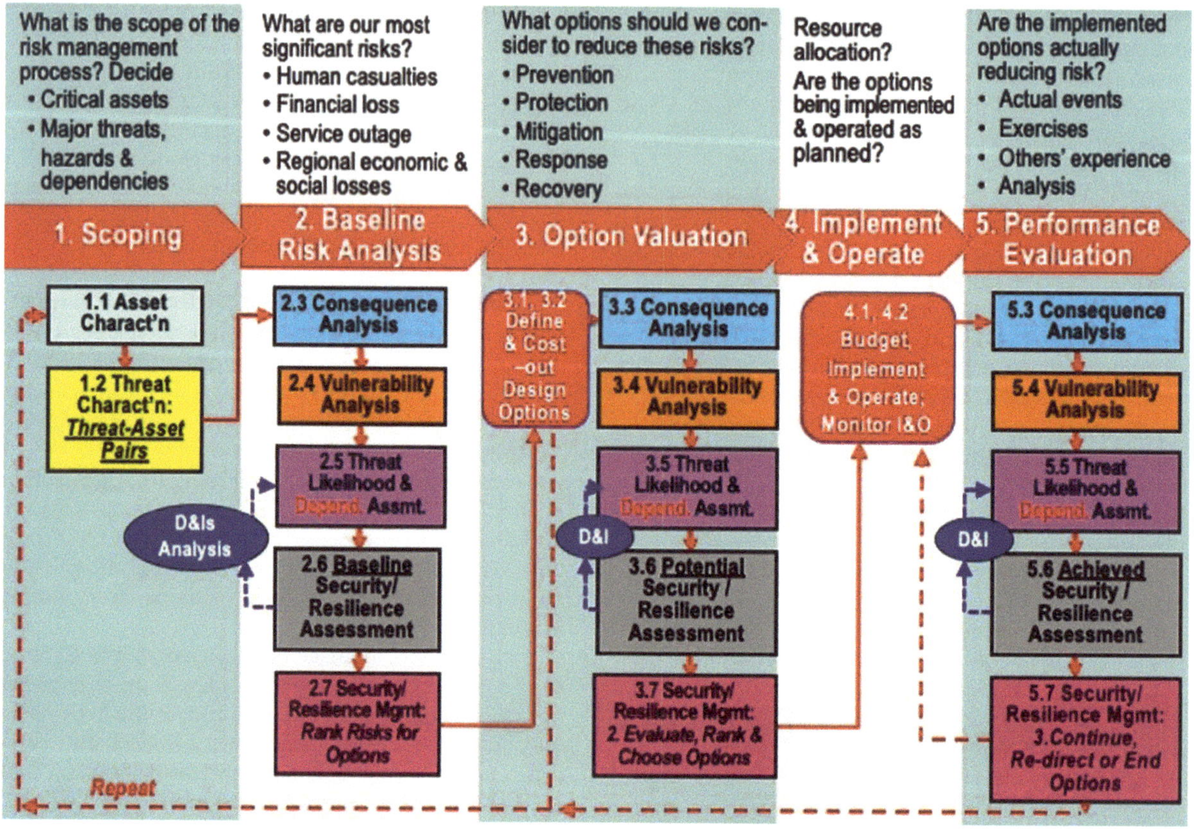

FIG. 3.2 Overall workflow of the CI risk management process.

highest-risk TAPs marked for risk-reduction options are taken to the next decision.

3. **Option Valuation:** *What risk-reduction and resilience-enhancement options are worth investing in?* Options are defined in enough engineering/operational detail to estimate how much they would reduce one or more of the terms in the risk equation for each of the TAPs. The terms of the risk equation are reestimated assuming the acceptance of the option. The gross benefit of each option is the difference in risk without the option (the above baseline) and the risk with it. Life-cycle and initial (first-year budget) costs are estimated for each option. *The value of each option is its net benefit*—the gross benefit less the option's life-cycle cost, both in present values if the option lasts more than one year. The most promising options are taken to the next decision. Note that the benefit/cost ratio or return on investment are not used in this decision (other than as a "tie-breaker") because they measure the *efficiency* of the option in generating benefits per dollar, not the total value, which the RMP seeks to maximize. If there were no budget constraint, these ratios could be used because all options that produce more benefits than they cost would be moved to the next decision.
4. **Implementation and Operation:** *Which options will actually be budgeted, implemented, executed and managed?* This is the executive decision to select the options that warrant inclusion in the organization's operating and/or capital budgets, competing with all other proposals. The options that are funded are implemented, managed, and monitored. Options not accepted in this decision might have significant benefits but could not be funded due to the CI's required rate of return and budget or bond constraints. Some of these unfunded options could enhance the security and/or resilience of organizations dependent on the CI or for the region as a whole. These are retained for reanalysis in the dependencies and regional analyses (Sections Dependencies and Interdependencies (D&Is) and Regional Risk and Resilience). The value of doing so

is that sound, technically feasible options with significant benefits to others are identified for possible incentives or outside funding.

5. **Performance Evaluation:** *Are the implemented options achieving their objectives by actually reducing risk and enhancing resilience?* All terms of the risk equation are reestimated based on *actual* conditions and *actual* risks calculated for each TAP. Data for this reestimation are drawn from actual events (to this or to other organizations by analogy), full-scale exercises and peer reviews, and reestimation in light of changed circumstances and new insights. This supports the decisions to continue, expand, redirect, or terminate the implemented options. Because the implemented options changed the original situation (and time has passed), *this analysis also serves as part of the baseline analysis for the TAPs for which options have been chosen.* New or previously deferred TAPs are added for a more comprehensive baseline and the process iterates, starting with the first decision, scoping for the second full pass through the process.

FIVE PHASES TO ADDRESS THE FIVE CRITICAL DECISIONS

Decision 1: Scoping

Fig. 3.3 highlights how the first of these decisions are supported. The highest priority assets (or subsystems of assets) are those most essential to carrying out the mission(s) of the organization, broken down as to function, subsystems, and assets.

The most important threats are those that have the potential to cause great damage to the critical assets, that is, those that most compromise the mission(s) of the CI. An initial threat set is common across organizations in the region that are conducting risks assessments at a given point in time. Analysts must consider all the initial threats but may eliminate all that do not apply or are not high priority and may add threats that are locally significant but not on the common list, for example, mudslides in certain parts of California. Having all participating organizations address a common initial list allows later analyses of dependencies due to common events, aggregation at a regional scale, and any comparisons needed in making the critical decisions.

Each TAP is considered briefly and summarily ranked as to severity of its consequences. The highest ranked TAPs are retained for the analysis, while all others may be deferred. Deferred TAPs may be included at any point if they are later determined to be potentially significant. Because the process is iterative, subsequent cycles will include previously deferred TAPs. Only the higher priority pairs in each cycle are analyzed to minimize analyst fatigue that could distort the analysis.

Decision 2: Baseline Risk Analysis

The baseline risk is the current level of risk posed to the critical assets by the identified threats assuming no change in security or resilience features already in place. It serves to alert the organization of the specific risks it faces and as the point of comparison for valuing options for improving security and/or resilience.

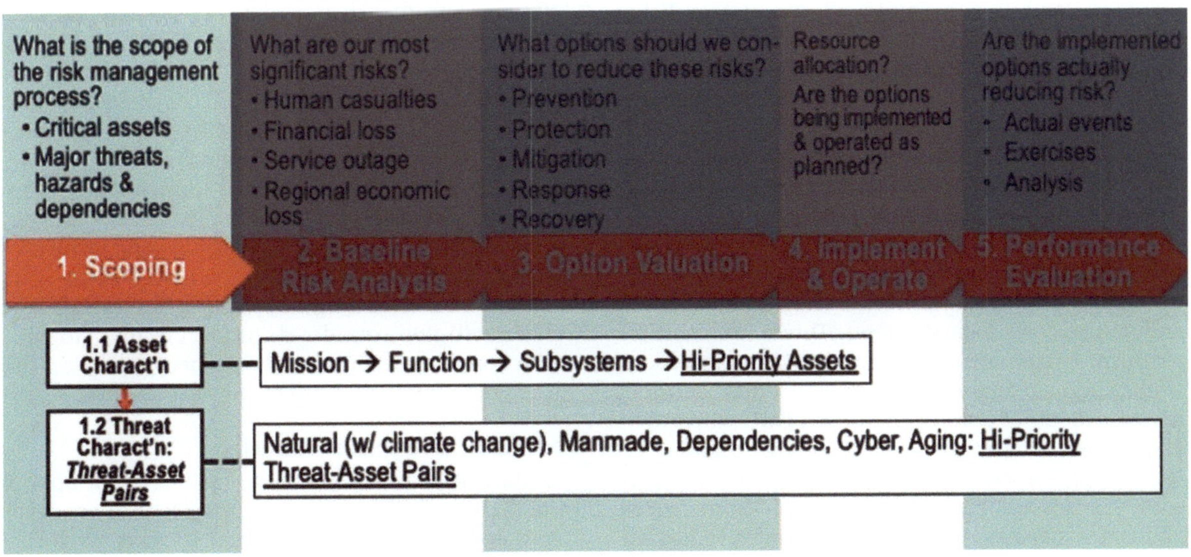

FIG. 3.3 The scoping decision produces ranked threat-asset pairs (TAPs).

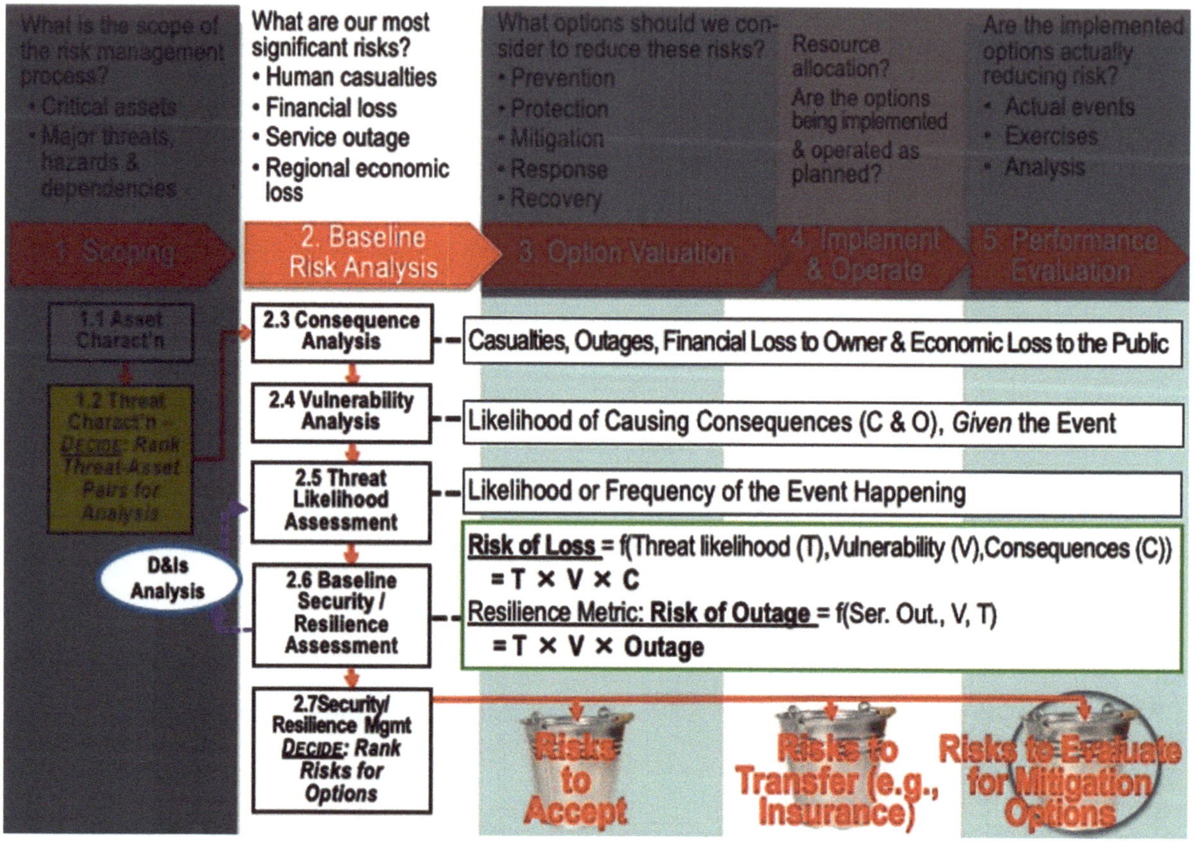

FIG. 3.4 The current baseline risk determines which TAPs to try to mitigate.

A team of operators, engineers, security/emergency managers, and line managers of the system being analyzed estimate the consequences (including outages), vulnerability, and likelihood for each threat–asset pair (Fig. 3.4). Engineering and management consulting firms also offer these services but are best engaged with significant participation by CI personnel. The estimates are generally discussed among the participants until a general consensus is reached.

Consequence estimates are made by closely considering the TAP scenario. Vulnerability is estimated through processes like event tree analysis, path analysis, and vulnerability logic diagrams. Threat likelihood of terrorism is estimated by federal intelligence agencies (if they begin to provide them) or by a "proxy" method, a simple model of terrorist choice of a TAP relative to its alternatives, based on published information on terrorist preferences. Threat likelihood of natural hazards is based on historical frequencies from federal and state databases. Where events are massive in scale relative to the region (e.g., earthquakes or hurricanes), the TAPs involving that threat are estimated at the same time so that they may be consistently assessed, and the respective TAP-level results may be added together as needed. These estimates are combined to calculate risks for each TAP, which are ranked by their risks, using any or all of the respective risks—casualties, financial losses, outages, or regional economic loss—individually or in their combined form. Massive threat TAPs are included as a single event for this ranking. Based on these rankings, the TAPs are sorted by which will be accepted by the organization if they occur, which will be transferred (e.g., by buying insurance), and which will be considered further relative to risk-reduction and resilience-enhancement options. This third group passes forward to the next decision.

Decision 3: Option Valuation

For the risks to be mitigated, the analysts define options and their life cycle and initial costs. Options are developed by systematically considering ways to reduce each term in the risk equation—threat likelihood,

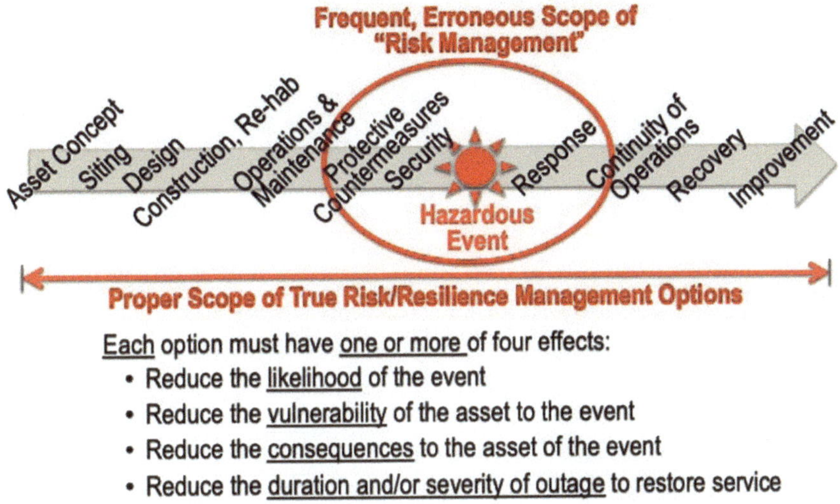

FIG. 3.5 The full scope of potential options to reduce risk.

vulnerability, and/or consequences (including outages). The scope of the types of options to be considered, illustrated in Fig. 3.5 covers the whole life cycle of an asset, from initial concept for an asset or program through all preevent phases (initial conception, siting, design, construction, operations and maintenance, protective countermeasures, and standing security arrangements) to postevent (emergency response, continuity of operations, recovery, and possibly improvement over the preevent asset). Rebuilding or replacing a damaged asset may present an opportunity to employ newer or less risky technologies, for example, replacing chlorine gas with an alternative disinfectant in a water treatment plant. Regardless of where along this continuum an option lies, it must reduce the threat likelihood, vulnerability, consequences, and/or the length or severity of service outages. Only these effects will reduce risk or enhance resilience, so they are the only way benefits can be generated by the options.

Life-cycle cost is the present value of the sum of all current and future, after-tax cash outlays and may include anything from initial design and siting studies, through construction and operations, maintenance, periodic rehabilitation, decommissioning, and environmental restoration. If these extend over more than 1 year, the present value is calculated using the weighted average cost of capital to the organization. The initial cost is the commitment in the first budgetary period.

Having established which TAP risks appear to justify consideration of mitigation and developed options for doing so, the next critical decisions determine which options should be accepted for implementation.[4] The process for doing this (Fig. 3.6) is to reestimate the consequences (including casualties, financial loss to the organization, service outage and regional economic impacts), vulnerability, and threat likelihood given the implementation of the option.

The gross benefits of the options are the difference between the baseline risk and the risk with the option. If an option is durable (lasting more than 1 year), the stream of future benefits is discounted by the same rate as used for discounting life-cycle costs. Benefits may be stated in terms of any or all of the respective consequences, for example, reduction in casualties, direct losses to the organization, outages or economic losses to the regional economy or in terms of the respective combined values. The primary terms that may be arrayed for decision-makers are shown in Table 3.2.

Decision-makers may consider each risk individually (they may or may not be correlated), or in the combined, or "inclusive" forms. The combined forms are most useful when the number of options is large and/or they must be considered in the context of competition with other, nonrisk-related investment options as

[4]Because the estimates used in the RMP are based on the judgments of a group of experienced security, technical and operating personnel, they are acceptable for ranking options for the next step, valuation, and choice for implementation. In cases where a major investment is to be made in a durable asset or subsystem, these quick estimates may be judged as too hasty and ad hoc to justify massive cash outlays. In such cases, the user is advised to follow this decision with an engineering feasibility study based on a detailed design of the option.

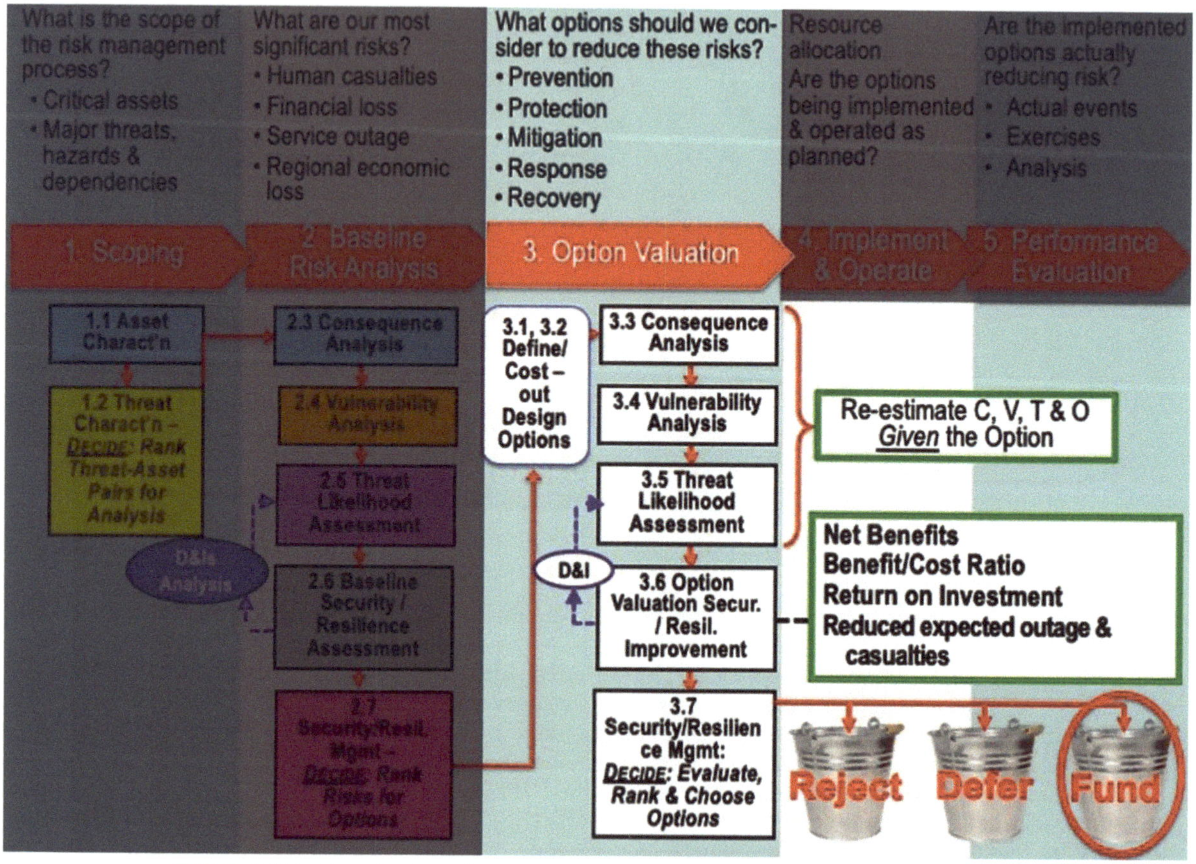

FIG. 3.6 Valuation of options by re-estimating risk and funding high net benefit options.

in a budgeting process or in trade-offs associated with asset management strategies. In such cases, a single dollars-to-dollars comparison can be most useful. Net benefit, as opposed to other metrics of merit, is the preferred decision criterion because the objective is to seek the maximum possible risk reduction under budget and other constraints. As noted, ratios like benefit/cost ratio and return on investment measure the efficiency of the options in reducing risk, not the amount of risk reduced, introducing the possibility of selecting a portfolio of efficient options that collectively produce less benefit than using net benefits.

Decision 4: Implement and Operate

At this point, the process transfers from the analysis team to the executive level (aided by the analytical staff) to decide which among the recommended options to include in the capital and operating budgets. Risk-reduction and resilience-enhancement options must compete for resources with all other organizational requirements, so it is necessary to justify them in the same terms as other budget requests are justified. This varies from organization to organization but usually is a minor algebraic variation from net benefits (preferred), benefit/cost ratio or return on investment (both after tax for taxable entities). Once selected, the funded options are implemented, operated and managed by the appropriate units of the organization. The process of implementation and operations should be tracked and recorded and compared to the plans for use in interpreting the results in the next decision.

Decision 5: Performance Evaluation

Different from most other risk management approaches, the RMP includes a structured performance evaluation phase to assure that the options selected and implemented are, in fact, reducing risk and/or enhancing resilience—the fifth and final critical decision. This decision is supported by a third pass through the process of estimating consequences (including outages), vulnerabilities, and threat likelihoods, but this time estimating the levels actually achieved. This is in addition to

TABLE 3.2
Criteria for Decision-Makers to Use to Select Options for Implementation

Decision Criteria: Both Options Valuation and Performance Evaluation	RISK OF:				COMPOSITE RISKS	
	Human Casualties	Financial Loss to the CI Organization	Service Outage	Economic Loss to the Region	CI Organization Composite (Note 1)	Regional Public Composite (Note 2)
Gross benefits	X	X	X	X	X	X
Net benefits	X	X	X	X	X	X
Benefit/Cost ratio	X	X	X	X	X	X
Return on investment	X	X	X	X	X	X

Notes: 1. Organization Inclusive Risk = Σ (Direct Financial Loss to Organization Risk) + Σ (Casualties Risk × "Human Capital" Cost). 2. Regional Public Inclusive Risk = Σ (Economic Loss to the Region Risk) + Σ (Casualties Risk x Statistical Value of Life) + Σ (CI Financial Losses Risk).

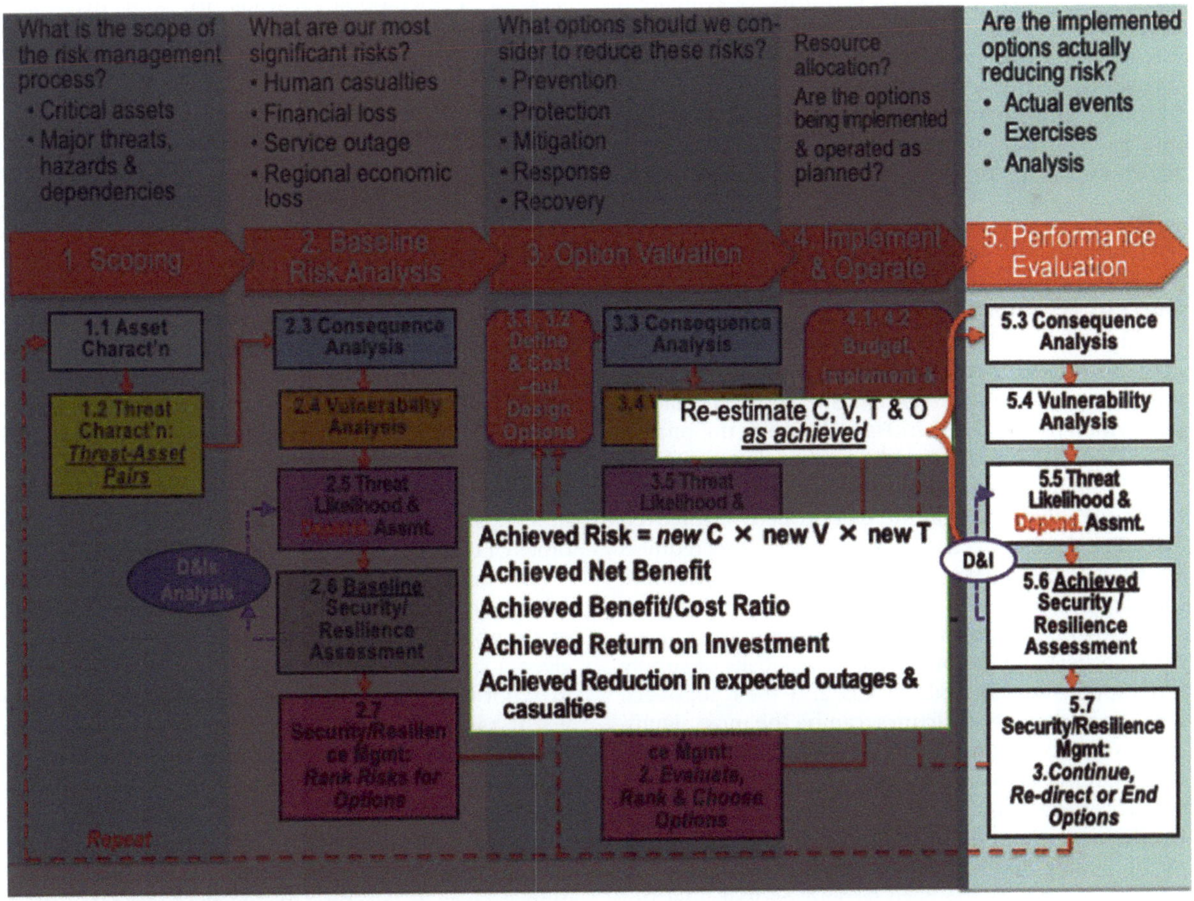

FIG. 3.7 Performance evaluation calculates how much risk has been reduced and drives continuations and redirection by results.

appraising whether the options have been implemented and operated as planned. Fig. 3.7 shows this process and the achieved levels of risk. These new estimates are informed by actual events both locally and by other organizations using similar risk-reduction options, full-scale and tabletop exercises, and analysis of the current situation for the high-priority threat–asset pairs.

Because performance evaluation could present opportunities for self-serving, the RMP suggests an independent review or audit of the analytic process, estimates and results. Experts from a specialized risk management unit of the organization or from outside the organization could carry out the audits. Similar organizations in the same industry could conduct peer reviews of one another. The decisions based on this evaluation are whether to continue, grow, diminish, redirect, or terminate the options based on whether they are cost-effectively reducing risk and/or enhancing resilience.

Implementation of the selected options has altered the baseline situation, so this performance evaluation also serves as a portion of an updated baseline risk analysis for the threat–asset pairs previously included. In this sense, the performance evaluation and the new baseline risk analysis can be conducted at the same time to iteratively and continuously improve the security and resilience of the organization.

DEPENDENCIES AND INTERDEPENDENCIES

Current Situation in Managing D&Is

No CI-level risk management process is complete if it does not explicitly analyze and manage dependency, interdependency, and proximity risks. The ovals marked "D&I Analysis" or "D&I" in Fig. 3.1 signify the process of analyzing D&Is. These risks may be larger risks to the CI's ability to continue or restore service than any of the risks that occur on the CI's property. Moreover, extending the CI-level RMP to the regional level requires addressing the challenges posed by dependencies and interdependencies.

CIs are both suppliers of essential inputs (e.g., electricity, water) and customers for them. Risks of supply interruption due to D&Is of CIs and their suppliers, including other infrastructures, can be the most significant risks an individual CI faces and can lead to cascades of CI failures across whole regions and beyond with massive social and economic disruption, e.g., the Northeast Blackout of 2003 affected 50 million people in the United States and Canada and cost $6 to $10 billion [23]. It has long been recognized that interruption of supply of some CI services can be a major threat, one to be actively protected against—to the level that "feels right." Many CIs maintain back-up electricity generators at significant cost, for example.

Such "soft" criteria as "feels right" are used because CIs and regions lack the necessary information, methods, and models for true D&I risk analysis. Several investigators at National Laboratories have undertaken efforts to describe and model D&I behavior [24] for purposes other than regional CI risk management. So far, these are too aggregated, require too much highly sensitive information, and are too reliant on assumptions to guide investments by CIs. Although the Laboratories continue to work in this area on projects that appear promising, the tools currently available to CIs simply do not provide the information and modeling needed for CIs to analyze their D&I risks, especially at the level needed for the RMP. No one has yet produced an acceptable method for CIs to use in analyzing D&I threats at a level where concrete risk-mitigation options can be valued specifically. This lack of a methodology leads to both over- and underinvestment.

Using RMP for Dependencies and Interdependencies Analysis

The emergence of two new capabilities fundamentally improves the ability to measure and manage D&I risks. The first is the engineering/operational threat–asset RMP described in this chapter. The fine-grain TAP approach allows CIs to analyze D&Is to specific assets and subsystems under specific threat conditions and to develop and evaluate concrete options for reducing these specific risks, often in collaboration with the supplier CIs (background developments are described in several works [e.g., 8,14,18,19]). When this method is coupled with the RCISR Critical Infrastructure Security and Resilience InfoXchange described in the text box, the second new capability, CI users are able to communicate with other CI users to share RMP data in full confidence [25].

The necessary information a customer CI needs but does not have for analyzing its D&I risks is the likelihood of a supplier CI's outage, with its severity and duration, given a specific threat event. Although the customer CI may not have this information, the supplier does have it if it has carried out a competent risk analysis on its own assets using a common threat set. Assuming that the CIs and localities of a region conduct risk analyses using logically consistent definitions and processes that include outage risk, a common initial set of threats, and a secure and controlled information-sharing capability such as RCISR InfoXchange, the basic core dependency questions can be asked by customers and answered by suppliers:

At this specific geographic point (or defined area) on a map—the location of one or more of the customer's critical assets—what is the likelihood the supplier will be unable to provide service under this specific threat event? And, if that likelihood is not zero, how long will the supplier's service be out?

Answers to these simple questions allow the customer CI to complete its RMP dependency assessment for an asset at that geographic point or area under that specific threat because it has already estimated its own vulnerability and consequences of that asset under that threat.

The Regional Critical Infrastructure Security and Resilience InfoXchange

The Regional Critical Infrastructure Security and Resilience (RCISR) InfoXchange [25] is a "system of systems" that provides confidential communications, data (including from GIS), information, analytical tools and methods and "dashboard" displays to screened and qualified members. Members may include CIs, local and state jurisdictions, businesses and business associations, healthcare and public health organizations, and other stakeholders concerned with regional risk/resilience and incident management. It orchestrates and shares detailed data among vetted members to facilitate cooperative action before, during, and after major disasters.

RCISR InfoXchange is based on the integration of the Single Automated Business Exchange for Reporting (SABER)/XchangeCore data orchestration technology [26] and the Alliance for Community Solutions (ACS) [27]. SABER/XchangeCore, originally designed to facilitate coordinated, rapid recovery of businesses after a disaster, was originally developed by the US Department of Homeland Security (DHS) as the Unified Incident Command and Decision Support. It has matured to independence and is now privately offered as a set of web services that translate data formats in a two-way exchange, transform non-web data formats into web services, orchestrate common operational data into useful data packages, and facilitates secure communications. The ACS i-INFO Suite develops and maintains web-based communities of interest to facilitate secure information sharing and coordination. Both technologies have their own approaches to increasing community resilience.

The RCISR InfoXchange complements other incident management systems by providing user-defined standard information as well as controlled access sharing of highly sensitive data and information under rigorous information sharing protocols. One of its primary objectives is to facilitate data-sharing and analysis of D&Is and regional community decision-making, along with coordinating emergency management across CIs and public agencies.

Both customer and supplier CIs will be concerned about the confidentiality of the locations of their critical assets. It is not uncommon in certain businesses and industries that large customers require their suppliers to conduct analyses of the risk they will be unable to provide specific goods and services. The customer CI must list the locations to be queried, implying it has critical assets there, but can add additional, even random points, to disguise those that are actually critical.

The supplier CI will be able to answer because it has competed its own, logically and methodologically equivalent risk analysis using the same threats in which the probability and duration of outages under that event have been estimated. Note that the customer does not need to know the location or criticality of the supplier's assets that provide the service or the route by which the service is delivered—the specific, sensitive information that has impeded the willingness to share detailed data. That is, in addition to being shared under highly secure conditions, the information requested is of little or no use to malign actors seeking to damage the supplier CI or the customer CI, or in any liability or negligence suits, even if it were to be hacked. Furthermore, the initial supplier CI is also most likely a critical customer of the initial customer, for example, a water purification plant and distribution pumps need electricity to operate, while the process control computer of the electric utility needs water for cooling. This reciprocity — i.e., interdependency — once demonstrated, could also encourage collaboration among CIs in a region.

Such data may take the form of spreadsheets of asset locations, polygons on a geographic information system (GIS) layer, or other ways. With experience, this information exchange can be made automated and efficient. The focus is on the most critical assets of the customer CI, but over time may be extended to major community assets like hospitals and emergency command centers. To keep the analysis manageable, it should not include all assets of any system and not all systems.

These answers can be linked together using a network model based on a GIS for specific threat events and, by simulations, traced throughout the region to estimate the total outage for all analysis participants in the regional analysis, including those in any cascading failure across infrastructures. Once using a GIS map, potentially dangerous neighboring and nearby facilities are noted and analyzed as proximity risks—risks that would impact the CI if it had a major disruption. Integrated together, these answers provide the basis for each CI to complete its own risk analysis, including D&I and proximity risks. It also provides an overall regional analysis of dependencies and interdependencies that supports a regional analysis of risks and benefits to the public.

Once the capability for capturing and modeling dependencies and interdependencies exists, it can be used in the five critical risk management decisions to complete the respective CIs' analyses and to begin accumulating the data needed to address the regional public risk issues.

REGIONAL RISK AND RESILIENCE

Regional Community Benefits Defined and Calculated

To this point, the discussion has focused exclusively on the risks, costs, and benefits that directly affect the CIs in question. This is necessary and customary to maintain fiscal integrity, as in any other business. However, CIs are fundamentally different from other businesses and industries in three important ways. First, they have critically important public missions and obligations that transcend simply keeping revenues greater than costs because their production is essential to life, health and livelihoods of whole regions. Second, water, electricity, and sometimes other lifeline CI services are priced to recoup operating and capital costs, debt service, and a negotiated profit (for investor-owned CIs)—notably the cost of the service, not its value.

And, third, partly due to the first two, CIs characteristically generate enormous indirect "spin-offs" or "ripple effects" as their output is used by their customers, whose output is used by the customers' customers, and so forth down the line, giving rise to "multiplier effects." With CIs, especially the lifelines, the external costs of CI risks and the external benefits of CI risk reduction are huge—up to and including the total economic and human well-being of the entire regional community. At each level, value is added that would not be added if the CI's outputs were inoperable. These indirect effects do not enter into the decision-making of the CI because CIs receive no revenue past the first transaction. Indirect effects are experienced as the chain of customers adds value, but are "external" to the initial sale of services, called "externalities" by economists. Externalities may be positive (e.g., honey bees pollinate orchards) or negative (e.g., air polluters impose health costs on others). Externalities indicate that the service's price equilibrium cannot reflect the true costs and benefits of the services, a "market failure" because the price signals do not result in the socially optimal level of production [6]. In the case of risk reduction and resilience enhancement, these "products" are systematically underproduced relative to the socially desirable level. That is, the overall well-being of the region can be improved by increasing the investment in risk/resilience improvements. The CIs cannot be expected to undertake these investments beyond the level where direct benefits equal or exceed direct costs of the investments—unless incentives increase the benefits to the CI or reduce the CI's costs. Many government programs are designed to correct market failures. In the case of managing risk and resilience of CIs, many regions potentially have ways to provide such incentives through various stakeholder collaborations in addition to governments, as discussed in the sections on Regional Community Process and Incremental Funding, below.

Once the detailed, bottom-up engineering/operational risk analysis of the specific risks of individual TAPs and the additional risks of outages due to D&Is are included, the CIs initially determine which risk-reduction options they will undertake with their own resources. The analysis of regional risks and benefits to the community can then be undertaken. The objective of regional risk analysis is to minimize total regional risks by exploiting high-public-benefit options that otherwise would not be selected or would be undersized by the CIs because the direct benefit to the CI alone is insufficient to justify the cost or because of the CI's budget constraints. This situation arises when the direct benefits to the CI are modest but the benefits to the regional community are significant due to externalities. Most risk management processes proposed or used for CIs include only direct risks, costs, and benefits, ignoring their larger roles where externalities dominate.

The challenge is to find ways to identify and calculate externalities and to bring them more into the risk and resilience decision-making of the CIs and/or the regional community, as both have a rightful say in these outcomes.

Regional risk management may be initiated and performed by any organization or coalition of organizations that has access to the necessary data. This may well involve state and/or local governments in collaboration with CIs and other local economic organizations (e.g., chambers of commerce, industry-specific local coalitions, or ad hoc cooperation). Regional risk analysis allows identification of cases where significant benefits to the regional population will be foregone if the region relies solely on the CIs to address options from their own direct, internal benefits and costs. This knowledge allows the CI and these stakeholders to collaborate to provide incentives, subsidies or to take other actions that capture a portion of these benefits to the regional public.

Estimation of Regional Economic Losses

Given the estimated direct and dependency-based outages of all CI services, conventional input–output modeling can be modified to estimate the overall loss to the regional economy due to CI failures in substantial detail. The work by Haimes, Rose, Santos, and their collaborators [28–37] on risk and economic analysis of disasters extends conventional input–output (IO) and computable general equilibrium modeling to estimate loss in gross regional product (GRP) as a function of service interruptions of the respective CIs. GRP is analogous to gross national product—the sum of all value-added economic activity based on outputs of economic sectors less intermediate consumption. One group of such models is inoperability input–output models (IIM) and dynamic IIM (DIIM). These specialized models differ from conventional IO models in that they account for the natural resilience inherent in a regional economy. For example, an intended purchase of capital equipment may be deferred due to a regional scale disaster, but will most likely not be lost, whereas restaurant meals not eaten during a disaster are less likely to be made up later.

Establishing a regional baseline (no incidents and no changes from present), then re-estimating the GRP under the outage calculations of the TAPs sets a baseline for estimating regional economic risks and risk-reduction benefits that result from options that reduce outages. These simulations account for dependencies, interdependencies, and cascading failures. Individual CIs estimate their service interruptions, if any, for each TAP due to both direct and D&I impacts of the threat event. The total of the direct outages and D&I outages in each CI sector are the essential inputs to the IO model, which estimates the total lost GRP and the losses to each major economic sector, including households and governments. Lost jobs, wages, local income taxes, etc., can be estimated from the losses suffered by each economic sector in the region. Gross regional benefits are the difference in regional GRP without options and the GRP with the options.

Several government agencies, universities, and consulting firms have recently collaborated on an update of the widely cited 2005 report of the National Institute Building Sciences Multihazard Mitigation Council, which documented how every federal dollar spent on risk mitigation saves society an average of four dollars [38]. The updated version, *Natural Hazard Mitigation Saves: 2018 Interim Report* [39], showed that the ratio varied substantially across federal programs (from an unacceptable 0.2:1 for a particular bridge reconstruction to 11:1 for another bridge/road project and 31:1 for protecting water treatment facilities), with an average of four dollars saved for every dollar invested in infrastructure-focused risk mitigation.

The modeling for these studies is the same as the modeling required to enable the estimation of regional economic losses from threat events and the economic benefits of risk mitigation and resilience improvements. The required underlying regional economic data are available for purchase at reasonable cost from at least two sources. Many state, local, and regional jurisdictions already subscribe to these data and use them in

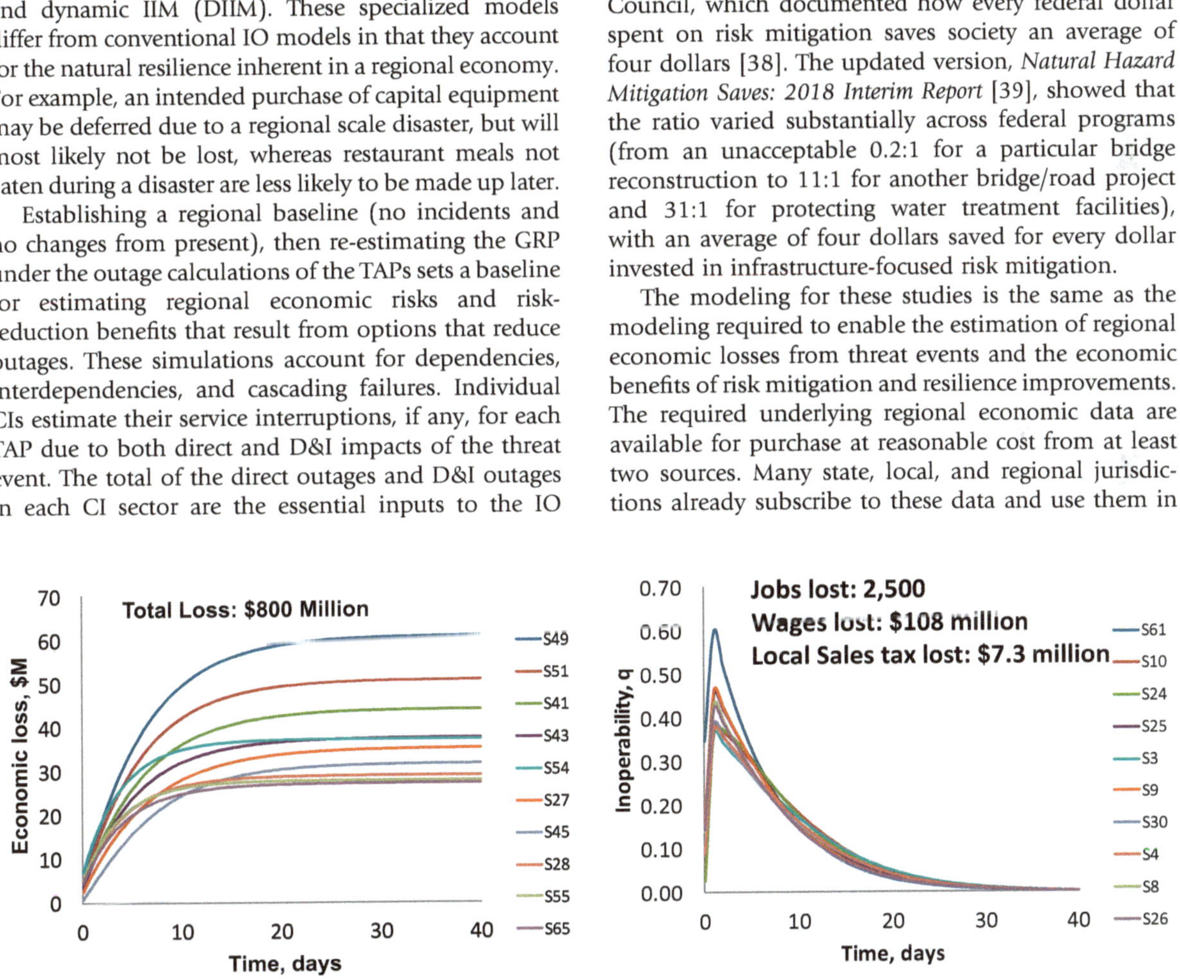

S49	Computer systems design and related services
S51	Administrative and support services
S41	Federal Reserve banks and credit intermediation
S43	Insurance carriers and related activities
S54	Ambulatory health care services
S27	Wholesale trade
S45	Real estate
S28	Retail trade
S55	Hospitals and nursing and residential care facilities
S65	State and local general government

S61	Other services, except government
S10	Apparel and leather and allied products
S24	Other transportation equipment
S25	Furniture and related products
S3	Oil and gas extraction
S9	Textile mills and textile product mills
S30	Rail transportation
S4	Mining, except oil and gas
S8	Food and beverage and tobacco products
S26	Miscellaneous manufacturing

FIG. 3.8 Loss of regional economic activity, total by industry and related impacts. (Source: [8].)

land use, economic and developmental planning, using very similar models.

Fig. 3.8 exemplifies the results of a DIIM as applied for a series of outages among interdependent infrastructures in a mid-sized US metropolitan region. The overall loss of $809 million in economic activity is made up of diverse sectors and represents a major negative event for this community due to interruptions initiated by outages in water and wastewater services alone. D&Is were not included in this estimated economic loss but would have increased it materially. Mitigation projects to reduce this loss could have significant benefits in avoided losses.

This loss, combined with the number of casualties multiplied by the value of a statistical life and health (human suffering) and the aggregate direct losses to the CIs and localities (because they will be recovered through the tax base or rate base), would serve as the combined estimate of regional consequences to the public of the initiating threat event. When risk-reduction options are being evaluated, this estimate allows consideration of the benefits to the regional public by analyzing the options that reduce outages. This, with the estimated consequences to the respective CIs, allows balancing of the interests of the CIs and the public, respectively. Such a clear statement of the economic and human dimensommunity Process

As with the CI and D&I level analyses, the regional level process deals with all five of the key risk management decisions:

- *Decision 1: Scoping* at the regional level starts with formation of a suitable coalition or partnership if one does not already exist. It must, however, contain at least one organization that is trusted enough by all the other organizations to be authorized to handle their confidential and sensitive information at the most detailed level, as facilitated and protected by the RCISR InfoXchange or its equivalent. Once the coalition has been formed, this decision sets the goals, boundaries, common threat set, and other terms of reference for the subsequent work.
- *Decision 2: Baseline Risk Analysis* estimates the baseline risk for the region, incorporating the results of the CIs and D&I analyses using the DIIM. In addition to the regional baseline risk, individual regional risks of casualties, financial losses to the CIs, total service days interrupted, and the economic losses to the region are each calculated to complement the regional risk estimate. In this step, the benefits of a "No-New Options" case (only the options likely to be funded by the CIs based on their own internal analyses) is also calculated for these terms. Obviously, the highest level of trust and information protection is essential for the exchange about options to be shared. At first, this is probably best done in face-to-face meetings of the highest-level personnel involved in the actual risk management. Based on these analyses, the coalition designates which threat–asset pairs should be considered for further risk mitigation from the regional community's perspective.
- *Decision 3: Option Valuation* determines which risk mitigation options beyond those that the CIs have planned to implement should be recommended for funding. Some options will have been analyzed and rejected by a CI because its direct benefits were insufficient to justify the outlay or because it was not feasible due to overall budget constraints. These options are included in this phase, as well as additional options suggested by the coalition members' examination of the untreated risks. Truly new options may require additional analyses at the CI and/or D&I levels, so an iterative process is expected. The regional community's incremental benefit on options being implemented by the CIs are examined as externalities to determine whether the coalition or others should provide incentives to the respective CIs to reduce the costs of the options enough for the CIs to undertake them. Those that have significant community benefits that would otherwise be forgone are recommended for regional funding or incentives.
- *Decision 4: Implement and Operate* allocates regional resources and/or incentives to capture incremental regional community benefits by providing the CIs with incremental funding for specific options with high regional community net benefits that the CIs would not otherwise undertake. These incentives would require the CIs to implement and manage very specific, technically feasible options, defined in detail because they are based on the engineering/operational analysis of specific TAPs or groups of TAPs. If these have significant net benefits to the regional public, coalition members, or others (see next section) may pay an amount needed to lower the CI's cost enough for the CI to fund the rest. Because the funding would be just enough incentive to make the option attractive—at the margin—to the CI based on direct benefits alone, the benefit/cost ratios on the outside funding would be exceptionally high.
- *Decision 5: Performance Evaluation* decides whether individual options are actually reducing risk and/or enhancing resilience. These assessments build on the comparable CI-level assessments and may augment

that information with regional table-top and field scale exercises, simulations, etc. Each implemented option will be subject to the decision as to whether it should be continued, enhanced/expanded, modified or redirected, or terminated. As noted earlier, the analysis supporting these decisions also supports the new baseline risk for a new full cycle of the overall RMP.

It is clear that there are aspects of dependencies and interdependencies that are too complex and amorphous to be captured fully in this coarse, mechanical analysis. Moreover, there are significant factors that are not included in this analysis, for example, nonparticipation of major CIs or major local industries, future locational and relocational decisions for major industrial facilities, start-ups that become major new industries, major technology changes for the region's current industries, or loss in productivity due to a traumatized workforce. For all these reasons and more, the estimates that come from this process are best considered a lower bound of actual consequences. Although such externality effects are real, they are extremely difficult to measure or manage. A lower bound based on "hard" data, if recognized as such, can be effectively used in decision-making. If the lower bound estimates make the case for an investment in an option, the true number would only enhance the case.

Incremental Funding to Generate Significant Regional Community Net Benefits

The RMP identifies where it would be prudent for CIs to invest their own resources based on the direct benefits and areas where investments could be induced by incentives based on the value to the community. For options not justified by direct benefits to the CI but that have significant net benefits to the regional community, outside assistance can be justified as correcting the market failure of externalities. For example, if an investor-owned utility requires a 10% return in its capital budgeting and a risk-reduction project investment returns only 8% in reduced expected losses, an outside party could subsidize the cost of the project by enough to raise the 8% to 10% because the outside investment is justified by the outsider's or the community's benefits. This "buy-in" strategy is likely to be the most cost-effective of incentives because they complement, not supplant, the CI's own investments and the mitigation options are based on detailed engineering/operational level risk analysis. This sort of buy-in could be local and private or could be from outside, public sources. It is especially attractive when it results in "public goods"—goods whose use cannot be denied to anyone desiring to use them, so cannot be priced, and the use of which does not diminish the amount that can be used by others, for example, national security through military preparedness or a sea wall to mitigate the risk of sea level rise for the whole community.

This sets up the possibility of a series of sequential investment tranches, each based on the self-interests of different outsiders. Specific tranches, in order of likelihood of success, are as follows:

1. **The CIs, themselves**. Risk reduction and resilience enhancement are, in the first order, the respective CI entities' responsibility to the extent they can directly capture enough benefits to justify the cost and have enough budget (and often regulatory approval) to make the investment. The results of CI-level risk analyses should drive investments by each CI based on its own, legitimate self-interest to the limit it can justify and afford. These entities have fiduciary responsibilities that require that their investments be justified by the benefits they accrue directly. Completing the above analytical process may well cause reallocation of internal funds to reduce risk and enhancing resilience.
2. **Other CIs, especially lifelines**. After participating in this detailed analysis of their vulnerabilities to D&I threats, CIs that supply each other with essential services might be willing to collaborate on options that make them collectively more secure and resilient.
3. **Other major customers**. This is well precedented today, as many major industrial customers pay premiums for the additional security of redundant feeder lines and highest priority restoration after an outage. Some of this collaboration could directly result in cost-saving coordination, e.g., opening Main Street only once for the restoration of water, sewer, natural gas, and electricity instead of opening it three or four times.
4. **Major regional employers or their associations.** Infrastructure failure of any significant duration makes employees' residences uncomfortable to uninhabitable. Employers are aware of their reliance on their skilled and experienced workforce, so they could find significant benefits to reducing the likelihood and duration of CI failures.
5. **Rate-payers.** Where large-scale regional community net benefits might be foregone due to a CI's inability to justify the investments, rate-setting and bond-rating agencies might well be convinced that these investments justify rate increases or improved bond

ratings because of the greater resilience of the regional community.

6. **Regional public–private partnerships or coalitions.** Direct collaborative investment by coalitions of local jurisdictions, major employers, and local civic groups might well back options with major community net benefits, especially where they result in public goods. Risk analysis from the public's perspective would identify these opportunities. The local business and industry community could be willing to participate to assure their communities, employees, and suppliers (including infrastructures), and their own facilities will be reliably available in a potentially disastrous event that could otherwise cause costly downtime, human casualties, or business failures.
7. **Taxpayers at any level.** Correcting market failures such as those resulting from externalities and providing public goods are second only to national security as justification for governmental action. Localities could provide direct support to CIs as marginal incentives for specific options. An interesting variation of this idea is the recent passage of California Measure AA, "Clean and Healthy Bay." Nearly 70% of voters approved an initiative to pay a $12/year surtax per property for 20 years to restore wetlands around San Francisco Bay. Here, the public of a nine-county area voted to tax itself to raise $500 million to invest in a natural infrastructure system [40]. State and federal governments have programs to enhance security and resilience of major regions that could be directed to specific options identified by CIs. A sizable amount of funding could be accumulated by federal interagency cooperation, each with the local and regional facilities of their respective missions, for example, GSA for federal facilities, DOD for local mission assurance, HUD for housing and redevelopment, FEMA for mitigation and postevent management preparation, and the infrastructure bank (if it is ever passed and funded).

This approach works if, and only if, a standard, repeatable, auditable risk analysis and option valuation process provides transparent, trusted, and protected information to determine the true return to the owner, the amount of necessary subsidy and the benefits to the potential funding source and/or the regional community. Without that, any buy-in or incentive strategy is subject to "free-rider" abuse, whereby an unscrupulous CI owner arranges justifiable investments to seem to require subsidy from the outside party to do what the CI would do without the subsidy. This suggests some sort of technical oversight or peer review of the risk analyses to assure that there is no "gamesmanship." The peer review at the CI risk management level could provide the needed expertise and experience.

Previous approaches to improving regional community security and resilience have not had the benefit of the "bottom-up" engineering/operational risk analysis core that the RMP provides, so the regional level approach has never been tried, except for limited but successful testing in a pilot project that contributed to the development of the RMP [8]. Wholesale grants for broadly defined regional preparedness for generic threats, as used today, have seldom been used to enhance CIs' security and resilience.

The result of such sequenced, complementary analyses and funding decisions holds the potential for a maximum total risk reduction given the total funds available to the CIs and to the other interested stakeholders. In the absence of such a process, the allocation of funds by each entity and level based on their own direct benefits and budgets will produce materially less total beneficial investment because each successive share will be less sharply focused on the specific, diverse, highest priority needs and, even with cost-sharing, is likely to require a larger federal share to induce CI investments. Conventional "top-down" federal programs work by paying essentially the first tranche, even with cost-sharing. The RMP process will reduce more risk because it assists in funding highly specific investments that are both technically feasible based on the detailed engineering/operational analysis and are the CIs' and communities' highest unfunded priorities based on their own detailed analysis—and then, only the portion required to make it fundable.

The National Institute of Building Sciences' Multihazard Mitigation Council and the Council on Finance, Insurance, and Real Estate have developed a rich compilation of innovative incentives to stimulate investment in resilience, *Developing Pre-Disaster Resilience Based on Public and Private Incentivization* [41].

CONCLUSIONS AND IMPLICATIONS

Security and resilience of critical infrastructures is an issue of national and international significance, but of local and metropolitan solutions. The goal of this chapter is to reduce risks and enhance resilience of critical infrastructures and regional communities. It is necessary but not sufficient to this goal for CIs and communities to employ competent and effective risk management decision-support processes to assist in allocating constrained resources to achieve the greatest possible security and resilience improvement. This RMP must quantify baseline risks at a fine-grain engineering/operational level; establish the dollar value of options for reducing these risks; invest in, implement, and manage the preferred options; and evaluate how well the options have, in fact, reduced risk and improved resilience.

The complicating factors of dependencies and interdependencies and the risks they impose must be factored in to the extent available shared data will support. The perspectives of both the infrastructures' owners/operators and the community they serve are necessary to sustain fiscal integrity of the CIs while capturing maximum positive risk-reduction externalities for the full community. The RMP described in this chapter accomplishes this goal and meets these requirements. All elements of the process, except the D&I risk management, have been demonstrated several times and widely applied in related contexts, but have yet to be demonstrated as an integrated, multi-CI, full community-based system. The suggested D&I element is a simple, straight forward approach using the basic methodology and available modeling tools. It is expected to work smoothly after a few tests and refinements.

The most difficult aspect of employing RMP is the establishment of trust and committing the means to share sensitive information about the reliability of CIs' services during major, potentially disruptive events. Until this challenge is met, the critical analysis of dependencies and interdependencies cannot be completed. Building awareness of the community-wide implications of this shortcoming will bring pressure on those reluctant to participate in a controlled, secure exchange of the minimum amount of information needed to complete the D&I analyses and plan programs to reduce the attendant risks.

The quality and value of using the RMP increase with the level of collaboration among the CIs and between them and local and state governments, civic and business associations and community groups. Early on, this collaboration will improve the analysis of D&Is, while maturation may well bring additional resources for programs with greatest benefit to the community's well-being.

Current situations of most CIs lead them to systematically underinvest in security and resilience for several reasons. One of the most common is the systematic, significant underestimation of the value of risk-reduction and resilience-enhancement options, driven by an exclusively internal focus and valuation using CIs' pricing as value of their services. The value generated by critical "lifelines" of energy and water/wastewater lies in their externalities—the value added by all their customers, their customers' customers, and so on down the line, potentially including the whole regional community—a quantity that is virtually always orders of magnitude greater than price-based direct value. This increased value would justify significantly larger investments (e.g., Miami's $400 million bond program for resilience relative to sea level rise and flooding [42]).

The RMP addresses this challenge by requiring that benefits be valued in terms of both:

- Direct losses, casualties, and service outages avoided by the infrastructures, valued by an internal, price-based method; and
- The whole community's total direct and indirect avoided losses of lives, livelihoods, and economic activity valued by the contribution to overall regional economic activity.

The community's lost economic activity is based on adaptations of common regional economic models used in land-use and economic development planning in most major cities and metropolitan regions. These models estimate overall loss of GRP and by each of its business and industrial sectors, including households and government. Demonstrating that CI risks and resilience are everyone's risk and resilience—and can be concentrated in unexpected industries—can be used to mobilize community participation and potential financial contributions, such as incentives, targeted directly to the threat–asset pair(s) that the engineering/operational risk analysis method has demonstrated has the greatest potential for total risk reduction and resilience enhancement.

Thus, the risk management process described in this chapter is comprehensive in that it starts with an all-hazards risk analysis of threat–asset pairs at an engineering/operational level, includes the additional risks due to dependencies, interdependencies and proximities, estimates the external community risks and benefits and suggests an approach to finding significant new sources of funds—that can be targeted directly to specific, untreated, significant threat–asset pairs.

The RMP, in this sense, is a strategy for elevating infrastructure and regional community security and resilience by relaxing financial constraints and targeting incentives to the specific points where they can generate the maximum net benefits to both CIs and their communities. It is a matter of spending millions of dollars on security and resilience to save billions of dollars in losses while avoiding tragic levels of human casualties and devastated communities. This is a strategy worth consideration by all critical infrastructures and communities.

REFERENCES

[1] Public Health Security and Bioterrorism Preparedness and Response Act of 2002, Title IV, Drinking Water Security and Safety. Public Law 107–188, 42 U.S.C. 300i-2, 116 STAT. 682.

[2] 42 U.S.C. 300i-2, Section 143, as Amended by "America's Water Infrastructure Act of 2018.".

[3] R. Shumard and S. Schneider. Utility Security: Understanding NERC CIP 014 Requirements and Their Impact. EE Online. September/October 2014. https://electricenergyonline.com/energy/magazine/813/article/Utility-Security-Understanding-NERC-CIP-014-Requirements-and-Their-Impact.htm.

[4] B. Harrell, NERC CIP-014 Standard Explained, Curricula, 2019. https://www.getcurricula.com/nerc-cip-014-standard-explained/.

[5] Pub.L. 112-141, 126 Stat.405, "Moving Ahead for Progress in the 21st Century Act, MAP-21.".

[6] B.M. Frischmann, Infrastructure: The Social Value of Shared Resources, Oxford Univ. Press, 2012.

[7] E. Roe, P.R. Schulman, Reliability and Risk: The Challenge of Managing Interconnected Infrastructures, Stanford Univ. Press, Stanford, CA, 2016.

[8] J.P. Brashear, et al., Regional Resilience/Security Analysis Process for the Nation's Critical Infrastructure Systems, 2011. http://www.serri.org/publications/Documents/ASME%20Project%20-%20Final%20Report%20-%2020%20Dec%202011%20(Brashear).pdf.

[9] P.L. Bernstein, Against the Gods, Wiley., NY, NY, 1998.

[10] M. Lebeau, P. Duguay, The Costs of Occupational Injuries: A Review of the Literature, Institut de recherhe Robert-Sauve en santé et en securite du travail (IRSST), Quebec, Canada, 2013. Report R-787.

[11] National Institute of Occupational Safety and Health, Economic Burden of Occupational Fatal Injuries in the United States Based on the Census of Fatal Occupational Injuries, 2003–2010, Washington, DC, 2017, https://www.cdc.gov/niosh/data/datasets/sd-1002-2017-0/default.html. Dataset Number SD-1002-2017-0, p.3.

[12] M. Moran, C. Monje, Guidance on the Treatment of Economic Value of a Statistical Life (VSL) in U.S. Department of Transportation Analyses – 2016 Adjustment, U.S. DOT, August 8, 2016. https://cms.dot.gov/sites/dot.gov/files/docs/2016%20Revised%20Value%20of%20a%20Statistical%20Life%20Guidance.pdf.

[13] U.S. Environmental Protection Agency. Mortality Risk Valuation. https://www.epa.gov/environmental-economics/mortality-risk-valuation#whatvalue.

[14] J.P. Brashear, P. Scalingi, R. Colker, A Business Process Engineering Approach to Managing Security and Resilience of Lifeline Infrastructures, National Institute of Building Sciences (NIBS), Washington, DC, 2015. Submitted under Contract HSHQDC-14-00089 between NIBS and the U.S. Dept. of Homeland Security.

[15] U.S. Dept of Homeland Security. National Infrastructure Protection Plan (NIPP) 2013: Partnering for Critical Infrastructure Security and Resilience. Washington, D.C., Last Published Date: February 21, 2019.

[16] D.W. Hubbard, The Failure of Risk Management: Why It's Broken and How to Fix it, Wiley, 2009.

[17] American Water Works Association, ANSI/AWWA J100-10 Risk and Resilience Management of Water and Wastewater Systems, 2010.

[18] J.P. Brashear, J.W. Jones, Risk analysis and management for critical asset protection (RAMCAP Plus), in: J.G. Voeller (Ed.), Wiley Handbook of Science and Technology for Homeland Security, 2010.

[19] J.P. Brashear, et al., The RAMCAP Approach for the Water Sector, Overview & Ten Annexes, ASME Innovative Technology Institute, LLC, Washington, DC, September 2007.

[20] Subtitle G, Title VIII, Pub. L. 107–296, 116 Stat. 2238 (6 U.S.C. 441–444).

[21] Applied Engineering Management Corporation. Herndon, vol. A. https://www.aemcorp.com/.

[22] U.S. Environmental Protection Agency. Vulnerability Self Assessment Tool. Washington, DC. https://vsat.epa.gov/vsat/.

[23] Electric Consumers Resource Council (ELCON), The Economic Impacts of the August 2003 Blackout, 2004. https://elcon.org/wp-content/uploads/Economic20Impacts20of20August20200320Blackout1.pdf.

[24] National Infrastructure Simulation and Analysis Center, a Joint Program of Sandia National Laboratories, Los Alamos National Laboratory and Pacific Northwest National Laboratory; Argonne National Laboratory has an independent R&D program on D&Is.

[25] Bay Area Center for Disaster Resilience, Regional Critical Infrastructure Security and Resilience InfoXchange (RCISR InfoXchange), 2018. https://prod.i-info.com/dashboard/portal.aspx?gid=4380D6FBA4BA4861B42A9D14A4DCD1B0.

[26] Saber Space. XchangeCore. https://www.saberspace.org/xchangecore-home.html.

[27] Alliance for Community Solutions. https://www.youracs.org/our-mission.

[28] Y.Y. Haimes, Risk Modeling, Assessment and Management, fourth ed., Wiley, NY, 2016.

[29] J.R. Santos, Inoperability input-output modeling if disruptions to interdependent economic systems, Systems Engineering 9 (1) (2006).

[30] K. Barker, J.R. Santos, Measuring the efficacy of inventory with a dynamic input-output model, International Journal of Production Economics 126 (1) (2007) 130–143.

[31] J.R. Santos, K.D.S. Yu, S.A.T. Pagsuyoin, R.R. Tan, Time-varying disaster recovery model for interdependent economic systems using hybrid input-output and event-tree analysis, Economic Systems Research 26 (1) (2014) 60–80, https://doi.org/10.1080/09535314.2013.872602.

[32] J.R. Santos, S.T. Pagsuyoin, L.C. Herrera, R.T. Tan, K.D. Yu, Analysis of drought risk management strategies

using dynamic inoperability input-output modeling and event tree analysis, in: Environmental Systems and Decisions, Springer, 2014, https://doi.org/10.1007/s10669-01409514-5. ISSN 2194-5403.
[33] A.Z. Rose, A framework for analyzing and estimating the total economic impacts of a terrorist attack and natural disaster, Journal of Homeland Security and Emergency Management 6 (1) (2009). Article 4.
[34] A.Z. Rose, S.B. Blomberg, Total economic impacts of a terrorist attack: insights from 9/11, Peace Economics, Peace Science and Public Policy 16 (1) (2010). Article 2.
[35] A.Z. Rose, D. Wei, A. Wein, Economic impacts of the ShakeOut scenario, Earthquake Spectra 27 (No. 2) (2011) 529–557.
[36] I.S. Wing, A.Z. Rose, A.M. Wein, Economic Consequence Analysis of the ARkStorm Scenario. Natural Hazards Review, A40165002, American Society of Civil Engineers, 2015.
[37] A.Z. Rose, I.S. Wing, D. Wei, A. Wein, Economic impacts of a California tsunami, Natural Hazards Review 17 (2) (2016) 04016002. American Society of Civil Engineers.
[38] Multihazard Mitigation Council, Natural Hazard Mitigation Saves: An Independent Study to Assess Savings from Mitigation Activities, National Institute of Building Sciences, Washington, DC, 2005.
[39] Multihazard Mitigation Council, Natural Hazard Mitigation Saves: 2018 Interim Report, National Institute of Building Sciences, Washington, DC, 2019.
[40] Sierra Club, San Francisco Bay Chapter. June 2016 Election Results. https://www.sierraclub.org/san-francisco-bay/june-2016-election-results.
[41] Multihazard Mitigation Council and the Council on Finance, Insurance and Real Estate, Developing Pre-disaster Resilience Based on Public and Private Incentivization, National Institute of Building Sciences, Washington, DC, 2015. www.nibs.org/resource/resmgr/MMC/MMC_ResilienceIncentivesWP.pdf.
[42] B. Ki-moon, F. Suarez, Miami's Battle Plan for Rising Seas, New York Times, February 21, 2019, p. A27.

CHAPTER 4

Resilience of Electric Power Infrastructure

JEFF DAGLE, MSEE, PE

On August 14, 2003, four root causes coalesced to plunge 50 million people into the largest blackout in the history of the North American power system [1]. Customers and commerce were disrupted spanning an area that included Cleveland, Detroit, Toronto, and New York City. This blackout was not different from other large blackouts. Often, they occur during periods of high system stress. Because the grid is designed and operated to prevent a single failure from cascading into a large blackout, large-scale blackouts always result from the confluence of multiple causes. On this particular day, the four root causes were the following: (1) failure to adequately understand system limitations, (2) poor situational awareness, (3) inadequate vegetation management, and (4) improper reliability coordinator oversight.

Protection systems automatically and quickly clear faults (i.e., short circuits) using localized measurements to determine whether a device should be quickly deenergized by opening circuit breakers. Protection is ubiquitous in all power grids, protecting lines, generators, transformers, and other hardware. Overlapping zones of protection provide redundancy (for instance, when a circuit breaker fails to clear the fault in a timely manner). Protection is also selective. Selectivity means that it is designed to only clear faults within a well-defined zone of jurisdiction. However, when a cascading failure occurs, protection systems may operate unpredictably. During the sequence of events associated with the 2003 blackout, many independent decisions by various protection systems disconnected several transmission lines forming multiple islands (portions of the system separated from each other).

Loss of generation caused by their own protection will reduce the frequency. Automatic underfrequency load shedding (described in detail later in this chapter) may be inadequate or improperly timed to prevent a collapse within that particular island, or voltage problems may cause generators to trip. When multiple things go wrong at the same time, the uncontrolled cascading failure can result in the system failing in a very unpredictable manner. In the 2003 blackout, at least 265 power plants with more than 508 individual generating units were automatically disconnected from the grid and shut down [1].

Collectively, these actions to trip transmission lines and power plants formed multiple islands and disrupted power to millions of customers. Those areas that were importing electricity before the islands forming were much more prone to customer outages than those areas that were exporting power before the sequence of events began. Moreover, if generation is lost in a generation-deficient area to begin with, the consequences will be much more significant. This happens in nearly all blackouts, not just this blackout.

Delving into the sequence of events and analyzing the root causes, the investigation team produced dozens of detailed recommendations to minimize the likelihood of something similar from happening again because of these lessons learned. Events like this illustrate the importance of a reliable and resilient electric power system. The 2003 blackout was estimated to result in an economic cost between $7 and $10 billion in the United States [2]. In Canada, Ontario workers lost 18.9 million hours of employment, while manufacturing shipments dropped by $2.3 billion [3].

EXAMINING ELECTRIC POWER RESILIENCE

This chapter provides a perspective on electric power resilience, with a focus on the interconnected North American power system. It will explore how reliability and resilience, while inexorably intertwined, have separate and nuanced differences.

Reliability is a measure of how often and for how long power is disrupted to utility customers. Resilience

Optimizing Community Infrastructure. https://doi.org/10.1016/B978-0-12-816240-8.00004-5

addresses the preparedness of the electric power system and its ability to cope with various hazards that can befall the grid. This chapter is intended to provide a broad perspective of these issues with limited technical depth. For additional detailed information, the author invites the reader to pursue the technical reports provided in the references, particularly the 2017 report published by the National Academies Press, *Enhancing the Resilience of the Nation's Electricity System* [4].

Defining Electric Power Resilience

There are several widely accepted definitions of electric power resilience that are generally quite similar to each other. The National Infrastructure Advisory Council (NIAC) provided a comprehensive and concise definition in a 2010 study report recommending specific steps that could be undertaken to improve the resilience of the electric power and nuclear sectors [5]:

> *Infrastructure resilience is the ability to reduce the magnitude and/or duration of disruptive events. The effectiveness of a resilient infrastructure or enterprise depends upon its ability to anticipate, absorb, adapt to, and/or rapidly recover from a potentially disruptive event.*

The "disruptive events" in the above definition can come in various forms. They can be weather-induced or the result of other naturally occurring phenomena, such as earthquakes or solar storms. Or they can be human-induced, such as physical attack, cyberattack, or even an electromagnetic pulse. Each of these types of hazards represents a different threat to the safety and security of the system. Taken individually, it is important to understand the consequences associated with the likelihood to each, so that the overall risk can be appropriately managed. However, resilience is a deeper concept than risk management against specific threats. Resilient design concepts are rooted in the understanding that it is possible to design a system to withstand a wide range of hazards, even those that are not enumerated or even imagined.

Another appealing aspect of the 2010 NIAC study report is that it articulates a resilience lifecycle as illustrated in Fig. 4.1. This figure denotes the various stages of resilience planning and management, including a feedback loop to incorporate lessons learned from prior events. This illustration reinforces that resilience is more than hardening and is more than restoration. It involves planning for and mitigating these events before, during, and after they occur.

RELIABILITY METRICS AND THE CHALLENGE WITH DEVELOPING RESILIENCE METRICS

Industry-accepted standards for power system reliability metrics are used to measure the frequency and duration of power system disruptions. The IEEE Guide for Electric Power Distribution Reliability Indices (Standard 1366) defines several reliability indices, the most common of which are the System Average Interruption Frequency Index, System Average Interruption Duration Index, and Customer Average Interruption Duration Index [6]. These serve as an excellent basis for understanding past performance. These metrics also provide a means for benchmarking relative reliability between different systems in a way that supports meaningful communication between utilities, regulators, and customers. These metrics can also be used to provide trending information, determining whether reliability is improving or degrading over time. These reliability metrics are often used to justify capital improvement programs and other operational and maintenance expenditures. For example, more aggressive vegetation management (clearing trees and other brush away from transmission lines and distribution feeders) might be warranted in specific areas. However, these metrics

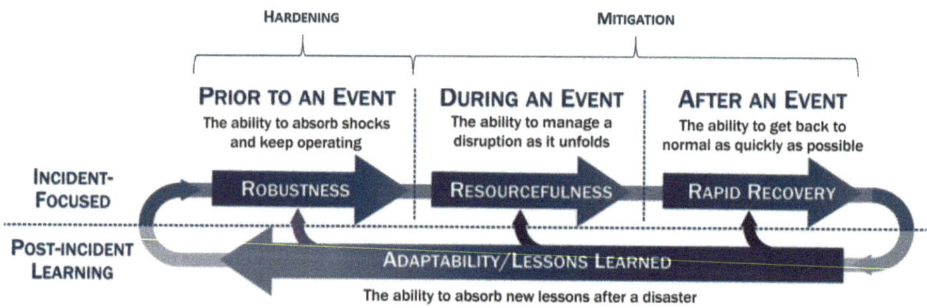

FIG. 4.1 The resilience lifecycle. (Source: National Infrastructure Advisory Council. A Framework for Establishing Critical Infrastructure Resilience Goals, 2010.)

are inherently lagging indicators of system performance. Therefore, although they are an excellent indicator of how the system performs associated with routine disruptions, they may not fully reflect how resilient the system might be in the future against different types of events.

Although the aforementioned reliability metrics do include provisions for how to measure extreme events (some of the metrics deliberately exclude them, because a large weather-induced event would mask other reliability concerns of interest, such as trending, etc.), they all still have one thing in common: they measure things that have already occurred. There is a need for a whole new class of resilience metrics that can become leading indicators of performance in response to future events.

As part of the US Department of Energy's Grid Modernization Initiative, there is a research project to develop metrics associated with resilience and other facets of grid modernization [7]. The goal of this research is to develop metrics that are able to proactively measure the resilience of the power system with enough rigor and standardization to provide a basis of benchmarking and investment, similar to the way traditional reliability metrics are used today.

To illustrate the challenges associated with developing proactive resilience metrics, consider the following thought exercise with two hypothetical scenarios. First, consider two identical utilities in every respect, including reliability statistics. If the first utility purchases and inventories more spare parts than the other, they likely become more resilient to potential future events. They would be able to more quickly recover from a catastrophic disruption than the other utility that would be dependent on supply chain delivery and logistics for more of their replacement parts. So, one aspect of measuring resilience should be the spare parts inventory. In a second hypothetical scenario, again with two identical utilities, the first one decides to cross-train all of their employees in emergency response skills and conducts more drills and exercises. Would they then be more prepared? How much of an impact would that have? Would that impact be measurable? Would the particular training given have more or less of a measurable impact than alternative training topics? Would this be more or less effective than some other aspect of resilience, such as the spare parts strategy, or something else altogether?

In a perfect world, robust resilience metrics would be able to evaluate the relative effectiveness of a myriad of potential resilience-enhancement alternatives. Then, the cost and benefit of these different options can be evaluated in a consistent manner. However, given that there are so many different things that could be done, which may or may not be effective to a wide variety of different scenarios, the challenge of defining these resilience metrics is immense. These metrics are aiming to measure the effectiveness of preparedness associated with events that have not occurred, or may never occur. The ultimate objective is encouraging the right actions and investments in architecting, designing, building, operating, and maintaining an infrastructure with heightened resilience. However, everything involves a financial and engineering tradeoff that requires balancing the benefit to minimize the overall risk.

MAKING THE SYSTEM MORE RESILIENT

Many things can be done to improve the overall resilience of the power system. These include hardening critical components, increasing the modularity and interoperability of components (making quick replacement and restoration more efficient), and changing the overall architecture of the system to reduce the criticality of individual components. For example, many of the hazards faced by the power system are related to severe weather events, and these are manifested by wind, water, or ice. Carefully choosing the location of substations and making that equipment and facilities resistant to extreme wind or flooding goes a long way toward improving the robustness of the infrastructure. Selecting poles and towers that can resist the mechanical forces exerted by high wind or heavy ice is also a way to harden against these conditions. Although it may be impractical to invest in upgrading all of a utility's assets in this manner, selective investment in critical stations and key paths could nevertheless represent a prudent resilience investment. These resilience-enhancement investment strategies should also focus on electric power infrastructure assets critical for the recovery phase. This will be discussed in greater depth later in the chapter.

Widespread devastating weather events also impact related critical infrastructure, such as transportation networks. Therefore, considering accessibility of critical stations is also a key factor as it relates to siting substations and power lines.

Some design tradeoffs may not be a simple matter of economic justification but may include other engineering trade-offs. For example, undergrounding transmission and distribution (T&D) lines (significantly more expensive than overhead infrastructure) is commonly cited as a means to eliminate the possibility of wind damage. If properly designed and installed, they can also be resistant to flooding. But this type of infrastructure may be vulnerable to widespread disruptions in the event of an earthquake. Furthermore, regardless of the hazard, damage assessment and restoration times for underground infrastructure are often significantly longer than overhead assets. Therefore, although there may be some resilience enhancement offered by undergrounding T&D infrastructure, there are additional tradeoffs that must be considered. These decisions can become quite complex.

Ideally, the system should be designed such that it has inherent properties of resilience: bend without breaking, fail with minimal disruption, and enable fast restoration and recovery. These principles should hold regardless of the nature of the event, even if it was unforeseen during the design phase. Fortunately, the power system is already replete with many of these design principles as the system evolved over its history. For example, careful thought has been given to minimizing the dependency on communication for automation systems that operate the system. Although communications are certainly important, particularly for voice communication between the control center, power plants, and field crews, automated controls are designed to be somewhat self-sufficient in the absence of communication. During normal operation, centralized controls optimize the system. However, it is often possible to operate the system in a degraded mode in the absence of communications. There is also a hierarchical separation of control functions, where fast-acting local controls operate to a set point that is provided by a slower-speed wide-area control scheme. Thus, if the wide-area communications are lost, the local controls continue to operate to the last setting it received. In addition, substations are often sited and designed so that they can backup loads being served by adjacent substations—a very handy design feature during a severe emergency. Overlapping zones of protection provide extensive redundancy to provide safety and reliability if there is a failure in the protection scheme, or breaker failure and/or misoperation.

FREQUENCY CONTROL SAFEGUARDS

The architecture and design of the bulk power system also includes other types of controls that provide additional defense against cascading failure. One notable example of these emergency controls is "under frequency load shedding." The power system frequency, nominally 60 Hz (in North America), is a balance between the generation and load. If there is less generation than load, the system will slow down; vice versa with more generation than load (system will speed up). Automatic controls will act to bring the system back to the nominal frequency by dispatching generation. Governors at the generators act more quickly, and this is called "primary frequency control." Secondary frequency control is undertaken by control centers, sending dispatch signals to generators automatically based on wide-area measurements and supervisory control systems. However, if the frequency mismatch becomes too great, there is a risk of an adverse outcome resulting from automatic controls associated with generators designed to protect them from damage.

For power generators that involve synchronous machines, there is a direct connection between the frequency of the grid and the rotational speed of the shaft. A two-pole machine operates at 3600 rpm corresponding to 60 Hz, a four-pole machine at 1800 rpm, and so on. Steam plants include large turbines on the same shaft and often have design tolerances of about 1 Hz above or below nominal speed. Automatic controls will trip the generator to protect itself from damage if the previously mentioned primary or secondary frequency controls are unable to maintain nominal frequency within the plant's acceptable operating range of system frequency, and hence rotational speed of the shaft.

In the very rare instance that the system frequency drops too low and risks tripping generators due to their low-speed protection requirements there is a coordinated scheme of automatic controls that will disconnect load from the system, otherwise called "under frequency load shedding." This is an emergency control that will help restore system balance of frequency under extreme conditions. It prevents tripping generators that would have resulted in a worsening cascading failure and will enable more rapid restoration of load (customers) when the system conditions have been stabilized.

This is an elegant control scheme because it does not rely on external communications to function. It measures system frequency locally, and trips based on a predefined threshold of frequency and duration. Any coordination considerations inherent in the design of the overall scheme are embedded in those set points. The scheme is highly robust and not prone to failure. The potential of misoperation is very low. These controls rarely operate because the normal means of controlling frequency are almost always adequate. Furthermore, the beauty of this control scheme is that shedding load when the frequency is low for a prolonged period of time should always benefit the overall system (because the low frequency condition is the result of more load than available generation at that instant in time), regardless of the details associated with specifically how the system got into that condition in the first place. This is why under frequency load shedding is such an excellent example of a resilience-enhancing emergency control.

MAKING INDIVIDUAL ASSETS LESS CRITICAL

There are different strategies for making individual assets less critical. One strategy is segmentation. For example, a bus at a substation can be split into sections connected with a tie breaker so that a fault or short circuit on one segment can be isolated by clearing the fault without deenergizing more than that single segment. The utility also may have multiple substations in an area each serving a portion of the load in a region or community with alternate pathways for bringing power to different portions of the system. Often there are overlapping transmission lines of multiple voltages in the same area. For example, a 500 kV backbone would not be the only way to transport power in a region because there would be lower voltage transmission infrastructure in the area, such as 230 kV. A single transformer would not be responsible for delivering power to distribution feeders; there would be alternate pathways with overlapping zones of different voltages to provide a robust selection of alternatives to delivering power in the event of a failure. Depending on the criticality of the load, there might be redundant feeders, and in some cases, redundant feeders from different substations.

There are also opportunities to reduce the criticality of generation assets. The most resilient options are when there are is diversity of fuel supply, including fuel-secure generation where the power production is not vulnerable to fuel supply disruptions. The prospect of distributed generation to reduce the dependency of centralized power generation can also enhance resilience, especially if that distributed generation is hardened to be able to reliably deliver power under a diverse range of extreme circumstances. There has been a significant shift over the past several years toward more distributed generation, including customer-sited solar generation in many regions. These trends are driven primarily by economics, but also have significant resiliency advantages if these generation assets can self-sufficiently operate without the voltage and frequency support from the interconnected system.

Energy storage is another nascent technology (in terms of cost-effectiveness for widespread deployment) that will dramatically improve resilience in the future. Primarily motivated by the need to support nondispatchable generation associated with solar or wind generation, energy storage can also provide greater flexibility during emergency events. To make these systems more valuable in emergency scenarios will require that the current state of the art (currently a few hours of cost-effective storage) would need to be extended to several days of storage capability.

The deployment of energy storage is largely an economic decision, in competition with either generation or demand response technologies. Although there are a variety of energy storage technologies, ranging from pumped hydro to batteries, they have cost considerations associated with their power discharge capacity, energy storage capability, and other costs associated with efficiency and maintenance. Over the next several years, we expect to see more energy storage deployed on the system to help "firm up" power from nondispatchable power generated from renewable resources. In addition, there will continue to be a niche for high-end customers requiring uninterrupted power. However, because the energy storage costs for most technologies are currently prohibitive for multiday storage, the deployment of widespread energy storage for resilience-enhancement benefit will be limited for the foreseeable future.

LIMITING THE CONSEQUENCES OF COMPONENT FAILURES

Component failures are inevitable. Multiple simultaneous failures are also likely under extreme conditions. One of the key hallmarks of resilience is limiting the consequences to the overall system in the face of many failures of individual components. This can be achieved through redundancy, preventing cascading failure sequences, and designing for graceful degradation of system operations when there are failures of multiple individual components.

Redundancy is the most common method of limiting the consequences of equipment failure, but this comes at a significant cost. Therefore, the utility industry has adopted rigorous planning standards that govern the appropriate level of redundancy for specific circumstances. The North American Electric Reliability Corporation (NERC) develops and enforces mandatory standards for bulk power transmission system planning performance requirements [8].

Well-established fault-tolerant design principles can be employed to design a system architecture that is less dependent on the continuous functionality of individual components.

FAILSAFE COMMUNICATIONS

Communications are an important and growing area of dependency for modern electric power systems. Historically, the electric power infrastructure was designed to limit the dependency on communications, which were deemed expensive and prone to reliability issues. As various forms of communication have become more ubiquitous over the past few decades, there has been greater dependency on these communications for various systems to operate. In many cases, it has migrated beyond measuring for enhanced situational awareness toward increased dependency on wide-area communications necessary for the reliable operation of the power grid. One notable example of this dependency on Internet-style communications are market systems involved in the provision of services such as generation dispatch, allocation of critical ancillary services, and other vital communications that would significantly degrade the reliable operation of the power system if they were compromised. Although this dependency has been mitigated through the provision of highly redundant and reliable communications infrastructure and designed redundancies and backup systems, there remains a dependency that could be compromised under extreme scenarios where the communication system functionality would be degraded. It is recommended that further effort be made to identify strategies to operate the system absent dependent infrastructure, such as communications. A well-designed and often practiced "fail-safe" approach to operating the system in the absence of communications is one such strategy.

ADAPTIVE ISLANDING

The interconnected power system derives much of its reliability through its immensity. With more inertia and shared reserves, operational benefits were obtained when smaller systems were formed into larger systems over the past several decades. In addition, there are also significant economic benefits that can be realized when power can be transferred over vast distances. However, when a cascading failure occurs, there is a risk to portions of the system that are far removed from the epicenter of the initial problem. The process of "islanding" the system, or breaking the interconnection into smaller portions, can result in significant outages if the islanding occurs haphazardly as the result of an uncontrolled cascading failure. However, if the islanding is preplanned and properly managed, the impact could be significantly reduced. Adaptive islanding is a concept that has been around for several years and includes well-known and implemented concepts such as remedial action schemes [9]. The goal is preserving the benefits of large-scale interconnected system operations during normal conditions while reducing the risk of cascading failure propagation during abnormal or emergency conditions.

Following the 2003 blackout, academic studies suggested that cascading failures would be less disruptive if the grid were divided into smaller systems. After all, should critical load centers like New York City be connected to a system as far away as Nebraska or Saskatchewan? Is such exposure necessary?

Through the first part of the 20th century, smaller power systems were interconnected in North America to reap efficiency gains and improve system reliability. These consolidations are still occurring throughout the world. Although the North American grid has not changed its fundamental structure in several decades, the systems in Europe, and more recently China and India, have been forming fewer and larger interconnections. The benefits are increased shared reserves, and a more stable frequency that is more resistant to change with additional inertia. Adaptive islanding is a concept that enables the benefits of a larger grid during normal conditions and containing disruptions during off-normal conditions.

Remedial Action Schemes

Adaptive islanding is an advanced form of modern remedial action schemes. Sometimes known as special protection systems, these schemes are used in conjunction with protection to maintain power system stability. For example, path ratings consider credible contingency violations of reliability criteria. Remedial action schemes enable increased flow (increasing asset utilization) while maintaining reliability by implementing fast-acting control to mitigate adverse conditions following an event.

One particularly well-known state-of-the-art remedial action scheme is associated with the western North American power system. Along portions of the west coast, there are several 500 kV lines in parallel with a large direct current line. This intertie enables seasonal energy exchange between the winter-peaking Northwest and summer-peaking southern California. When the scheme detects circuit breaker operation (caused by a faulted component) fast-acting control is triggered (armed in advance by operators based on the prevailing conditions). This includes tripping loads on the receiving end and generation runback on the sending end of the interface, among other things, to maintain overall system stability. These schemes have also included controlled separation of the system as appropriate. Future adaptive islanding schemes will leverage and build on this experience.

Implementing Adaptive Islanding

The prospect of limiting the disruption associated with an uncontrolled cascading failure is the benefit of adaptive islanding. The system would be implemented similarly to remedial action schemes today, where the islanding schemes would be preplanned and adjusted based on the patterns of generation and load. The boundaries of these planned islands would be periodically updated as changing conditions warrant. When needed, fast controls would cleave the system into predefined islands. Fast-acting load shedding would be necessary to maintain stability in areas that are generation deficient. Quickly matching the available generation with the real-time load will be crucial, while respecting dynamic transient response characteristics.

Several research papers have proposed adaptive islanding [10]. However, wide-scale implementation will require additional development and demonstration to be fully realized. In some areas, such as Alberta or Florida, adaptive islanding will be easier to implement than in more tightly interconnected regions such as Ohio. If these challenges can be worked out, adaptive islanding will result in a more resilient power system by minimizing the disruption caused by future cascading failures.

FLEXIBILITY

Another way to enhance resilience is by enhancing flexibility. Increased flexibility will more readily accommodate available resources, adapt to the available transmission pathways, and service as much load as possible under varying conditions. This will likely be accommodated in the future with more sophisticated control schemes. Care must be taken to ensure that these control schemes themselves are not brittle and do not by themselves introduce issues that might otherwise detract from a simple and resilient system. Inherently, complex systems may not be the most elegant solution during extreme emergencies. For example, if communications are unavailable, these controls need to revert to a "failsafe" and reliable mode of operation. These considerations need to be incorporated into their design.

Specifically in the distribution arena, there have been significant advancements in the cost-effective deployment of various distribution automation schemes. Generally referred to as "smart grid" technologies over the past several years, these are technologies that provide increased automation given advances in low-cost communications. Enabling the interconnection of various advanced technologies, the "smart grid" provides a framework for improved planning and operation of the power system, ranging from sensing and measurement to various control and automation technologies [11]. The smart grid will enable greater resilience because the power system operators will have more timely information about status of the infrastructure, and control technologies to more quickly react to emergency situations.

Under the general category for fault location, isolation, and service restoration, these technologies can automatically identify a faulted segment, deenergize it with circuit breakers, and then use automatic switching to restore portions of the distribution feeders that do not have permanent damage. These schemes can also be extremely helpful to aid in large-scale system restoration, when there are multiple failed segments that need to be repaired. The US Department of Energy, with funding through the American Recovery and Reinvestment Act of 2009, provided smart grid investment grants to support the deployment of several projects utilizing this technology [12]. Utilities are continuing to deploy these technologies today.

ENHANCING RESTORATION

No matter how robust and hardened the power system infrastructure has been built, extreme events will require repairing damage that will inevitably occur. One key dimension of resilience is how fast critical service can be restored. This can be enhanced through spare parts strategies, sufficient manpower that is trained and equipped, as well as close coordination spanning all levels of the restoration effort, including damage assessment, workforce prioritization, and dealing with logistics issues (such as housing and feeding the restoration workforce). The availability of critical backups, including mobile substations and transformers, can be very important if there is damage to that type of equipment.

Importance of Mutual Assistance

No matter how well prepared an individual utility may be, restoration time can often be significantly enhanced with additional manpower and resources (equipment such as bucket trucks, etc.). Especially in the case of distribution system restoration, the timeline for restoring poles and feeders is almost a linear function of the number of crews that are actively working on the restoration. Double the workforce and cut the recovery time in half (in round numbers). Therefore, the utility industry has evolved a very robust system of "mutual assistance programs" where utilities help each other when recovering from large events. This is particularly evident with hurricanes and similar events, where the magnitude of disruption is great, and there is advance time to plan and stage the recovery assets from other utility companies. Various trade organizations representing the utility industry work well together to provide coordination of this type of response. This is one aspect of resilience in which the electric power industry already excels. Each event represents an opportunity to gather lessons learned for future improvements and utilities have honed this type of response over the years.

One potential disruptor to mutual aid is when the affected utility is facing financial pressure associated with bankruptcy. Mutual assistance requires the financial backing provided by the host utility to reimburse the expenses of the participating utilities lending their crews and equipment. In cases where the affected utility is unable to pay for or extend lines of credit, mutual aid can be disrupted, and restoration times significantly extended. There are two notable examples where bankruptcy issues thwarted the rapid recovery from significant hurricanes: New Orleans in 2005 and Puerto Rico in 2017. In both cases, the level of devastation to the electrical infrastructure was severe, but also in both cases there were fewer outside crews brought in to help the restoration and recovery efforts. Therefore, in cases where the financial health of the affected utilities is compromised, it may be prudent to provide State- or Federally backed financial support to provide for mutual assistance.

Spare Parts and Logistics

The ability to quickly mobilize crews and equipment to begin restoration efforts is also vitally important to improve restoration of electrical services, particularly for scenarios where widespread destruction of physical assets has occurred. Knowing when and where to stage these restoration crews, their equipment such as bucket trucks, and the parts they will need to effect the restoration, such as poles, wire, transformers, and other supplies, are crucial. During certain large-scale events such as hurricanes, where there are updated forecasts of the storm track, it may be possible to optimize where to locate the crews so that they are out of harm's way during the initial event but have access to adequate transportation infrastructure to start mobilizing immediately when it is safe to do so. In other cases where there is little to no warning of the impending event (e.g., an earthquake or malicious attack), strategies to stage parts and equipment, and training to quickly mobilize the workforce are important.

CYBER RESILIENCE

As the utility industry has become increasingly dependent on cyber resources (computers, communications, etc.), they are also vulnerable to disruptions caused by breaches to those assets. Recovering from a large cyber event is somewhat different than recovering from an event where there is physical damage to the electricity infrastructure. The skillsets are different and the approach to incrementally recover system functionality will be unique to the targeted utility and the specific attack scenario. One approach to address this issue is planning for cyber recovery similarly to other response and recovery approaches where scenarios are drilled and response and recovery is practiced.

Cyber resilience is fundamentally different from cyber security. Cyber security is the necessary mindset of trying to keep the adversaries from compromising the functionality of critical cyber systems. The utility industry in general has taken great strides toward enhancing cyber security over the past several years. The mandatory NERC cyber security requirements apply to the bulk power system in North America [13]. As part of the American Recovery and Reinvestment Act of 2009, the US Department of Energy required grant recipients to implement cyber security measures for their projects [14]. There have been a number of additional initiatives and technologies aimed at enhancing the cyber security of critical control systems [15]. However, cyber resilience goes beyond cyber security. Cyber resilience is taking the mindset that a breach will occur and implementing the necessary preparedness to effectively deal with it when it does.

Therefore, utility cyber security programs should be enhanced to include a specific focus on cyber resilience. This includes the assumption that a major breach of critical cyber assets will occur at some point in the future and implement plans, recovery procedures, training, and practice drills to restore cyber functionality under worst-case scenario assumptions. Something akin to a "black start" restoration plan for the cyber systems would be very beneficial for utilities to enhance their ability to recover from a very serious breach to their cyber infrastructure.

PERSONNEL RESILIENCE

Properly trained, equipped, and motivated people are the most important ingredient for a highly resilient system. There is nothing better than a skilled brain that can troubleshoot problems and develop effective workarounds. This is particularly true if there is an extreme event unlike anything that has happened in the past. Therefore, it is important that utilities maintain appropriate staffing levels and provide opportunities for training and developing their staff so that they can be successful in dealing with extreme events when they occur. This also includes planning for how to help their families and other logistics considerations so that they can be most effective focusing on the restoration and recovery efforts.

Dwight D. Eisenhower is credited with saying "In preparing for battle, I have always found that plans are useless, but planning is indispensable [16]." This is a good perspective as it relates to preparing for a major unexpected event. The process of planning will reveal significant ways to improve the emergency response procedures. However, as the range of possible events is so diverse, it is unrealistic to expect that the plans themselves will completely address all possible permutations of possible scenarios.

REFERENCES

[1] U.S.-Canada Power System Outage Task Force, Final Report on the August 14, 2003 Blackout in the United States and Canada: Causes and Recommendations, April 2004. https://www.energy.gov/sites/prod/files/oeprod/DocumentsandMedia/BlackoutFinal-Web.pdf.

[2] ICF Consulting, "The Economic Cost of the Blackout: An Issue Paper on the Northeastern Blackout, August 14, 2003." http://www.solarstorms.org/ICFBlackout2003.pdf.

[3] U.S. – Canada Power System Outage Task Force. Final Report on the Implementation of Task Force Recommendations. https://www.energy.gov/oe/downloads /us canada-power-system-outage-task-force-final-report-implementation-task-force.

[4] National Academies of Sciences, Engineering, and Medicine, in: Enhancing the Resilience of the Nation's Electricity System, The National Academies Press, 2017. https://doi.org/10.17226/24836.

[5] National Infrastructure Advisory Council, A Framework for Establishing Critical Infrastructure Resilience Goals, October 19, 2010.

[6] IEEE, IEEE Standard 1366-2012: IEEE Guide for Electric Power Distribution Reliability Indices, May 2012. https://ieeexplore.ieee.org/document/6209381.

[7] U.S. Department of Energy. Grid Modernization Initiative. https://www.energy.gov/grid-modernization-initiative.

[8] North American Electric Reliability Corporation, Reliability Standards for the Bulk Electric Systems of North America, January 2019. https://www.nerc.com/pa/Stand/Reliability%20Standards%20Complete%20Set/RSCompleteSet.pdf.

[9] North American Electric Reliability Corporation, "Remedial Action Scheme" Definition Development: Background and Frequently Asked Questions, June 2014. https://www.nerc.com/pa/Stand/Prjct201005_2SpclPrtctnSstmPhs2/FAQ_RAS_Definition_0604_final.pdf.

[10] H. You, V. Vittal, Z. Yang, Self-healing in power systems: an approach using islanding and rate of frequency decline-based load shedding, in: IEEE Transactions on Power Systems, February 2003.

[11] U.S. Department of Energy. The Smart Grid: An Introduction. How a Smarter Grid Works as An Enabling Engine for Our Economy, Our Environment and Our Future. Brochure available at: https://www.energy.gov/oe/downloads/smart-grid-introduction-0.

[12] U.S. Department of Energy, fault location, isolation, and service restoration technologies to reduce outage impact and duration, in: Smart Grid Investment Grant Program of the American Recovery and Reinvestment Act of 2009, December 2014.

[13] North American Electric Reliability Corporation. Critical Infrastructure Protection Standards. https://www.nerc.com/pa/Stand/Pages/CIPStandards.aspx.

[14] U.S. Department of Energy, 2012 DOE Smart Grid Cybersecurity Information Exchange, December 2012. https://www.smartgrid.gov/files/2012_Cybersecurity_Information_Exchange_.pdf.

[15] U.S. Department of Energy, Roadmap to Achieve Energy Delivery Systems Cybersecurity, September 2011.

[16] Quote Investigator. https://quoteinvestigator.com/2017/11/18/planning/.

CHAPTER 5

Becoming a Resilient Water System: A Transformative Process

MICHAEL E. HOOKER, MBA • GEOFFREY G. MILLER, PE, BCEE • TIMOTHY P. TABER, PE, BCEE

INTRODUCTION

Few things in life are as important as access to clean water. The evolution from water delivered through individual wells to centralized treatment and distribution systems opened the door for the growth of suburbs, the improved management of water resources, and the protection of public health. With such an important role in the community they serve, water utilities must be prepared for the risks they may face.

Understanding how a water system has evolved over time, the components critical to delivery, and the management policies in place all contribute to understanding and expanding its resilience. This chapter examines how the Onondaga County Water Authority (OCWA) grew to the system it is today and how resilience became an important component of its day-to-day management.

THE EVOLUTION OF OCWA

To fully understand and appreciate the state of the OCWA in 2019 with respect to asset management and its vulnerability assessment, it is helpful to understand where the water system started and how it evolved over the course of the past 150 years. The construct of the Authority is similar to some systems in some ways and is widely different from the construct of most systems in many ways. At the onset, the water supply for Onondaga County, outside of the City of Syracuse, was fractured and comprised of both public systems and a private water supplier. Over time the private supply changed hands multiple times and eventually evolved into a public water supplier, the OCWA.

As the OCWA evolved, little by little during its first 40 years of operation, in addition to internal growth, smaller systems were acquired by lease or purchase. Then, in 1994, the pace quickened, and over the following 25 years, OCWA saw seven major wholesale customers turn their systems over to the Authority. During that same period, nine formerly independent systems connected and leased their systems to OCWA. Another 12 town water systems were newly constructed and subsequently leased to the Authority. Most recently, on January 1, 2017, the Onondaga County Water District, a major wholesale supplier to OCWA and the City of Syracuse, was consolidated under the OCWA umbrella.

In addition to the fast pace of consolidations, OCWA and every other water system in the country have been dealing with the fallout of the September 11, 2001 attacks and adjusting to changes in their local climate. September 11th led to the introduction of the vulnerability assessment process. Climate change introduced the all-hazards approach to risk management. More recently, risks associated with cybersecurity were introduced to the mix.

All these factors lead to OCWA seeking a comprehensive and timely way to manage its assets both with respect to dealing with aging infrastructure and the addition of new and acquired infrastructure and then in turn having the ability to access, rank, and address the risks all facilities are facing.

The Creation of the Authority

The OCWA was created by an act of the New York State Legislature in 1951; however, OCWA did not begin operations until December 29, 1955, when the Authority successfully acquired the New York Water Service's Syracuse plant that was supplied by Otisco Lake (See Fig. 5.1). How the Authority came about is a story worth being told, specifically when taking into account what the system entails today.

Optimizing Community Infrastructure. https://doi.org/10.1016/B978-0-12-816240-8.00005-7

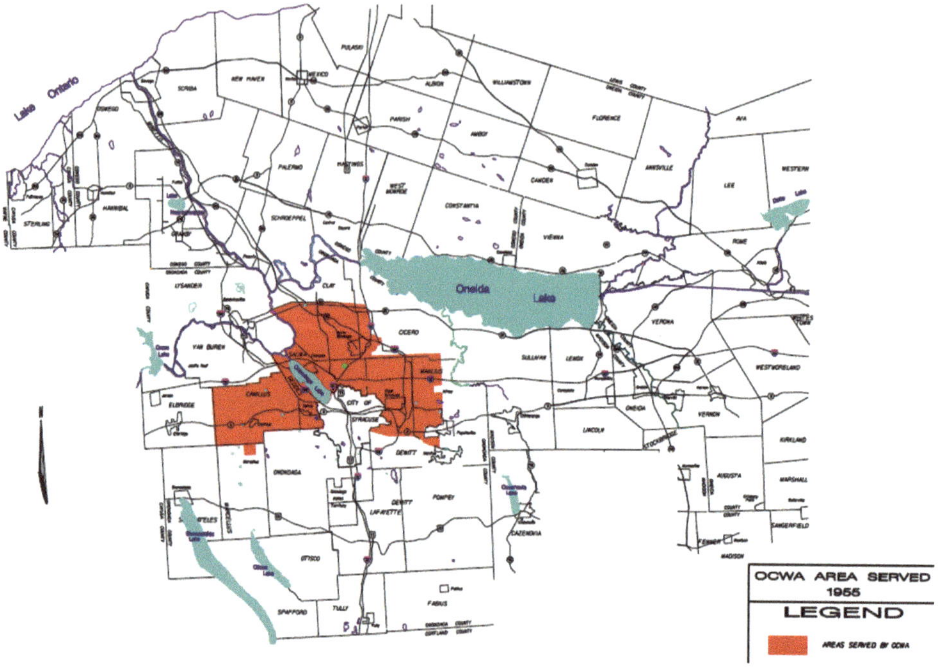

FIG. 5.1 OCWA service area in 1955. (Source: *Provided by OCWA.*)

OCWA's Predecessors

Syracuse Suburban Water System

The Syracuse Suburban Water System predated federal involvement in local development, and unlike many urban water systems, it was created primarily to meet the requirements of large-scale industry.

Syracuse Suburban's primary source of water, Otisco Lake, was first developed by the State of New York starting in 1868 as a federal system for the Erie Canal in times of drought.

American Pipe Manufacturing Company

In the early part of the 20th century, public water systems were not in the best position to compete for scarce investment funds. Enter the American Pipe Manufacturing Company, a Philadelphia firm that formed and partly financed several water companies, including the Onondaga County Suburban Water Company, which was incorporated on April 11, 1907. It is worth noting that this was an effort by American Pipe to assure a steady demand for their cast iron pipe.

The pipe company also drew up the plans for the construction of the new dam, intake, reservoir, and transmission line. At that time, the system was being designed to supply several large consumers who were finding their current supplies inadequate, including the nearby Village of Marcellus. Similarly, the villages of Camillus, Solvay, and DeWitt were also looking for a new water supply.

However, the most important potential customer was the New York Central and Hudson Railroad. At the time, the New York Central's repair shops and yards east of the city were supplied by the City of Syracuse through its line from Skaneateles Lake, but the City's single transmission line was incapable of meeting peak demands. During the summer of 1907 (and not for the first time), the New York Central had to cut way back on its water use to keep the City's Woodard reservoir from being drained. Solvay Process was also looking for a supplement to its existing supply.

Suburban Water Company

The Onondaga County Suburban Water Company was originally capitalized at $50,000 ($1.46 million in 2019 dollars), but this was not enough to start construction. After changing its name to the Syracuse Suburban Water Company in September of 1907, the company attempted to increase its capital to $600,000 ($17.6 million in 2019 dollars). However, few of their shares sold at first as the international financial panic of October of 1907 ushered in a severe economic depression. As a result, the water company could not raise enough money to begin construction until 1909.

Syracuse Suburban Water began operations with a new dam and intake at Otisco Lake, a reservoir in Fairmount, and a standpipe in Eastwood. The single transmission line was 20-in. in diameter from the lake through Fairmount to Solvay Process. From there to the DeWitt yards, they ran a 16-in. main; and from DeWitt to the Minoa shops and from Fairmount to Amboy, they ran 8-in. lines. Whenever possible, the mains were laid in the New York Central's right of way, presumably to avoid long and expensive negotiations with individual property owners and local governments to obtain their own right of way.

For more than a decade, the system ran uneventfully. The transmission line was shut down only three times before 1924, once to clean the Fairmount reservoir (July 16, 1913) and twice to make connections (June 21, 1913 and June 26, 1913). The first break in the transmission line, in May of 1924 at Marcellus Falls, was repaired before customers were inconvenienced. Two more cracks in the same bridge in February 1925 caused Syracuse Suburban's only other major shutdowns.

Federal Water Service

In December 1926, the Federal Water Service bought Syracuse Suburban and changed its name to the Onondaga Water Service. Shortly after, the system began expanding and became the center of controversy. In 1927, the 20-in. transmission line between the lake and Fairmount was augmented by a 24-in. line. The next year chlorination equipment at Otisco was replaced, and the Wolf Street booster station and service building were added. Then, anticipating heavy demand from the New York Central, the Onondaga Water Service increased capacity by building a new Otisco booster station in 1929.

Demand for water soon began to justify improvements in the system. Suburban Syracuse experienced a housing boom in the late 1920s. Developers either installed or paid the Onondaga Water Service to install mains and services in new tracts in Mattydale, Lyncourt, and Fairmount.

Additionally, supply to some older areas was upgraded. Both the Village of Lyncourt and Town of Salina had 2-, 2½-, and 3-in. mains replaced by 6- and 8-in. mains when they formed fire districts and paid for hydrants.

The Onondaga Water Service also added 400 customers when the Village of North Syracuse gave it a perpetual franchise in exchange for 6- and 12-in. mains, 90 hydrants, and an elevated tank in 1929. For several years, Village officials tried to form their own water systems, but voters repeatedly turned down bond issues.

The Onondaga Water Service began selling water wholesale to new water districts formed in already settled areas, including the Dooley tract in Fairmount in 1928, the Fairmount water district in 1929, and the Camillus and Cityview water districts in 1930. Finally, the Water Service picked up three new industrial customers: Fairmount State School (1929), Syracuse Airport (Amboy, 1929), and Will and Baumer Candle factory (1930).

A stock market crash in 1929 and the subsequent collapse of the economy by the end of 1930 sharply reduced demand for water and stopped expansion, with one exception—a main extension on Buckley Road in 1939 to serve several truck farmers. The only major improvement to the system in the 1930s was the installation of chlorination equipment at Fairmount reservoir to eliminate *Escherichia coli* from the lines east of the reservoir.

Conflict between the Onondaga Water Service and local mill operators increased in the 1920s and took off during the Depression. When Onondaga Water Service planned to expand their capacity, they went to the New York State Water Control Commission to obtain permission to draw 11 million gal/day (mgd) from Otisco Lake—an increase of 6 mgd.

In 1930, Onondaga Water Service officials tried to reduce flow to Nine Mile Creek to maintain water quality. Mill owners went to the State Commissioner of Waterways to try to force the Onondaga Water Service to increase the flow to the creek. Recognizing the impossibility of satisfying both sides, the State Commissioner of Canals and Waterways, Ralph O. Hayes, recommended that the state abandon its interest in the lake and sell its water rights. With all interested parties then objecting, the state Attorney General, John J. Bennett, Jr., ruled that the state could not pass on ownership of the waters of Otisco Lake and Nine Mile Creek, only certain riparian rights.

The Onondaga Water Service's reputation with the public through these episodes was not enhanced by the nature of its ownership. On December 12, 1926, the New York Water Service Corporation, which was in turn controlled by the Federal Water Service Corporation, bought a controlling interest in Syracuse Suburban Water Company and changed the name to the Onondaga Water Service. Then in April 1929, a holding company called United Power, Gas and Water company, or Tri Utilities obtained control of the Federal Water

Service and People's Light and Power Company, thus stacking a third holding company over Onondaga Water Service.

Utility holding companies, in general, were under attack at the time, and the Federal Water Service was investigated by a House of Representatives committee. Tri Utilities collapsed at the end of August 1931 and went into receivership. Federal Water Service was separated from the others in February 1932 when Chase National and Central Hanover Banks bought its voting stock. The banks then auctioned Federal Water Service stock. In the meantime, New York Water Service was reorganized. The Onondaga Water Service was merged into New York Water Service and functioned from then on as one plant in a centrally managed company.

Surprisingly, the Syracuse plant of the New York Water Service escaped public censure for termination of service for nonpayment during the Great Depression.

As the Water Service passed through the 1930s, time began exposing the physical weakness of the system. Low pressure, leaks, increasing demand, and, in one neighborhood, worms forced the Water Service to replace undersized mains. Transmission main breaks revealed the impossibility of maintaining regular service with a single transmission line and inadequate storage.

The first major break occurred on the 16-in. line feeding Liverpool in June 1930. It took more than 2 days to repair and caused the temporary closing of a major railroad line when the water weakened its bridge crossing. The second, in August 1931, was on a 24-in. line. Although it was repaired in less than 12 hours, hundreds were out of water or had low pressure, according to the *Herald*. When the line to Liverpool broke again in February 1936, the Mayor of Liverpool ordered the old village spring system reactivated; otherwise, 800 families would have been out of water during the 24 hours it took to make repairs.

New York Water

America's entry into World War II demonstrated that not only was New York Water incapable of maintaining service during emergency repairs, they also were not willing or able to meet heavy industrial demand. When Onondaga Water Service was absorbed by New York Water, it became part of a highly centralized organization. Everything but emergency repairs had to have the prior approval of the main office, and even emergencies had to be justified down to the penny.

Intercompany correspondence indicates that either overall profits were down or that money that should have been reinvested in replacements and improvements was siphoned off. When an ailing pump at Otisco was replaced shortly after the attack on Pearl Harbor, it was not junked or left in place as a standby. Instead it was sent to the Bayshore plant as their "new" pump.

The Otisco Lake dam site was fenced for the first time because of orders from the War (now Defense) Department. Correspondence dated 1942 over a supply line to a new industrial park (the Carrier line) indicates a determination to spend as little as possible. They argued that because the industrial park represented wartime demand and not a steady customer, they should rely on the City to make up shortfalls on a regular basis. By 1953, demand on this line was causing low-pressure problems on the eastern end of the system.

After the War, New York Water's financial condition appeared to be critical. Construction of a 36-in. main west of the Fairmount reservoir was halted when the contractor was not paid by the New York office. General Auditor, C. B. Myers, sent them a priority system for submitting vouchers; they were instructed to put red stickers on the bills that had to be paid on time.

Nor were they able to oblige water districts literally begging for their services. In 1940, the West Genesee Water Service approached them about supplying Westvale with water but ended up having to turn to the Village of Solvay. In 1945, Solvay notified them they could no longer supply them. Albert Korves of the Syracuse office of New York Water then began a series of memos urging New York to authorize taking it over. West Genesee Water Service served 458 customers directly and was linked to the Taunton Water District, which consumed 4½ million gal/yr. Serving them would cost New York Water a storage tank, a booster station, and 6000 ft of mains to connect them with Otisco water. This was more than the main office was willing to spend.

In 1946, New York Water went before the State Water Power Commission to fight, unsuccessfully, an order to supply Westvale on a wholesale basis. Even after they lost this case, they stalled on making adequate provision for service. In 1950, the New York office finally authorized replacing the 2-in. line supplying the district with an 8-in. line after pressure from the Health Department and the State Public Service Commission following numerous complaints from the public about water quality.

Supply to the housing tracts springing up after the War was also badly handled. Although the use of 2-in. mains was restricted by the Public Service Commission's requirement that all homes be within 500 ft of a hydrant, and although New York Water had problems with 2-in. mains in the 1930s, it began laying 2-in. galvanized pipes on a regular basis.

The post-War boom put a strain not just on equipment, but also on the supply of water. In 1945, New York Water got permission to raise the dam and draw up to 16 mgd from the lake. This was not enough. In 1948 and again in 1951 and 1952, New York Water began drilling test wells—first by the dam and later along Nine Mile Creek. In every case, the water proved to be too hard to be useable.

All of New York Water's shortcomings came to the surface when they filed for a 15.72% rate increase with the Public Service Commission in 1950. Eight communities served by the New York Water Service filed objections and sent representatives to the hearings that began in March. The Villages of Solvay, Liverpool, North Syracuse, and the towns of Geddes, Salina, Cicero, Clay, and Fire District #2 in the Town of Salina were mainly concerned because, although New York Water pumped an average of 10 mgd, it had storage for less than 6 million gallons. In the event of a break on the single line between Marcellus and the Fairmount reservoir, all of those communities would be out of water within 12 hours. Although New York Water had three connections with the City of Syracuse, their agreement limited the water company to 1 mgd from that source.

The construction of the New York State Thruway in 1951 made the lack of storage facilities especially critical. Transmission lines had to be relocated in several places, leading to heavy dependence on City connections and low pressure. New York Water dealt with the relocations and major scheduled repairs by running notices in the newspapers listing schedules and areas to be shut off or put on low pressure.

By this time, New York Water Service's operation in Onondaga County was running on borrowed time. Soon after New York Water lost its rate increase petition in 1951, the Onondaga County Water Authority was formed. After ordering a series of studies looking for a new source of water, the Authority settled on Otisco Lake and began condemnation proceedings. New York Water Service's Syracuse plant was formally acquired by the Authority on December 29, 1955.

Onondaga County Water Authority (December 29, 1955 to Present)

On December 29, 1955, when the Onondaga County Water Authority finally acquired the Onondaga County–based assets of the New York Water Supply Corporation, OCWA began operations of a system with 134 miles of water mains that served 9119 accounts spread throughout 12 towns in the county with an average daily demand of 14.26 mgd.

Because of limited funding, in 1958, OCWA sold its system components located in the Town of Dewitt to the Town, which remained a wholesale customer, purchasing water from both the City of Syracuse and the Authority.

Proceeds from the sale of the system to Dewitt allowed OCWA to undertake its first round of multiple improvement projects, including raising the Otisco Lake Dam to its present level.

Onondaga County experienced post-World War II growth starting in the 1950s that carried on into the 1970s. By 1965, OCWA had 12,161 retail accounts and 219 miles of mains with a daily demand of 21.79 mgd.

By 1975, those numbers more than doubled to 26,450 retail accounts and 561 miles of mains with daily demand reaching 32.69 mgd, and by 1985, OCWA was serving 44,603 retail accounts via 875 miles of mains with a daily demand reaching a peak of 45.77 mgd.

It should be noted that although OCWA's retail base was growing during this 30-year period, the customer base of OCWA's numerous wholesale customers was growing in parallel. Additionally, industrial water demand was also increasing.

OCWA's Otisco Lake daily withdrawal was and is still limited to 20 mgd, and until the mid-1980s, it was an unfiltered water source. In 1985, OCWA built a filtration plant north of Otisco Lake in the Town of Marcellus next to the gravity-fed transmission mains.

Initially the plant was designed as a slow sand, direct filtration plant with four filters capable of treating 24 mgd. The plant was also built with the ability to add powdered activated carbon to address taste and odor issues as they arose. In the 1990s, the filter profiles were changed from sand and gravel to mostly sand with a layer of granular activated carbon (GAC).

In 2010, the Otisco Lake plant was completely renovated, and two additional filters were added to enhance water treatment capabilities but not to add capacity. At that time, the filters were completely rebuilt, with new low-profile underdrains to accommodate multimedia filter materials, including 36 in. of GAC.

In the 1960s, OCWA's ability to meet daily customer demand was being taxed and exceeding the allowable 20 mgd withdrawal limits for Otisco Lake. On top of the increasing growth-driven daily demand, Central New York experienced a drought in the mid-1960s, which reduced the water flow from Otisco significantly and reduced the Lake to more akin to a large creek.

Forward thinking community leaders worked with Onondaga County to search out and develop an additional water supply for Onondaga County not only for the present but also well into the future. After extensive review of available water sources in the region, it was determined that Lake Ontario was the best and most reliable source of supply. At the time, the City of Oswego already had an intake tunnel in place, drawing water from the Lake, and by agreement a deal was struck to allow Onondaga County to jointly use the intake to supply a water filtration plant that would be built in the Town of Oswego in Oswego County.

Ultimately, the Onondaga County Water District was formed by the Onondaga County Board of Supervisors and created the Metropolitan Water Board (MWB) to oversee the construction and eventual operation of the Lake Ontario water treatment plant. Construction of the 60 mgd plant was completed and placed in service on June 1, 1967. The County still owns the Onondaga County Water District, and the Metropolitan Water Board remains in place. However, on January 1, 2017, by agreement with Onondaga County, the operations of the water plant and all Lake Ontario–related facilities were turned over to OCWA for operations and maintenance.

The agreement also requires OCWA to undertake upgrades and improvements of the 52-year-old treatment plant, including the long overdue addition of auxiliary power generation for the Water Plant and the Raw Water and Clear Water pumping facilities. The estimated cost for the combined projects is in the $50 million range.

Presently, the Lake Ontario water treatment plant pumps water from Lake Ontario through an 8-foot diameter intake it shares with the City of Oswego.

Water for the plant is pumped from the Lake Ontario raw water pump station to the plant 2 miles away. There the water is processed and filtered. Water leaving the Lake Ontario plant is pumped via the Clear Water Pumping Station to the Terminal Reservoirs via 22 miles of 54-in. prestressed concrete cylinder pipe transmission main. Because of the size of the line and pumping constraints, water delivery to Onondaga County is limited to 50 mgd, leaving excess plant capacity available for delivery to customers in Oswego County that are near the treatment plant.

From the Terminal Reservoir, which was initially a 30-million-gallon open reservoir, water is pumped by the Farrell Road pumping facility via transmission mains to reservoirs in the eastern and western ends of Onondaga County. Initially, the Eastern Reservoir was an open 30-million-gallon storage facility that includes a pumping station that pumps water to higher elevations in Onondaga County and allows for water to be conveyed to customers that were eventually added in Madison and Oneida Counties to the east of Onondaga County. Water pumped to the west end of the County initially fed into a 100-million-gallon open reservoir that was designed to address projected growth in the western portion of the County as well as supply large industrial customers, including breweries.

With the Lake Ontario Plant on line by 1967 and OCWA being its lone full-time customer, water availability for Authority customers jumped from 20 to 70 mgd once the plant came on line. Correspondingly, OCWA daily water demand increased from 21.79 mgd in 1965 to 32.69 mgd in 10 short years. From 1975 to 1985, daily demand increased to OCWA's historical peak of 45.77 mgd. It is worth noting that, in 1985, OCWA was supplying over 50 industrial accounts, including two breweries, which, at their peak, drew a combined total of 11 mgd. Over that same 20-year span, OCWA's customer accounts total rose from 12,641 in 1965 to 26,450 in 1975 and totaled 44,603 accounts by the end of 1985. Furthermore, breakdown of OCWA sales was roughly 29.9% residential and commercial, 35.5% industrial, and 34.6% wholesale.

The mix of sales began to slowly change in the 1980s when the Town of Cicero leased its system to OCWA.

Then in 1995, a parade of water system conversions from wholesale to retail began, starting with the Town of Salina, which turned one wholesale water system into 7500 retail customers. Over the course of the next 20 years, seven other wholesale customers leased their systems to OCWA, adding a grand total of 20,293 retail accounts. By the end of 2018, wholesale water sales dropped to 20.3%, industrial sales represented 22.3% of total sales, and residential/commercial sales totaled 57.4% of sales.

In addition to the conversion of wholesale customers, increasing water regulations, degrading water supplies, and overall cost increases drove another 21 towns and villages to come to OCWA for water supply. In some instances, the municipality chose to abandon its existing supply. In other cases, degrading private well supplies led towns to form water districts that in turn constructed a new water system that would be leased to OCWA for ongoing operations and maintenance. Through town and village actions, the Authority added an additional 9967 retail accounts between 1995 and 2018.

By 2018, OCWA's customer base increased, from 44,603 accounts in 1985 to 103,330 accounts, an increase of 59,727 accounts. Of the overall increase in total customers, 36,260 (60%) were existing users that were on private wells or supplied by their municipality. Furthermore, during the same 33-year time frame, OCWA added 23,467 retail customers through organic growth See Fig. 5.2.

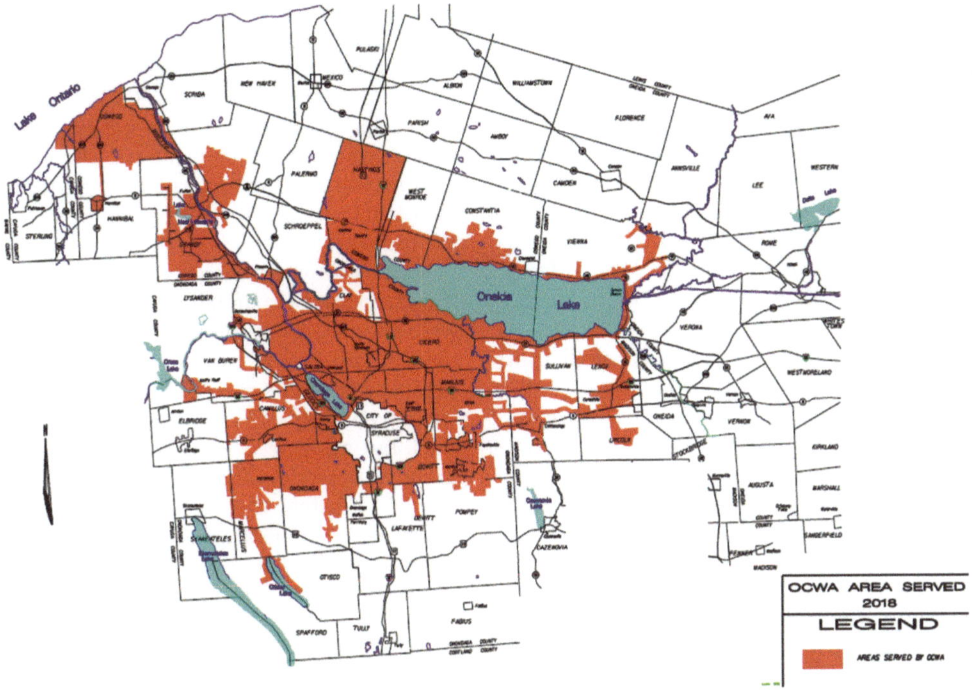

FIG. 5.2 OCWA area served, 2018. (Source: *Provided by OCWA.*)

From 1985, when average daily demand peaked at 45.7 mgd, daily demand slowly began to decline, even in the face of added customers. The Central New York economy began to slow dramatically in the 1980s and has yet to fully recover. A major blow to the region and OCWA came about when the Authority's largest customer at the time, Miller Brewing, announced it would close their facility in 1994. Along with the loss of jobs, the loss of water sales revenue also took its toll on both OCWA and the Metropolitan Water Board. By 1995, average daily demand fell 3.49 mgd. The one saving grace, albeit a short one, was the addition of several cogeneration facilities that were coming on line at the same time and somewhat offset the loss brought about by the closing of the Miller Brewery.

From 1995 on, the mix of OCWA's customer base was in flux. At the same time, the Authority began to see the impacts of climate change impacting the region, particularly from 2006 and beyond. As opposed to many areas in the country that were experiencing dryer periods and prolonged droughts, Central New York started experiencing wetter summers that resulted in seeing the drop-off of summer peak demands. Coinciding with the changing weather, favorable pricing rules related to cogeneration facilities ended, and in turn, the cogeneration facilities were shut down completely or turned into peaking facilities, resulting in greatly reduced water demand. By the end of 2006, daily demand had declined to 38.52 mgd, all due to a declining industrial base and a major decline in peak residential summer use.

Since 2006, daily demand has hovered around 38 mgd, in spite of the addition of customers. By the end of 2018, average daily demand was 37.58 mgd for 103,330 accounts served by a system that includes 2140 miles of mains, 13,400 hydrants, 61 water storage facilities, 47 pump stations, and 2 water treatment plants.

Over time the growth projected in the 1960s did not occur, particularly in the western end of the County. However, OCWA did experience growth in the eastern end of the County. OCWA also expanded outside Onondaga County and saw expansion take place as former stand-alone water systems in Oswego, Madison, and Oneida Counties connected to OCWA. More recently, OCWA began serving a small area in Cayuga County where existing homes had failing private wells. In most cases, the newly connected systems abandoned their existing groundwater source and leased their systems to OCWA, leaving the operations, maintenance, and future capital improvements in the hands of the Authority.

OCWA has a third source of supply, the City of Syracuse and its water department. The City's source of supply is Skaneateles Lake, located in the southwestern corner of Onondaga County. The Skaneateles supply is a gravity-fed system that is unfiltered but maintains a filtration avoidance permit. OCWA has an agreement that allows for the Authority to buy up to 3 mgd from the City to service OCWA customers that border the City.

It should be noted that the Lake Ontario supply can supply up to 10 mgd into the City of Syracuse via the Central Branch pipeline construction by MWB. Additionally, MWB built a Southern Branch system that comprises two large pumping stations and three water storage tanks that can supply Skaneateles water to both City and OCWA customers in and around the southern portions of Syracuse. Although MWB built the system, it was initially operated by the City of Syracuse Water Department. In November of 2008, the water system, which was in need of substantial rehabilitation, was leased by Onondaga County to OCWA.

EVOLUTION OF ASSET MANAGEMENT AT OCWA

Prior to the 1990s, asset management was not even on the radar of the Onondaga County Water Authority. Annual capital reinvestment was minimal and relatively sufficient for the size of the system. Keep in mind, in most cases the mains and service were less than 100 years old, with most assets being less than 40 years old. The Otisco Lake water treatment plant was barely 10 years old, and the number of pump stations and storage tanks in service was minimal, primarily because of the system being gravity fed by Otisco Lake and the plant.

Emergency response planning at the time was centered around dealing with water main breaks and operating remote facilities manually should the telemetry fail. Power failures were recognized as a potential problem. However, because power failures were infrequent on a gravity-fed system and recognizing there was more than adequate storage for the system, power failures were not of major concern.

The thought process began to change as the pace of consolidation picked up in 1995. It became apparent to OCWA that the mix of systems now being operated and maintained brought a new set of challenges. Although OCWA, and its predecessor's, policy specified cast iron and eventually ductile iron pipe for all main installations, many of the systems being acquired did not. Furthermore, until 1995, OCWA specified a minimum pipe size of 6-in. and then 8-in. mains thereafter. OCWA also specified copper tubing for all customer service lines, and all customers had to be metered. Services over 100 ft in length required the meter be installed at the property line.

On the converse, the systems being added to OCWA specified various materials and sizes, and not all services were metered. The Authority discovered every type of service line imaginable, including wooden mains still in service. The type of material used for services has varied as well. Copper was the most predominant material used, but galvanized lines have been found as well as some lead service lines. With respect to lead service lines, fortunately the overall number of lead lines is relatively small, as a recent survey and records review has indicated.

The number of pump stations and water storage facilities increased, as older systems were added to the OCWA system. The same held true for newly constructed town systems that wanted to be served by OCWA, as the Authority required the town to provide adequate storage and pumping facilities as part of the service agreement.

As the pace of consolidation was picking up in the 1990s, the need to reinvest in the overall water system came into focus, prompting the OCWA Board and Executive team to develop its first long-range (20-year) capital budget and operations and maintenance budget.

As the asset management planning process evolved, several major events that transpired had a firm hand in shaping the overall planning process. The first major event took place on Labor Day, September 7, 1998 when several major thunderstorms consolidated creating a huge derecho (a straight-line wind storm) that had winds up to 115 mph hit central New York along with 10 to 20 lightning strikes per minute. By the time the storm was over, OCWA lost all power and more than a quarter million people were left in the dark. At the time, OCWA had minimal auxiliary power systems. The Otisco Lake Plant could operate, and portions of the main headquarters building were operational. As for the major pumping station, there were no generators in place and the existing emergency response plan (ERP) did not address an outage of this magnitude. On the fly, operating personnel identified a supplier of portable generators and in the middle of the night obtained three large generators needed to keep the major pumping facilities on line. As it turned out, acting quickly and obtaining the generators in the middle of the night proved to be fortunate, as it turned out the Authority had just rented the last three generators available in New York. OCWA was able to meet all customers' water needs in the aftermath of the storm, and the Authority also was able to supply water to adjacent water systems that had lost power.

Once the dust had settled, more than 2 weeks after the storm, OCWA conducted an after-action review and took an introspective look at its ERP and overall system operations. The review led to a major overhaul of the ERP and resulted in the executive team recommending that future capital budgets include retrofitting the existing pump stations with auxiliary generators. The recommendation was taken under advisement by the Board, until the following winter brought a major ice storm to the region.

The ice storm in the winter of 1999 had a similar impact on central New York, although not as widespread. Much of the damage was limited to the northern portions of the OCWA system, and it forced the need to rent generators and to go to manual operations for a good portion of the system.

It was also around this time that the discourse on climate change impacts was revving up. Although talk for much of the country centered around hotter, dryer weather, and prolonged droughts, the predictions for upstate New York centered around more frequent and intense storms both in the winter and in the summer, much like what was experienced during the Labor Day storm of 1998 and the ice storm of the winter of 1999.

This leads the OCWA Board to go beyond taking the executive staff's recommendation under advisement. As part of the 2000 budget process, the Board adopted a policy whereby OCWA would systemically retrofit every major facility with auxiliary power. Smaller facilities were retrofitted to be able to accommodate a portable generator, and in turn, OCWA acquired two large portable generators. The Board took the policy a step further, mandating that all future facilities would be built with auxiliary power. This policy extended to new systems that towns were constructing with the intent to lease to OCWA. It was at this time that the OCWA management team set out to update the Authority's ERP and make it more relevant to the changes going on related to climate change and system growth.

The defining event, for the entire country, took place on September 11, 2001. The events of that day do not need to be recapped here but suffice it to say they forever changed the way OCWA operated and those changes are ever evolving.

On the heels of the September 11th attack water systems across the country were tasked with developing a vulnerability assessment that addressed potential outside threats. For many utilities, this was a one and done exercise. However, New York State has required its water systems to update their vulnerability assessments and their corresponding ERPs every 5 years. Over time, New York's vulnerability assessment process has evolved to become an all-hazards assessment, and with the last cycle of submittals, a cybersecurity component was added to the program.

As is all too often the case with 5-year submittals, the tendency is to complete the update and then let it sit on the shelf for 5 years and then embark on a mad scramble to complete the updates. OCWA's Board and executive team did not want to fall into that routine and in turn decided to continually update both the vulnerability assessment (VA) and the ERPs.

OCWA's initial VA was developed with the assistance of a local security consultant, AMRIC Associates, which employed former FBI and Department of Defense investigators. OCWA's part-time safety manager was also retired military and had some security training and was assigned to work with AMRIC on developing the Authority's first VA. To facilitate proper development, OCWA sent the AMRIC agent and OCWA's safety manager to an American Water Works Association (AWWA) Risk Analysis and Management for Critical Asset Protection (RAMCAP) [1] training program to learn how to properly develop an assessment for the Authority. The initial assessment was a spreadsheet-based application that listed vulnerabilities and recommended countermeasures and, over time, tracked the implementation of the countermeasures.

Correspondingly, during this same time frame, OCWA set out to upgrade all its business enterprise systems (billing, accounting, and finance) and introduced the Authority to IBM Maximo for work and asset management. By the end of the project, all the formerly separate systems were fully integrated with each other and were also integrated with Maximo and OCWA's geographic information system (GIS) in one database.

Furthermore, as OCWA continued to grow and the demands related to safety and security evolved, a full-time, trained professional was brought on board to manage safety and security programs. Additionally, the safety and security manager is supported by the executive team, department heads, and an employee-led safety committee.

In the following years, with the introduction of ANSI/AWWA Standard J-100 [2] and as the requirements of New York's VA increased, OCWA looked for automated solutions. The US Environmental Protection Agency (EPA) Vulnerability Self-Assessment Tool (V-SAT) program was considered at one point, and OCWA personnel were sent to training on how to use the system. As the history of the evolution of the Authority system indicates, the makeup of the system is complicated and it turned out that the V-SAT tool was not a good fit for OCWA, particularly where the Authority wanted to go with respect to ongoing updates and improvement to both the VA and the ERP.

During a group review of the V-SAT program with the safety/security manager and the engineer that attended the training session, it was agreed that V-SAT was not the right tool for OCWA, but all agreed we needed to do something to make the process easier and timelier. On the heels of that meeting, OCWA contacted its Maximo consultant at B&L Engineers and inquired about Maximo and its capabilities with respect to developing a Maximo-based VA program, one that would be J-100 compliant. In turn, our consultant set out to familiarize himself/herself with New York's VA requirements and the J-100 Standard.

OCWA'S ALL-HAZARDS APPROACH

OCWA's VA risk analysis process is conducted on its asset/threat pairs. OCWA's assets are prioritized based on the types of consequences anticipated under worst-case conditions. The quantitative risk analysis includes an assessment of consequences regarding public health impacts and safety. The risk determination accounts for the likelihood of damages based on countermeasure capabilities to detect, delay, and respond to the threat against a specific asset.

The process and tools utilized by OCWA were designed around the J100-10 Standard and follows the seven steps outlined in Fig. 5.3.

The J100-10 standard is an American National Standard, so designated by the American National Standards Institute (ANSI), and falls under the joint jurisdiction of the American Water Works Association (AWWA) and ASME Innovative Technologies Institute, LLC (ASM-ITI). This process analyzes and manages risks associated with malevolent attacks and naturally occurring hazards against critical infrastructure.

Management System

OCWA owns and utilizes IBM's Maximo system for managing all activities related to asset management. The system is used for project management and inventory control and tightly integrated with the financial management system. It was decided to utilize this system to manage the data and analysis related to vulnerability assessments. Benefits of this approach include the following:

- A single system to track all asset data, risk, and vulnerabilities

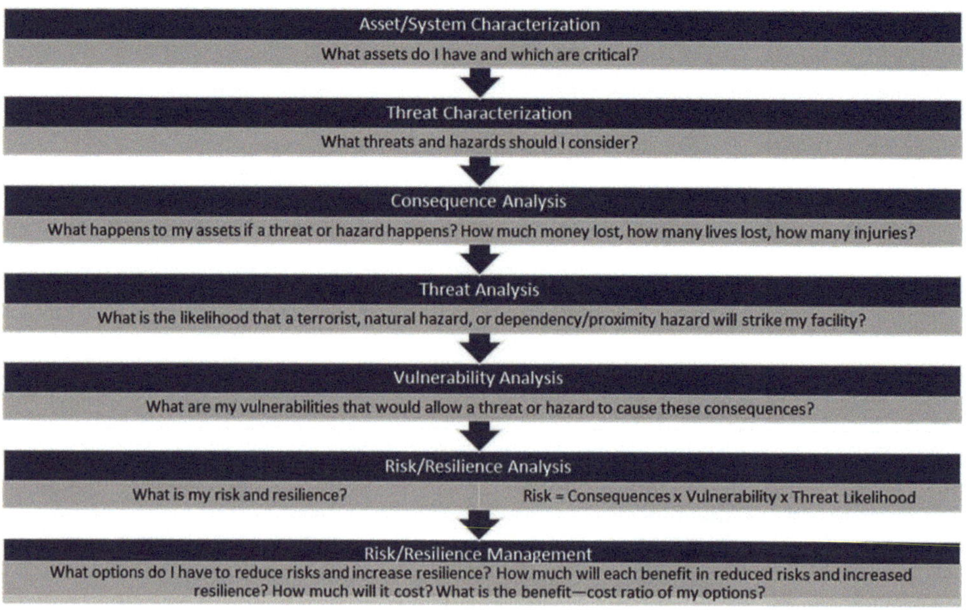

FIG. 5.3 OCWA vulnerability assessment risk analysis process. (Source: *Provided by OCWA, Courtesy of Barton & Loguidice*.)

- Detailed tracking of all follow-up activities identified in the vulnerability evaluations
- Comprehensive cost tracking of implementing countermeasures and other activities to reduce risk
- Full prioritization and scheduling abilities for follow-up work
- Putting all information into a single database so that multiple data sources do not need to be managed and synchronized

The following sections describe each of the seven steps in more detail.

System/Asset Characterization

The purpose of asset characterization is to determine the assets that, if compromised by malevolent, accidental, or natural hazards, could result in prolonged or widespread service interruption or degradation, injuries, fatalities, detrimental economic impact, or any combination thereof. Ultimately, asset characterization produces a list of critical assets that must be considered in subsequent steps (e.g., threat, consequence, and vulnerability analyses, and estimation of risks and resilience).

The first analysis step OCWA performed was to define the assets. OCWA maintains an inventory of assets within Maximo; verification of the data was performed as well as GIS coordination for the location of the sites. OCWA maintains a more detailed inventory of assets in Maximo, for example, individual pumps, motors, etc. The vulnerability assessments are performed at a higher level, which OCWA defines as a "site." Each site is categorized into one of these categories, which help to define the countermeasures that should exist to help protect it.

Threat Characterization

OCWA, referencing the J100-10 standard, utilizes an all-hazards approach when considering the threats to the organization. The following types of threats were being considered: man-made hazards or accidents, natural hazards, and dependency hazards (interruptions of supply chains or proximity to dangerous sites). Threat characterization was performed to identify general and specific threat scenarios to serve as reference threats for the remainder of the vulnerability assessment process. These threat scenarios characterize the events or combination of events that produce harm. Malevolent threats include various modes of attack (e.g., air, land, and water) and various magnitudes of attack elements. Attacks by both insiders (e.g., current or past employees, suppliers with access to facilities) and outsiders (e.g., adversaries, criminals, vandals) are also considered. Natural hazards include hurricanes, floods, tornadoes, earthquakes, snow storms, ice storms, and wildfires, and dependency hazards shall include interruptions of utilities, suppliers, employees, customers, transportation, and proximity to dangerous neighboring facilities.

To facilitate cross-asset and cross-sector comparison of risk and resilience values, a uniform and consistent set of reference threats (a set of threats to be used to evaluate vulnerability and consequence) are applied to all assets under evaluation. For each asset-threat pair, descriptions and notes are captured, which detail the range of magnitude that a threat can pose to an asset. OCWA used a uniform set of threats; decision-making prioritizes risk-reduction and resilience-enhancement options. The ability to prioritize assets is based on estimates of the worst reasonable consequences resulting from the destruction or loss of each asset, without regard to the threat.

Consequence Analysis

This step identifies the worst reasonable consequences that can be caused by the specific threats to the assets identified in the prior step. The consequence analysis estimates the results of threat scenarios using common quantitative metrics that include the J100-10 required criteria:

- Number of fatalities
- Number of serious injuries
- Financial loss to the owners of the facility
- Economic losses to the community (i.e., standard metropolitan or micropolitan area) in which it operates

In addition to these criteria, OCWA has added two additional criteria they feel are more representative to the true consequence of losing an asset, regardless of the cause of the event. The following criteria are included in the consequence analysis for OCWA:

1. Impacts to the ability to provide service:
   a. Adequate supply of safe water
   b. Adequate supply of safe water to meet emergency needs
   c. Parts of system without water
   d. Contaminated water available for firefighting, sanitary purposes
   e. No water available
2. Size of the service area
   a. <2500 customers
   b. 2500–5000 customers
   c. 5000–10,000 customers
   d. 10,000–25,000 customers
   e. >25,000 customers

OCWA assumed that for each threat, the consequences simulate a complete loss of the site/asset. All assumptions and notes associated with each asset/threat pair are maintained and reported from the Maximo system. Each asset/threat pair was assigned a combined consequence score in the range of 1 to 5.

Threat Analysis

This step estimates the likelihood of a malevolent event, dependency/proximity hazard, or natural hazard. OCWA has estimated the likelihood or frequency of all hazards and threats at each of their assets/sites.

OCWA took all this guidance into consideration when assessing the likelihood of each characterized threat at each site. Fig. 5.4 summarizes the scores applied to each asset/threat combination.

For each asset/threat pair, OCWA maintains a likelihood of occurrence score on the scale of 0 to 5. A score of 0 identifies with a threat that will never occur against a specific asset, an example of this is a marine attack against a site that is not located on a body of water.

Vulnerability Analysis

In this step, OCWA analyzed the ability of each critical asset and its protective systems to withstand each specified threat. OCWA performed their vulnerability analysis using the following four-step procedure, in compliance with the J100-10 standard:

1. Reviewed pertinent details of the facility construction, systems, and layout. Include countermeasures, mitigation measures, and other impediments to threats, such as topographic, design, and equipment features that provide deterrence, detection systems, and delay features, and local and supporting response measures. Include information on interdependencies, personnel interactions, and process flows within the facility. Identify vulnerabilities or weaknesses in the protection system.
2. Analyze the vulnerability of each critical asset or system to estimate the likelihood that, given the occurrence of a threat, the consequences estimated in step 1 result. The utility may use fault- or event-tree analysis, path analysis, vulnerability logic diagrams, computer simulation methods, or expert judgment rules-of-thumb that can be used consistently across all relevant assets.
3. Document the method used for performing the vulnerability analysis, the worst-reasonable-case assumptions, and the results of the vulnerability analysis.
4. Record the vulnerability estimates as point estimates. The likelihood of attack success may be expressed as a fraction, probability, or the number of successes among attempts.

An important step to the vulnerability analysis for OCWA was to define the countermeasures to assess for each site.

Likelihood Score	Criteria – Frequency of Occurrence
0	Will Never Occur
1	every 100 years
2	Every 25 Years
3	Every 5 Years
4	Every Year
5	Multiple Occurrences Per Year

FIG. 5.4 Likelihood of threat scores. (Source: *Provided by OCWA, Courtesy of Barton & Loguidice.*)

COUNTERMEASURES AND ASSIGNMENT

Countermeasures are those actions, processes, devices, or systems that can prevent or mitigate the effects of threats to a facility. When assessing a facility, OCWA considers two principal kinds of countermeasures:

- **Existing countermeasures** are those already present, defined as part of the baseline or current situation. In addition to physical countermeasures, OCWA includes any policies, procedures, and operating practices, which serve to protect their asset base.
- **Potential countermeasures** are those OCWA is considering adding to its utility in the future. When OCWA performs improvement analyses, they can add potential countermeasures to determine their effect on the outcome. Cost is an important consideration when defining potential countermeasures, and these costs are applied during the comparison of implementation costs to the assessed risk reduction.

OCWA includes all countermeasures that provide capabilities to reduce the likelihood of consequences from included threats. Countermeasure capabilities are categorized based on the effectiveness of each action to reduce consequences or limit the likelihood of damage or threat.

Countermeasures' Effectiveness Against Threats

OCWA's examination of the J100-10 standard revealed that the standard does not address how effective countermeasures are against specific threats, but common sense exposes the fact that countermeasures are only effective against specific threats. For example, having bollards at a site is not effective to protect the site from an ice storm. For this reason, OCWA established a rating system for each countermeasure/threat pair to identify which countermeasures are effective to protect against specific threats.

Risk Assessment

Risk (R) is the result of the analysis and aggregation of the likelihood of an undesirable event, the vulnerability of a specific asset to that event, and the consequences of that event:

$$R = C \times V \times T,$$

where

Consequence (C) is the result of an event occurrence, including immediate-, short-, and long-term, direct and indirect losses and effects. The loss may include human fatalities and serious injuries, monetary and economic damages, and environmental impact. It may also include less tangible, and therefore, less quantifiable effects, including loss of public confidence, political ramifications, decreased morale, reductions in operational effectiveness, or other impacts.

Vulnerability (V) is viewed as a function of countermeasure capabilities, also referred to as the likelihood of damages. Countermeasure capabilities reflect assets or processes, which effectively detect, delay, and respond to threats. They also consider any weakness in an asset's design or operation that can be exploited by an adversary or disabling natural event. Such weaknesses can occur in building characteristics, equipment properties, operational policies, or locations of people.

Threat likelihood (T) is the probability of the event occurring. OCWA can choose to score likelihood based on their qualitative level (based on historical patterns or consultation with authorities) or quantitative probability to evaluate all asset/threat pairs with a consistent likelihood.

Consequences Determination

The Maximo configuration allows the user to assess the consequences of failure resulting from each of the asset/threat pairs identified during the previous step by evaluating the severity of the impact on selected consequence attributes.

For each asset/threat pair, the risk is calculated every 24 hours in Maximo. As new data are entered and identified deficiencies are addressed, the risks to OCWA change, and the data are recalibrated.

RISK/RESILIENCE MANAGEMENT

The key element of OCWA's vulnerability assessment program is the continuous analysis of each of the more than 200 sites for the existence and condition of countermeasures. All identified vulnerabilities are documented, creating work orders in OCWA's Maximo system to address. All work orders are assigned and tracked through the workflow process in Fig. 5.5.

All data for the vulnerability assessment program are stored in OCWA's Maximo system and are regularly monitored and can be reported in a variety of formats (Fig. 5.6).

EMERGENCY RESPONSE PLAN ENHANCEMENTS

Concurrent with the development of its own J-100 compliant risk and vulnerability assessment program utilizing its existing Maximo system, Authority staff embarked on improving the ERP. Rather than developing an ERP and letting it sit on a shelf until the next 5-year regulatory period arrived, it was determined that the ERP should be a living, usable document. The ERP was also expanded to include emergency contact information for emergency response contractors, major customers, schools, regulatory personnel, highway departments, public officials, and more. The ERP is updated continuously, and at a minimum of four times per year, the updates are distributed to employees and regulatory personnel.

Using this approach, OCWA remains continually prepared to address emergencies as they arise, and management is not worried about going through a mad scramble at the last minute, every 5 years, attempting to update its risk assessment and ERPs. The risk assessment is updated continuously and is never more than a day or two old, and the ERP is never more the 90 days beyond its last update. In essence, OCWA's risk manager should be able to take both plans "off the shelf" and provide a copy to a regulator for review, and the reports should be in compliance.

As an all-hazard approach has been adopted as the driver behind updates to the risk assessment and the ERP, both plans continue to evolve and expand. Climate change equates to different needs and responses across the country, from state to state and region to region. In Central New York (CNY), climate change impacts the electric grid and in turn drives the need for backup power and more water storage facilities. The climate change–induced increase in frequency

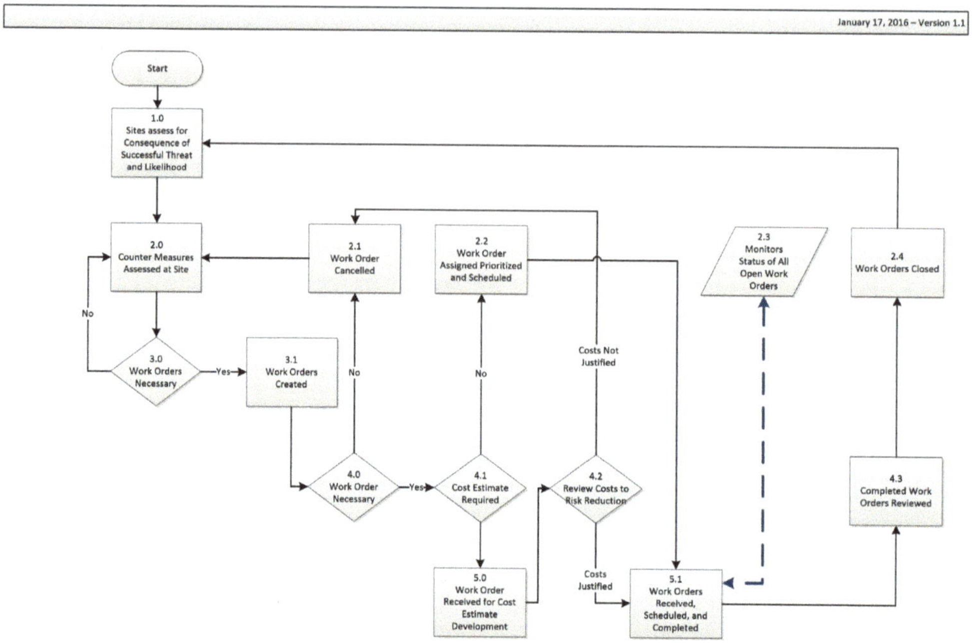

FIG. 5.5 Work order processing flow. (Source: *Provided by OCWA, Courtesy of Barton & Loguidice.*)

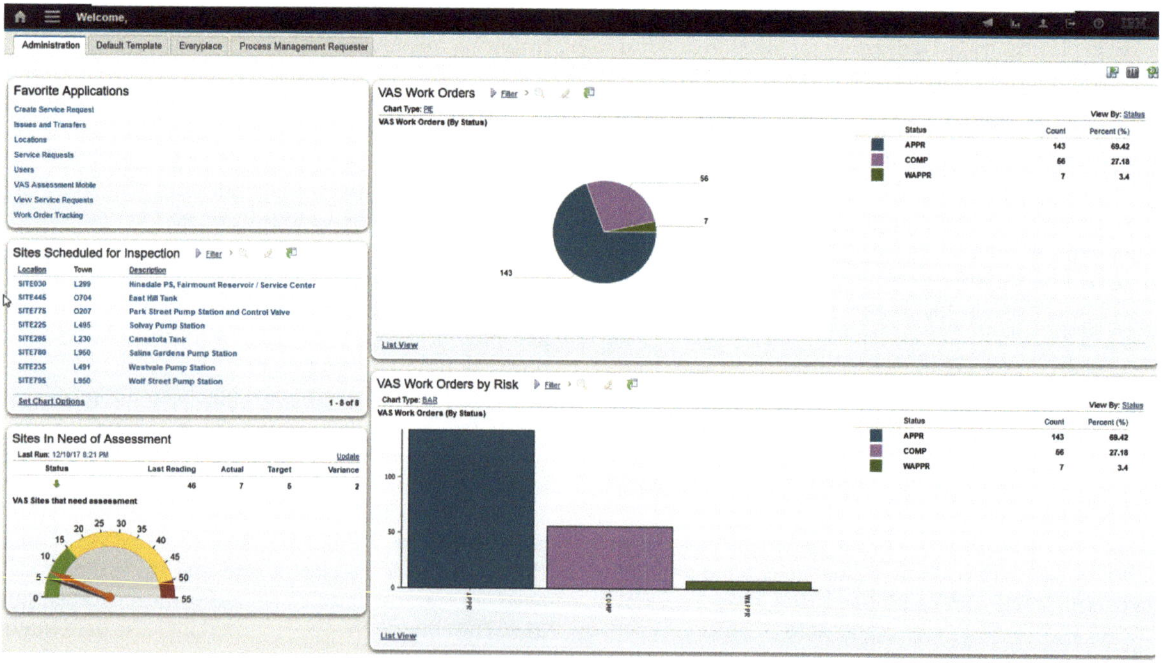

FIG. 5.6 Dashboard for OCWA's Maximo system. (Source: *Provided by OCWA, Courtesy of Barton & Loguidice.*)

and intensity of storms caused increased run-off into streams and lakes bringing with the run-off increased nutrients and turbidity. These increases have the need for increased awareness of the issues they bring, resulting in the expansion of the ERP, as well as adding additional components to day-to-day operating protocols employed by OCWA's two water treatment plants and daily distribution operations.

Increased sediment in the runoff in turn increases turbidity events. The Otisco Lake source and Skaneateles Lake source (maintained by the City of Syracuse) are more vulnerable to impacts than the Lake Ontario source. Water plant operators at the Otisco Lake treatment facility are more cognizant of storm events and are better prepared in case the intake turbidity increases. The same goes for the City of Syracuse water staff, who monitor Skaneateles Lake water. The Skaneateles supply is unfiltered and presents a whole realm of additional challenges. OCWA purchases around 3% of its daily demand from the City, but the lack of a filter plant limits options and presents a unique set of challenges all on its own. Accordingly, ongoing communication is vital to protect public health.

Once it has been determined there is a health safety issue, OCWA staff must be prepared to cease flow from Skaneateles, increase utilization of storage and, where possible, transfer water from OCWA's treated sources. Should those efforts not suffice, then plans need to be in place for bottled and/or bulk water distribution to impacted Authority customers. In these instances, coordination with the Onondaga County Health Department and Onondaga County Emergency Services is critical. To be successful, a plan must be in place, and it has been helpful to have gone through a table-top exercise or two before a real crisis takes place.

Another potential result from increased runoff is the addition of nutrients in the lakes that lead to the growth of algae blooms and bacterial growth, including blue-green algae and cyanotoxins. Skaneateles Lake, which has long been thought to be immune to such events due to its pristine water quality, suffered blue-green algae blooms in the past two summers. The potential impact on the City water system, which operates without a water treatment plant, could be severe. Should the blue-green algae release toxins, such as cyanotoxins and anotoxins, at levels above EPA and Health department limits, the entire population of the City of Syracuse and other communities supplied by the City would be significantly impacted. Although both Otisco Lake and Lake Ontario facilities have treatment facilities with activated carbon, additional steps would be taken to mitigate a blue-green algae event in either lake. These potential impacts have been recognized and are now included in OCWA's ERP, and the plan includes identification of a partial back-up water source for Syracuse.

In addition to physical and natural hazards being evaluated and addressed in the risk assessment and the ERP, OCWA, like every utility, must now recognize the potential of cyberattacks from criminals and foreign entities. Industrial control systems and business enterprise systems alike are computer based and vulnerable. Malicious actors, foreign and domestic, have displayed they are intent on disrupting government facilities and utilities by attacking any operating systems. Accordingly, OCWA included cyber assessments as part of its ongoing risk assessment process.

The Authority has formed strategic partnerships with external cybersecurity contractors, IT consultants, and cyberinsurance underwriters. OCWA has developed an ongoing program that includes continuing cybersecurity training for all Authority personnel. The program also involves ongoing review to ensure the OCWA is complying with guidance from its insurance carrier. Additionally, OCWA has used cybersecurity experts to conduct cyberattacks from outside the organization and from within the organization. In turn, weaknesses are identified and corrected. This is an ongoing process that has no end. OCWA has learned that threats evolve quicker than the patches, and only a foolish organization would rest on its laurels and believe they are protected and safe.

The same hold true for the Authority's risk assessment and ERPs. None will ever be complete, all will always be continually updated, and the Board, management, and staff will always worry about what is next or what lies ahead.

WHAT LIES AHEAD

OCWA has undertaken many actions since the 1998 and 1999 storms and the 2001 terrorist attacks. As OCWA responds to these events and prepares for the next, the Authority must be nimble and prepared for what else might be coming for the water system. OCWA has increased its resilience by beefing up physical security, installing generators for backup power, improving water storage facilities to have an ample supply of water, increasing water quality monitoring, and preparing for lake source water quality events, and it is now protecting itself from potential cyber criminals.

What is next, what potential hazard and threat will come? Could it be invasive species affecting water quality, something that affects the pipes, a pandemic impacting employees and their families? Or could it be something else? Something that cannot even be imagined yet? How should the OCWA prepare? The answer is a continued need to be vigilant, maintain our facilities in top condition, and secure, be nimble and able to change, and just plain be prepared.

By maintaining OCWA's risk assessment program and ERP as living documents, both have become integral components of day-to-day operations. All employees are trained and aware of potential threats and their consequences, and employee training is ongoing and continuous. Accordingly, everyone strives to protect components daily as they go about their duties. Strategic partnerships with outside stakeholders have been formed to enhance compliance and protection of OCWA facilities.

For OCWA, and any utility for that matter, the organization cannot rest on its laurels. None should ever believe a risk assessment, or an ERP is a "one and done" exercise. A utility must believe it is or may be a target of bad actors or bad weather and must always remain prepared. OCWA continually tests, retests, reviews, and adapts, and still the Authority is never satisfied. OCWA continues to evaluate threats and add new components when new or different threats are identified. Development of new partnerships is ongoing, and employee training is ongoing as are the evaluation, planning, improving, and preparation components of the process. In the end, it is all about protecting the 500,000 people being provided water every day.

ACKNOWLEDGMENT

A special thank you to retired OCWA employee, Meg Welch, who developed the OCWA history from inception through 1955.

REFERENCES

[1] Risk Analysis and Management for Critical Asset Protection (RAMCAP®). The American Society of Mechanical Engineers - ASME Innovative Technologies Institute, The American Water Works Association.

[2] ANSI/AWWA Standard J-100-10. The American Society of Mechanical Engineers - ASME Innovative Technologies Institute, The American Water Works Association.

Introduction

Implementing resilience at any appreciable scope and scale requires access to capital. In most cases, resilience improvements require additional funding up front, but as demonstrated by the National Institute of Building Sciences, the benefits can be significant [1]. Assuring that the benefits resilient infrastructure provides are captured and provide a measurable return on investment is essential to progress. New financing mechanisms or the evolution of existing mechanisms are emerging as potential solutions. The chapters that follow examine some of these mechanisms.

Concern is also growing among traditional investors and their regulators that the risks presented by disasters and climate risk may impact returns. The chapters in this part examine the risks and opportunities resilience provides within the finance sector. Specific to real estate investments, the dialogue continues in Chapter 10.

In the context of risk management, the line between financing and insurance offerings is blurring. Risk transfer products are developing at the intersections of insurance and finance. The risk transfer mechanisms can include insurance and reinsurance, catastrophe bonds, contingent credit facilities, and reserve funds.

Although this book includes discussion on financial strategies to support resilience, it does not examine the traditional role of insurance. Insurance plays many roles in advancing resilience—both as a signal of risk and a mechanism to support "bouncing back." When based on actuarial rates, the price and availability of insurance can send strong signals to consumers on the level of risk associated with a decision. When insurance is available, it functions as a form of risk transfer—the consumer (who often cannot or would not bear the cost of a catastrophe all at once) can shift the financial aspect of an adverse event to an insurer. Premiums are paid to the insurer on a regular basis to support this arrangement.

Despite these potential benefits, overall penetration rates for insurance are low. EY puts the nonlife insurance penetration rate between 4% and 5% in 2016 [2]. Even in high-risk areas, the numbers are particularly low. In California, fewer than 10% of residential buildings are insured against earthquakes in many counties, and in highly exposed urban areas coverage only approaches 25%. The percentage is higher for commercial risks, ranging between 30% and 40% in major cities [3]. The Insurance Information Institute found that 15% of American homeowners had a flood insurance policy in 2017, up from 12% who had the coverage in 2016 [4].

The Federal Emergency Management Agency (FEMA) has identified the need to increase insurance coverage as a major priority in its 2018–22 strategic plan. FEMA has set a "moonshot" goal of doubling flood insurance coverage by 2023 through the increased sale of policies under the National Flood Insurance Program and through private policies [5].

In addition to providing financial tools to reduce the impacts of events, the insurance industry has undertaken efforts to reduce physical risks as well. The Insurance Institute for Business and Home Safety (IBHS) has an active research program to study the impacts of hazards on buildings and test strategies for reducing such hazards. IBHS's FORTIFIED program provides guidance for both residential and small commercial structures to help reduce the damage caused by hurricanes and other high-wind events [6]. In some states, meeting the FORTIFIED criteria can result in an insurance premium reduction. The existence of a public-facing rating tool such as FORTIFIED has been shown to increase the desirability of more resilient properties, resulting in a 7% price premium in coastal Alabama [7].

The insurance industry has access to a variety of tools to support its understanding of risk and the underwriting process. Catastrophe modeling or cat modeling uses computers to estimate potential losses due to a particular event. Insurers also look to specific community characteristics such as their level of code adoption and enforcement. The Building Code Effectiveness Grading Schedule (BCEGS) developed by the Insurance Services

Office grades communities based on three main criteria: administration including adoption and enforcement of the latest code and staff training; plan review; and field inspections including staffing levels, workload, and items inspected [8].

The discussion on insurance and resilience can be the topic of a whole book on its own (although surprisingly few contemporary publications exist in the context of developed countries). However, the literature in this space is growing. Kousky and Shabman examine insurance design features that enhance resilience [9]. The American Bar Association has specifically focused on how the emerging risks of climate change intersect with insurance although more from the perspective of liability than the potential product offerings [10].

FINANCING RESILIENT INFRASTRUCTURE

As this book was being prepared, the United States was amidst a multiyear debate on how to address the significant backlog of projects to enhance and repair the nation's infrastructure. The need for such improvements is largely stipulated. The debate centers around how to fund these needs.

According to the American Society of Civil Engineers (ASCE), Americans will lose $3,400 per household annually through 2025 due to the current state of infrastructure. Household losses would increase to $5,100 per year through 2040, resulting in a cumulative per household loss of about $111,000. At a national scale if left unaddressed through 2025, the economy could see a $4 trillion loss in the gross domestic product and a loss of 2.5 million jobs in 2025 alone [11] See Fig. 1. This analysis does not take into account any of the losses associated with disasters or what it would take to enhance the resilience of infrastructure beyond current minimum requirements.

New financing mechanisms are needed to fund necessary improvements in infrastructure—particularly the incremental costs associated with resilience. The Centre for Global Disaster Protection and Lloyd's of London identified four innovative finance strategies that marry features of insurance and investment to support enhanced resilience [12]. These strategies parallel some of the thinking embodied in the concept of Incentivization.

- Insurance-Linked Loan Packages. This approach incorporates risk-transfer mechanisms with financing into a coordinated product offering. Loans would be given for projects where resilience measures have been built in and accompanying insurance would reflect premium reductions based on the maintenance of those measures.
- Resilience Impact Bond. In a Resilience Impact Bond, investors provide capital for infrastructure improvements where the return is based on the ongoing provision of resilience services by the project (or the impact it has).
- Resilience Bond. The Resilience Bond is a variant of the catastrophe bond where implemented resilience measures buy down the impact of a potential catastrophe resulting in reduced interest payments and lower investor risk.
- Resilience Service Company. A Resilience Service Company pays for measures up front in exchange for a share of the future insurance premium savings. This parallels offerings by energy service companies or ESCOs for energy efficiency investments.

In Chapter 6, Coffee outlines new or evolved financing mechanisms that support resilient infrastructure and recognize the benefits that accrue to communities and investors. As the infrastructure funding debate moves forward, these strategies should be part of the discussion. Investing in infrastructure without an eye toward the risks it could face over its life cycle may mean earlier than anticipated obsolescence or a need for further investment.

CLIMATE RISK IN FINANCIAL DECISION MAKING

As with most of the discussion in this book, the changes in risk posed by climate change present an additional layer to resilient finance. Efforts such as the Task Force on Climate-Related Financial Disclosure (TCFD) [13] and the Global Adaptation and Resilience Investment (GARI) work group [14,15] have focused on key metrics important for financial decision making in the face of hazards. The TCFD was established by the G20's Financial Stability Board and chaired by Michael Bloomberg. GARI was launched at the Paris COP21 global climate talks in conjunction with the United Nations Secretary General's A2R (Anticipate, Absorb, Reshape) Climate Resilience Initiative.

Common metrics and risk analysis methods across companies and their suppliers allow the holistic consideration of risk, informing investors, lenders, insurers, underwriters, and even regulators with tools to evaluate risks and exposures [13].

TCFD acknowledges the risks associated with a transition to a lower-carbon economy and the physical impacts of climate change. Transition risks include policy and legal, technology, market, and reputation risks. Physical risks include acute and chronic risks. Efforts around both adaptation and mitigation will cause shifts in how various sectors and industries will

	Surface Transportation	Water/ Wastewater	Electricity	Airports	Inland Waterways & Marine Ports	Aggregate Economic Impact of All Sectors
Business Sales						
2016–2025	$2,212	$896	$1,399	$625	$1,252	$7,038
2026–2040	$8,152	$5,907	$2,024	$2,397	$4,239	$29,292
GDP						
2016–2025	$1,167	$508	$819	$337	$784	$3,955
2026–2040	$1,981	$3,215	$1,071	$1,073	$2,003	$14,201
Jobs						
2025	1,052,000	489,000	102,000	257,000	440,000	2,546,000
2040	473,000	956,000	242,000	494,000	1,153,000	5,809,000
Investment Funding Gap—2016 through 2025						
Total Needs	$2,042	$150	$934	$157	$37	$3,320
Funded	$941	$45	$757	$115	$22	$1,880
Funding Gap	$1,101	$105	$177	$42	$15	$1,440
Investment Funding Gap—2016 through 2040						
Total Needs	$7,646	$204	$2,458	$376	$112	$10,796
Funded	$3,312	$52	$1,893	$288	$69	$5,614
Funding Gap	$4,334	$152	$565	$88	$43	$5,182

NOTE The total economic impacts caused by the gap are listed by sector. Note that the economic impacts are based on each specific sector and the research and modeling by sector developed in the initial *Failure To Act* reports and adjusted based on the gap from 2016–2040. As these impacts are not related to the investment gaps across infrastructure systems, they are not totaled. However, projected cumulative economic effects of the gaps in all sectors are presented in the aggregate section. All year totals in constant 2015 value. However, job totals are a single year impacts for 2025, not cumulative totals.

FIG. 1 Losses to the national economy due to infrastructure investment gaps. (Source: Reproduced from, American Society of Civil Engineers, Update to Failure to Act: The Impact of Infrastructure Investment on America's Economic Future, Prepared by Economic Development Research Group, 2016. https://www.infrastructurereportcard.org/wp-content/uploads/2016/10/ASCE-Failure-to-Act-2016-FINAL.pdf.)

fare in the future. The relationship between risks, opportunities, and finance are illustrated in Fig. 2 [16].

To build awareness and foster investor engagement, TCFD recommends that climate-related financial disclosures be incorporated into public annual filings. To be effective, the disclosures should meet seven principles:

1. Represent relevant information;
2. Be specific and complete;
3. Be clear, balanced, and understandable;
4. Be consistent over time;
5. Be comparable among companies within a sector, industry or portfolio;
6. Be reliable, verifiable, and objective; and
7. Be provided on a timely basis [16].

The Task Force structured its recommendations for climate-related financial disclosures around four themes that capture the typical operation of organizations: governance, strategy, risk management, and metrics and targets [16]. The content of recommended disclosures within each of these themes are captured in Fig. 3.

GARI's initial effort focused on identifying approaches to measure physical climate risk and identify examples of current climate-resilient investments. The work group identified six approaches to climate risk measurement and recognized the limitations presented by data consistency and coverage, scenario planning, implications of analysis, and the diversity of use cases

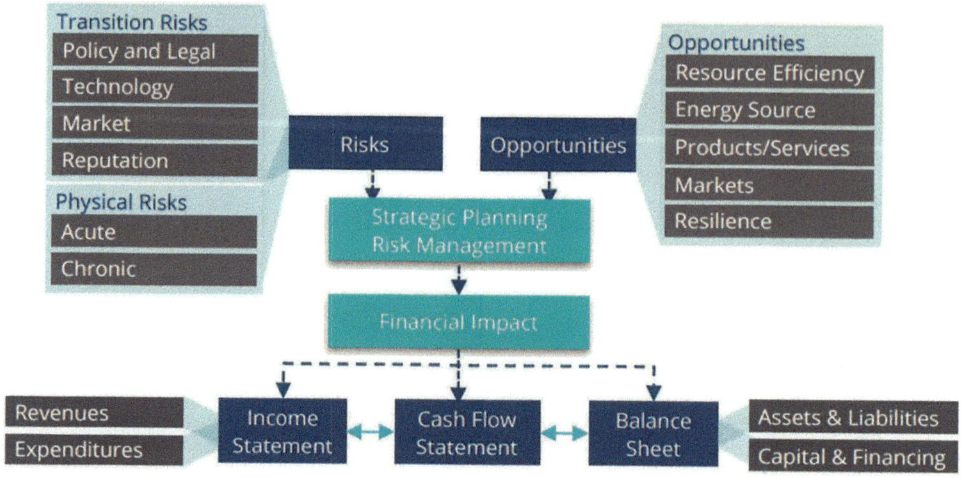

FIG. 2 Climate-related risks, opportunities, and financial impact. (Source: Reproduced from, Task Force on Climate-Related Financial Disclosures, Recommendations of the Task Force on Climate-Related Financial Disclosures, June 2017. https://www.fsb-tcfd.org/wp-content/uploads/2017/06/FINAL-2017-TCFD-Report-11052018.pdf.)

Governance	Strategy	Risk Management	Metrics and Targets
Disclose the organization's governance around climate-related risks and opportunities.	Disclose the actual and potential impacts of climate-related risks and opportunities on the organization's businesses, strategy, and financial planning where such information is material.	Disclose how the organization identifies, assesses, and manages climate-related risks.	Disclose the metrics and targets used to assess and manage relevant climate-related risks and opportunities where such information is material.
Recommended Disclosures	**Recommended Disclosures**	**Recommended Disclosures**	**Recommended Disclosures**
a) Describe the board's oversight of climate-related risks and opportunities.	a) Describe the climate-related risks and opportunities the organization has identified over the short, medium, and long term.	a) Describe the organization's processes for identifying and assessing climate-related risks.	a) Disclose the metrics used by the organization to assess climate-related risks and opportunities in line with its strategy and risk management process.
b) Describe management's role in assessing and managing climate-related risks and opportunities.	b) Describe the impact of climate-related risks and opportunities on the organization's businesses, strategy, and financial planning.	b) Describe the organization's processes for managing climate-related risks.	b) Disclose Scope 1, Scope 2, and, if appropriate, Scope 3 greenhouse gas (GHG) emissions, and the related risks.
	c) Describe the resilience of the organization's strategy, taking into consideration different climate-related scenarios, including a 2°C or lower scenario.	c) Describe how processes for identifying, assessing, and managing climate-related risks are integrated into the organization's overall risk management.	c) Describe the targets used by the organization to manage climate-related risks and opportunities and performance against targets.

FIG. 3 Recommendations and supporting recommended disclosures for climate-related financial disclosure. (Source: Reproduced from, Task Force on Climate-Related Financial Disclosures, Recommendations of the Task Force on Climate-Related Financial Disclosures, June 2017. https://www.fsb-tcfd.org/wp-content/uploads/2017/06/FINAL-2017-TCFD-Report-11052018.pdf.)

and physical risks [14]. The six identified approaches for measuring climate risk include the following:

1. Government indices and rankings;
2. Insurance risk ratings;
3. Corporate use data;
4. Project/portfolio risk screening tools;
5. Project scorecards; and
6. Engineering due diligence and design analysis.

Based on their work, GARI participants identified recommendations for key stakeholders. Regulators should establish standards and methodologies on how to assess climate impacts. Policy makers and public finance institutions should invest in resilient infrastructure and fixed assets that support resilience and support blended finance vehicles or other instruments to mobilize finance into resilience. Industry groups and think-tanks should develop guidance, methodologies, and standards for stress testing and risk screening and support development of investable products to deploy capital in adaptation and resilience. Investors and financiers should include climate risk in investment pricing and promote disclosure practices.

The Climate Policy Initiative has mapped the sources and uses of current funding for climate adaptation and mitigation. Between 2015 and 2016, the international funding reached an average of $463 billion from a level of about $360 billion in 2012 [17]. See Fig. 4.

The availability of consistent approaches to disclosure of climate risk can help the finance community incorporate climate risk considerations across investment decisions. In Chapter 7, Ambrosio, Kim, Swann, and Wang break down the infrastructure finance process to identify when and how climate-based resilience decisions impact funding streams. They set a path forward for increased focus on these important issues from both a societal and individual investor perspective. Through case studies, they clearly illustrate the need to understand where infrastructure sits, and how choices made by the broader community impact the resilience of individual facilities.

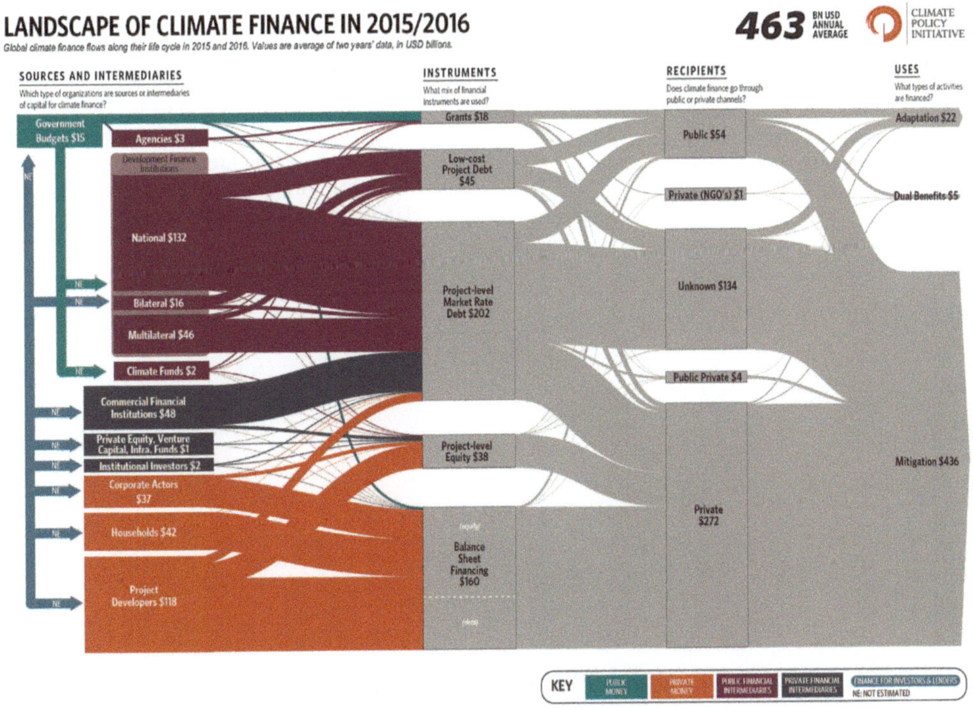

FIG. 4 Landscape of climate finance in 2015/2016. (Source: Courtesy, P. Oliver, A. Clark, C. Meattle, Global Climate Finance: An Updated View 2018. Climate Policy Initiative. November 2018. https://climatepolicyinitiative.org/wp-content/uploads/2018/11/Global-Climate-Finance-An-Updated-View-2018.pdf.)

REFERENCES

[1] Multihazard Mitigation Council, Natural Hazard Mitigation Saves: 2018 Interim Report. K. Principal Investigator Porter, C. co-Principal Investigators Scawthorn, C. Huyck, Investigators: R. Eguchi, Z. Hu, A. Reeder, P. Schneider, Director, MMC, National Institute of Building Sciences, Washington, D.C. https://www.nibs.org/resource/resmgr/mmc/NIBS_MSv2-2018_Interim-Repor.pdf.

[2] C. Crawford, L. Russignan, N. Kumar, Global Insurance Trends Analysis 2018, EY, June 2018. https://www.ey.com/Publication/vwLUAssets/ey-global-insurance-trends-analysis-2018/$File/ey-global-insurance-trends-analysis-2018.pdf.

[3] T. Jeworrek, The "Big One": Severe Earthquake in California Is Only a Matter of Time: Low Insurance Penetration a Risk for the Financial Sector, Munich Re, October 12, 2018. https://www.munichre.com/topics-online/en/climate-change-and-natural-disasters/natural-disasters/earthquakes/earthquake-risk-in-california.html.

[4] Insurance Information Institute, Facts + Statistics: Flood Insurance. https://www.iii.org/fact-statistic/facts-statistics-flood-insurance.

[5] Federal Emergency Management Agency, 2018-2022 Strategic Plan, 2018. https://www.fema.gov/strategic-plan.

[6] Insurance Institute for Business and Home Safety, FORTIFIED Construction Standards. https://ibhs.org/guidance/fortified-construction-standards/.

[7] S. Awando, H. Hollans, L. Powell, C. Wade, Estimating the Value of FORTIFIED Home Construction of Home Resale Value, Alabama Center for Insurance Information and Research, University of Alabama.

[8] Insurance Services Office, National Building Code Assessment Report: Building Code Effectiveness Grading Scale, 2019 ed.

[9] C. Kousky, L. Shabman, The role of insurance in promoting resilience, Resources (June 29, 2016). https://www.resourcesmag.org/common-resources/the-role-of-insurance-in-promoting-resilience/.

[10] C. Carroll, J.R. Evans, L. Patton, J. Zimolzak, Climate Change and Insurance, American Bar Association, 2012.

[11] American Society of Civil Engineers, Update to Failure to Act: The Impact of Infrastructure Investment on America's Economic Future, Prepared by Economic Development Research Group, 2016, https://www.infrastructurereportcard.org/wp-content/uploads/2016/10/ASCE-Failure-to-Act-2016-FINAL.pdf.

[12] Centre for Global Disaster Protection and Lloyd's of London, Innovative Finance for Resilient Infrastructure: Preliminary Findings, 2018. https://www.lloyds.com/~/media/files/news-and-insight/risk-insight/2018/innovative-finance-for-resilient-infrastructure.pdf.

[13] Task Force on Climate-Related Financial Disclosures, About the Task Force. https://www.fsb-tcfd.org/about.

[14] J. Koh, E. Mazzacurati, S. Swann, Bridging the Adaptation Gap: Approaches to Measurement of Physical Climate Risks and Examples of Investment in Climate Adaptation and Resilience, 2016. http://garigroup.com/discussion-paper.

[15] J. Koh, E. Mazzacurati, C. Trabacci, An Investor Guide to Physical Climate Risk and Resilience: An Introduction, 2017. https://garigroup.com/investor-guide.

[16] Task Force on Climate-Related Financial Disclosures, Recommendations of the Task Force on Climate-Related Financial Disclosures, June 2017. https://www.fsb-tcfd.org/wp-content/uploads/2017/06/FINAL-2017-TCFD-Report-11052018.pdf.

[17] P. Oliver, A. Clark, C. Meattle, Global Climate Finance: An Updated View 2018. Climate Policy Initiative, November 2018. https://climatepolicyinitiative.org/wp-content/uploads/2018/11/Global-Climate-Finance-An-Updated-View-2018.pdf.

CHAPTER 6

Financing Resilient Infrastructure

JOYCE COFFEE MCP, LEED AP

INTRODUCTION

Infrastructure underpins quality of lives and livelihoods and the productivity and vitality of cities. Furthermore, the success of global agreements such as the Paris Agreement [1], the United Nations' Sendai Framework for Disaster Risk Reduction [2], and the Sustainable Development Goals [3] requires resilient infrastructure. For purposes of this chapter, resilient infrastructure can withstand stresses and shocks, maintain operations, recover/bounce back from disasters, and be successfully adapted to future operational challenges. Given that most infrastructure has a useful life well beyond 30 years, decisions made in the next 10 years will determine many foundational aspects of the future of communities.

In its *2017 Infrastructure Report Card*, the American Society of Civil Engineers (ASCE) rated the state of the United States' existing infrastructure a D+ and estimated that the United States will have a $2.0 trillion investment gap by 2025 [4]. Closing this domestic investment gap will require that the government and private sector invest between 2.5% and 3.5% of gross domestic product (GDP) by 2025 [4]. ASCE identified the most vulnerable sectors as surface transportation, water/wastewater, electricity, airports, and inland waterways and marine ports [5]. Water systems are leaking two trillion gallons of treated drinking water annually, and at over 2000 dams the failure risk is high [6]. Studies show that 1 foot of sea-level rise will effect 60 wastewater treatment plants serving over 4 million Americans [7]. ASCE estimates that bringing the US infrastructure report card grade to a B- from a D+ will require a $4.59 trillion investment by 2025 [6]. Failure to address the infrastructure investment gap could trigger economic consequences for the US States in 2025 including the following:

- $3.9 trillion in losses to the US GDP.
- $7 trillion in lost business sales.
- $2.5 million lost American jobs.
- Substandard infrastructure costs per individual families estimated at $3400 in disposable income annually [5].

One of the resilience sector's major challenges is finding ways to fund resilient infrastructure projects—both modernizing existing infrastructure and building much needed additional infrastructure. The increased frequency and severity of climate change impacts (e.g., severe weather, drought, and extreme heat) place an additional toll on already strained infrastructure, exacerbating the need for reinvestment in critical infrastructure.

Increasing the number of resilient infrastructure projects that are financed will help to close the investment gap, improve the efficiency of existing infrastructure, accommodate a shift to new technologies, and support land use and other policies that create more resilient communities. Governments are key to closing this gap [8]. Leaders should harness additional capital for infrastructure projects by mobilizing private resources through creating projects that have more value for the market and that have better assessed risks. Leveraging private finance may create benefits beyond reducing the burden on public money, including the following:

- Engaging outside expertise and knowledge that fosters innovation.
- Improving performance efficiency.
- Establishing greater project buy-in.
- Creating additional local jobs [9–11].

Infrastructure investments represent an opportunity to leverage funding not only to address the crumbling infrastructure in the United States but also to improve resilience to climate impacts.

- Over the next 15 years, more infrastructure will be built than all now in place [12].
- Less than 1% of available private capital is being made available for infrastructure or sustainable infrastructure financing [12].

Optimizing Community Infrastructure. https://doi.org/10.1016/B978-0-12-816240-8.00006-9

- Infrastructure-oriented companies have outperformed in the new millennium with less risk [13].
- Increased investment in the next decade could raise global GDP by 0.6% and US GDP by 1.3% [14].
- Increasing investment by 1% GDP could generate an additional 1.5 million direct and indirect jobs in the United States [14].

This chapter seeks to describe finance sector trends related to resilient infrastructure and to increase understanding of existing resilient infrastructure finance options, instruments, and enablers, including identifying deal-making's challenges and solutions for resilient infrastructure projects.

FINANCE SECTOR TRENDS AFFECTING RESILIENT INFRASTRUCTURE DEMAND

The resilience sector has grown significantly in recent years in response to the emerging physical risks from climate change [15]. Public and private entities are developing plans to increase climate resilience, but these plans are only useful if they are deployed successfully, which often requires resilient infrastructure finance.

In the best resilient infrastructure projects, investors benefit because resilient infrastructure projects:

- Provide portfolio diversification as stocks and bonds tend to move in noncorrelated directions, so a diversified portfolio can be expected in the long run to earn higher returns with less volatility than a pure stocks or pure bonds portfolio.
- Mitigate the future climate risks that traditional infrastructure gets exposed to as long-term assets.
- Enable more efficient projects from design through approval.
- Ensure operational continuity and, thus, ongoing revenue generation following shocks and stresses.
- Provide additional environmental and social impact cobenefits.

Several finance sector trends could create more supply of resilient infrastructure finance capital: credit rating agencies including factors impacted by physical climate risks in their assessments of default potential; insurance rates rising in particularly vulnerable geographies; big data predicting the costs of future losses; the related growth of the catastrophe bond market, the potential for less federal financial assistance for postdisaster recovery efforts in the future; the increasing availability of privately financed risk management instruments; innovative finance tools like parametric insurance and resilience bonds; and finance industry guidelines recommending assessment of climate risk.

Municipal Credit Ratings Include the Physical Risks from Climate Change

Credit ratings are an important factor for finance and insurance of public and private entities. Credit risk represents the potential that a bond issuer will fail to make timely interest and principal payments to its bondholders. Credit ratings are given to governments that issue bonds as well as corporate bond issuers by three main agencies: Moody's, Standard & Poor's, and Fitch [16]. Cities, counties, and states generally are rated on their ability to pay bondholders back.

Recently, Moody's and Standard & Poor's announced that the physical risks of climate change could contribute to factors that may impact a government's ability to pay back debt, and the media announced that the credit rating agencies would be factoring physical risks into municipal credit ratings [17–20] as both shock and stress events can impact tax and fee revenue. When infrastructure fails and services are not delivered, rates are not assessed. These failures can even slow economic transactions, thus depressing tax revenue generation.

Practitioners worried about their credit rating going on the rating agencies' "watch" lists or being downgraded should increase their climate change-related governance, identifying the vulnerabilities and hazards that are likely to increase because of climate change and adopting policies and projects that proactively address them. This is especially important because when rating agencies assess that existing infrastructure is at risk, they may lower the bond rating of a municipality, therefore, increasing the cost of borrowing and leaving less funds available to invest in infrastructure upkeep and development.

Credit ratings also create a tension in the pursuit of resilient infrastructure projects. As more debt increases the chance that a bond issuer will not be able to repay its investors, municipal governments may hesitate to increase their general obligation bond to fund resilient infrastructure projects. However, cities that fail to act to improve resilience may see their credit rating lowered. Practitioners should examine their credit rating reports carefully and be transparent about the climate change-related risks their projects intend to mitigate as well as the fiscal prudence they use to ensure all existing and future bonds address these risks.

Insurance Premiums Rise

Insurance is the finance sector's means of risk transfer. Insured losses are growing with climate-related risks, reflecting a triad of factors: growing risks as the climate changes, more people buying insurance and a greater proportion of buildings located in high-risk areas, for example, as development expands into forested land prone to wildfire and coastal areas prone to coastal storms.

Several key US markets have experienced slightly higher insurance premiums: Houston's insurance costs have increased 9% and Miami's insurance costs have increased 5% in the last 15 years [21]. The reason most insurers have not yet increased prices is likely due to the ability to aggregate risks across so many different geographies and shocks and avoid overall losses. In other instances, state insurance regulators may not want to allow insurers to increase their premiums. However, as risk increases, insurers may abandon or decrease their exposure in certain high-risk markets or, in an extreme scenario, insurers may find themselves unable to cover the payment of damage claims.

Insurance companies are not transferring this risk through the private sector alone, however. Through the federal Robert T. Stafford Disaster Relief and Emergency Assistance Act, when a federal disaster is declared, the US government is, in some ways, the insurer of last resort. When a federal disaster is declared by the President, states and local governments gain access to federal disaster assistance, which generally covers 75% or more of disaster costs including debris cleanup, emergency repairs, rebuilding public buildings, housing, and infrastructure; provides assistance to individuals for meeting basic needs such as temporary housing; and hazard mitigation assistance projects, for addressing vulnerabilities that a disaster may have made all too apparent. Congress will also make additional funds available through special appropriations. In 2017, Congress made $120 billion in additional disaster aid available to areas affected by Hurricanes Harvey, Irma, and Maria, and wildfires in the Western United States [22]. Thus, all taxpayers pay for a proportion of losses due to federally declared flood, fire, and other natural disasters. For example, Florida has received almost $500 billion in federal disaster relief over the last 30 years [18].

If the National Flood Insurance Program (NFIP) grows to use private sector risk tools—a trend often discussed in both the private sector and government and in motion as the Federal Emergency Management Agency (FEMA) purchased both reinsurance and catastrophe bonds for NFIP in 2018 to help offset costs—insurance companies may increase insurance costs to signal risks adequately to property owners [23]. The existence of below market rate insurance in NFIP arguably leads to bad decisions around building locations and construction materials and distorts the market for insurance prices that reflect true risk. In addition, billions of dollars of NFIP debt is being forgiven at a rate similar to NFIP funds approved off-budget annually after disaster declarations, thus contributing to a federal budget deficit.

Big Data Inform Decision Making

Analysis of big data allows for better assessments of the costs and benefits of avoided risks. Firms such as RMS, AIR Worldwide, and CoreLogic, as well as in-house analysts in large financial institutions including insurers can now model the impact of hazards on existing assets. Other companies are also entering this field, like Jupiter Intel and The Climate Service who are extending beyond today's risks and also examining how climate change may exacerbate risks of flooding, drought, extreme heat, and other hazards. These technology and data-enabled insights inspire decision making to increase loss avoidance and risk transfer.

For instance, the availability of these data and models in the market inspired the catastrophe bond market as the models can ascribe a value to scenarios of future loss. In addition, the insurance industry uses these data and models to "stress-test" policies against, for instance, a 1-in-100 chance event. Especially, as insurance actuaries use 18 months of historic data rather than climate change scenarios to price insurance, this stress testing is key for both informing the industry of the risks of the policies it issues and, through price, signaling to policyholders the extent of their risk [24].

Of course, the most ubiquitous flood-related data set in the United States is the FEMA flood maps. FEMA's maps have historically been backward looking, not forward looking. In some communities such as New York City, FEMA maps are beginning to include forward looking climate projections, but in general, there is a discrepancy between the climate models, FEMA maps, and the private sector models. Reflecting on this discontinuity, Harvard's John Macomber notes: "At best, they will eventually converge into one open data set. At worst, some financial players with access to better information will take advantage of other financial players with worse information, with potentially big consequences for property owners left holding assets with risks that were historically improperly priced [25]."

Other big data sets, such as the US Government's National Oceanic and Atmospheric Administration's (NOAA) sea-level rise scenarios [26] and Climate Central's flood project maps [27], informed Zillow's easily consumed assessment of the number of homes predicted to be underwater at the end of the century and the value of those losses. Combined with its real estate database, Zillow anticipates that 6 feet of sea-level rise by century's end will result in 1.9 million homes under water with $900 billion in losses [28]. Millions are projected at risk from sea-level rise in the continental United States. Coastal cities are at particular risk of losing real estate value—contributing to decreasing property tax income and, subsequently, utility ratepayer income if people leave these at-risk residence [28].

Although the NOAA data are free, the large catastrophe models can be expensive to build and maintain. Groups such as the Oasis Loss Modeling framework offer the scaffolding on which decision makers can attach data for free [29], and practitioners should attempt to stress-test long-term resilient infrastructure projects against the 1% chance event.

International Regulatory Approaches to Addressing Climate Risk

Beyond guidelines, legislation, particularly in the European Union and the United Kingdom, addresses climate risk. French Article 175, promulgated immediately following the Paris Climate talks in 2016, requires public companies listed on the public stock exchange in France to divulge their physical risks and the potential impact on the issuers [31]. This law forces top management in companies to consider climate risk. For over 5 years, the United Kingdom Adaptation Reporting Power has guided corporations to report their adaptation actions. The program has helped move resilience management from sustainability practitioners to other departments, such as risk management, business continuity planning, and legal liability departments [32]. The European Union Water Framework Directive [33] requires anyone seeking money from European Union entities to illustrate their watershed approach. In the United States, no such comprehensive legislation exists to address climate-related risks. Legislative and policy precedent can inspire U.S. government practitioners to consider similar initiatives.

Investor Guidance Recommends Assessing Climate Risk

In mid-2017, the G-20's Financial Stability Board's Task Force on Climate-related Financial Disclosure (TCFD) [71a] published guidelines recommending how financial institutions should assess and disclose their climate change-related risk [30]. The guidelines are organized into several categories: governance, strategy, risk management, and metrics and targets. It is unlikely the United States will mandate the TCFD recommendations like France has done and the European Union is considering. Although US companies are embracing the recommendations—major sovereign wealth funds, asset managers, and other institutional investors as well as rating agencies have all said they will adopt the recommendations. TCFD guidance recommends that financial institutions assess the political risk of climate action, such as legislation for cap and trade programs or carbon taxes; the transition risk associated with the market disinvesting from high carbon intensity activities; and the impacts of physical risks from climate change on holdings within investment portfolios.

Helpful for increasing resilient infrastructure investment, TCFD emphasizes not only financial risk mitigation but also the opportunity for financial markets to invest in climate change action.

FACTORS BEYOND THE FINANCE SECTOR MAY INCREASE DEMAND FOR RESILIENT INFRASTRUCTURE

Trends beyond the finance sector also may affect demand for resilient infrastructure. Liability, experienced supply and value chain disruption, and geographic changes in demand for land could spark an increase in municipal, consumer goods, and real estate sector demand for resilient infrastructure.

Liability Grows

There is growing liability for decision makers as awareness of predicted climate change hazards grows and decreases the applicability of "force majeure" (i.e., acts of God) in case and constitutional law. The Conservation Law Foundation and Boston Green Ribbon Commission published a report, *Climate Adaptation and Liability: A Legal Primer and Workshop Summary Report*, which discusses how case law and constitutional law can shape potential liability in a climate-changed future [34]. Resilient infrastructure projects proactively address this liability, making it a strategic move to invest in these projects to avoid liability issues down the line.

Supply Chains Experience Climate Change-Related Impacts

Typically, corporate supply chain risks have been evaluated in terms of reputational and continuity risk [35]. Recently, these risks have been considered in the context of climate change. An Acclimatise case study examined the impact of the 2011 floods in Thailand, concluding that the floods resulted in $45 billion in losses for the global economy from supply chain interruption [36]. As climate change impacts worsen, the supply chains will continue to be disrupted by both climate change-related shocks and stresses, straining the economy and raising the attention of financiers.

Anecdotal evidence from more recent climate change-intensified storms suggests that gritty resilience to secure supply chains is now de rigueur. A major pharmaceutical appliance company with a manufacturing facility in San Juan, Puerto Rico, used its corporate jet to fly cash to its workforce so they could pay for their families' immediate needs and more swiftly return to work. A distillery in the Caribbean islands is doubling down on risk insurance and considering alternative distillers, given growing losses in the last decade.

Still, the cascading failures—often initiated by either a lack of access to the grid or the failure of the backup plan—suggest that distributed renewables like solar and wind generation play a role in protecting value chains.

Overall value chain impacts—including in supply chains—are changing leaders' risk perception. In 2017, the World Economic Forum released results from its annual Global Risk Perception Survey, which asked hundreds of corporate and government leaders what they believed to be their biggest risk in terms of likelihood and impact [37]. Over the years, climate risks have grown in importance, and the 2017 data indicate the top risks in terms of impact include water crisis, major natural disasters, extreme weather events, and the failure of climate change mitigation and adaptation.

These data suggest that global leaders have climate change risk solutions, such as resilient infrastructure, top of mind.

Risks May Change Real Estate Markets

In recent years, some real estate investors have begun to determine the likely losers to climate risk. The National Real Estate Advisors published data on the percentage contribution to risk by the largest markets in the United States. For their national portfolio, Miami is expected to contribute 38% of total risk; Houston, 14%; and New York City, 11%, among others [21]. The group predicts that lower risk areas will receive greater investments in the future.

As climate shifts, investment also is likely to shift. For example, Illinois' climate is expected to be more similar to that of New Orleans by the middle-to-end of the century. Therefore, based on climate risk analysis alone, Illinois may become more popular for tourism, real estate headquarters, and housing developments [38].

Population migration also may increase after extreme events, changing the geographic demand for new developments. Hurricane Katrina's diaspora included 800,000 people who left their homes after the hurricane and never returned [21]. Studies show that 4 to 13 million Americans are likely to be displaced by sea-level rise [39] and their movement will change existing cities with an influx of new residents [40]. Other examples include the increase in migration to Chicago from Puerto Rico after the recent hurricanes as Chicago's Puerto Rican community welcomed the newcomers.

Practitioners can consider how future market shifts will create opportunities to show investors better long-term returns on investments.

UNDERSTANDING FINANCE OPTIONS

Resilience professionals must become more familiar with finance instruments and enablers to find additional funding sources and reduce barriers to resilient infrastructure investments. As funds are being spent today on long-term infrastructure, there is no time to waste to ensure that all infrastructure investments relate to *resilient* infrastructure. Although innovative mechanisms for funding resilient infrastructure certainly will be deployed, traditional finance must be the funding engine for resilient infrastructure, given the scale of infrastructure needs and climate change risks. Thus, to understand resilience finance, it proves helpful to review infrastructure project finance tools: How money flows into the market and projects from public and private sources and how money flows from projects and the market to finance infrastructure. Fig. 6.1 shows how money flows into and out of the system.

Where the Money Comes From: Public Revenue Sources

A major source of public funding is federal grants for disaster assistance, including FEMA's Hazard Mitigation Grant Program, Pre-Disaster Mitigation Grants program, and the Department of Housing and Urban Development's (HUD) Community Development Block Grant-Disaster Recovery (CDBG-DR). Although these funds are not explicitly for resilience, they could

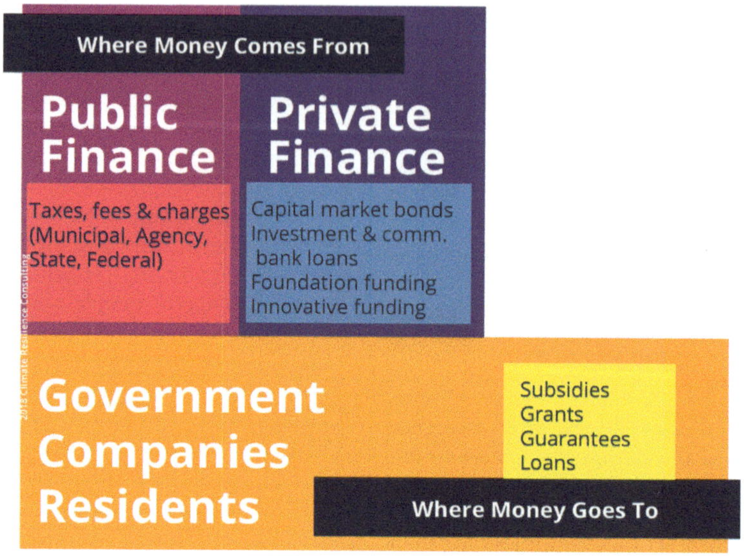

FIG. 6.1 Flow of money in and out of the resilience finance system. (Source: Joyce Coffee)

be used for that, and in 2018, an $89.3 billion emergency spending disaster supplemental included $28 billion for HUD's CDGB program and funds dedicated to mitigating risks due to sea-level risk [41].

In addition, public revenue sources to fund resilient infrastructure primarily comprise taxes and fees from municipal, utility, regional, and federal sources. Fees consist primarily of utility service fees, impact fees, business fees, tolls and user charges and, in a few cases, carbon pricing revenue. Taxes include such sources as general property, general sales, personal income, utility, real estate property transfer, mortgage recording, business fees, and tax audit revenues. Of these sources, property taxes are a significant source of local public revenue, while income and corporate tax are the greatest source of revenue for the federal government.

Public revenue also can be generated through project and service delivery cost savings and tax increment financing (addressed below). In certain regional, state, and municipal jurisdictions such as the Northeast Regional Greenhouse Gas Initiative, California's Cap and Trade System and the Boulder Climate Action Tax, carbon cap and trade and carbon tax systems also provide revenue.

Annual city and public utility revenue budgets are generally public documents available online, and they indicate sources of funds. Annual operating and capital budgets also provide information about the revenue flows. Although capital budgets provide information about longer-term investments for infrastructure, revenue budgets are useful for understanding money flow for day-to-day and annual infrastructure operations.

Practitioners should examine revenue budgets because they can provide ideas for funds to harness for resilient projects, combining a typically nonrevenue-generating infrastructure with a process or product that can generate fees or taxes. San Francisco's sea wall is one such pairing, applying revenue from motor vehicle fees to debt service for capital project bonds. By creating an investment payback mechanism, this pairing increases project "bankability," which is key for attracting resilience finance.

Most infrastructure projects are not self-funded exclusively from city service and utility-related revenue. Rather, the government or utility entity uses these revenues as leverage to issue a bond. Bonds are described in the following section.

Case Study: San Francisco Sea Wall

The California legislature passed AB-2578: Seawall Improvements to provide means to finance improvements to the San Francisco Seawall that address seismic and flood risk. The legislation enables the generation of $330 million in revenue over 45 years. Thus, it pairs the sea wall improvements with revenue generation from the state's share of tax increment from the Educational Revenue Augmentation Fund and motor vehicle in-lieu fees from a port infrastructure finance district.

The legislation authorizes use of these revenues for debt service on revenue bonds issued for sea wall improvements, given its "community-wide significance" and land-secured financing to finance capital costs [42].

From the perspective of bankability, this is interesting because the sources of long-term revenue to service the bonds (the Educational Fund and the In-Lieu fees) do not directly relate to revenues from the asset (sea wall). The sea wall itself does not create revenues, as nobody is tolled or taxed to use it.

Where Money Comes From: Private Investment Instruments

Debt and equity are the two major private sources of funds. Debt is borrowed funds that leverage a tax or fee revenue stream to pay back funds owed. Debt providers include banks, bonds, and federal revolving funds from programs like those under the Water Infrastructure Finance and Innovation Act [43]. On the other hand, equity is owned capital, generally stocks or direct investments. Equity providers include contractors, operators, and public pension funds. Both debt and equity can be used in the capital stack for infrastructure projects, although debt, most often from banks, accounts for the majority of financing for most infrastructure projects [44].

Although equity and debt are the primary funding sources for resilient infrastructure projects, other less common sources of funding include grants (including from philanthropy), which can fund technical assistance or advisory services auxiliary to the actual resilient infrastructure investment. They serve as a catalyst for creating resilient infrastructure.

Given debt's role in financing infrastructure, this chapter emphasizes debt and project bankability.

RESILIENT INFRASTRUCTURE INVESTMENT INSTRUMENTS: DEBT

To finance infrastructure capital projects with private funds, governments and other utilities ("the issuer") issue bonds. When investors invest in a bond, they are loaning money to the issuer in exchange for interest and the return of the original investment, or "principal." Because these funds are a loan and are a financial instrument, they are called debt securities. In part because the interest on municipal bonds is exempt from federal income tax in the United States, municipal bonds are a key part of many investors' diversification strategy to balance stock assets. The interest may also be exempt from state and local taxes if the investor resides in the state where the bond is issued.

Debt requires payback or "service," and funds for this debt service come from the public revenue sources mentioned earlier—taxes and fees. The two most common types of municipal bonds are general obligation (GO) bonds and revenue bonds, with funds from the issuer's fees paying bondholders' principal and interest.

General Obligation Bonds

The GO bond is the most used of any type of funding for public works, including toll-free transportation infrastructure, sea walls, and social infrastructure (public facilities such as hospitals, schools, and parks). The bond issuer uses the initial influx of money for the capital project and pays it back over time through taxes. Many of these projects are considered "unbankable" because they do not have a nontax revenue stream [45].

Especially, as GO bonds fund so much infrastructure modernization and new construction, practitioners should make a significant effort to work with their finance counterparts to ensure that the bonds are investing in resilient rather than traditional projects.

Revenue Bonds

Revenue bonds are the second type of bonds used for private finance of public infrastructure. They are supported by the operating revenue of a specific project that generates income, such as fees from a tollway or water supply. Issuers pay back investors' interest and principal with the revenue generated by the services delivered by the projects they fund.

Creating Bankability: Applying Public Revenue to Service Debt

Publicly generated funds, including taxes and fees, are the revenue that services, or pays back debt. Infrastructure projects with no revenue generation potential are typically funded by GO bonds, which are paid back from general tax revenue.

On the other hand, revenue bonds are issued for infrastructure projects that generate fees through the amenities they deliver. These fees are key to creating an "investable" or "bankable" project by providing monetary returns that can be generated and captured for investors.

To increase the number of funded resilient infrastructure projects and to ensure investors do not presume that all resilient infrastructure projects are unbankable, practitioners should be practical and creative in finding ways to generate revenue streams in their resilient infrastructure projects. One approach is to consider what other benefits or cobenefits, aside from the infrastructure's primary service, the project could create. Cobenefits could include serving as a tourist attraction, parking, or commercial amenity. Another opportunity is to examine other sectors that could be involved in the project, including transportation, water supply, or energy generation [46]. Ultimately, integrating multibenefit, multisector goals into the project's vision and design from the beginning will inform procurement requests and encourage designers to build in these multiples, fulfilling resilience objectives, even sometimes with no additional costs.

Green, environmental, or climate bonds

One type of revenue bond is the green bond, or climate bond—a category of investment of particular interest to social impact investors. Typically, green bonds are more sustainability-oriented—focusing on projects that reduce greenhouse gas emissions—than resilience-oriented—focusing on projects that provide collateral societal benefits and rebounding from shocks and stresses. Green bonds have their own marketplace where sponsors and investors can form partnerships to issue the bonds. External players such as green banks (see below) often coordinate these partnerships.

Issuance of green bonds has grown significantly since the market commenced in 2007 (with offerings by the European Investment Bank and World Bank). In 2017, total labeled green bond issuance amounted to $221 billion in debt outstanding [47]. Although only 3% to 5% of the issuances focus on climate change adaptation, green bonds may become a major source of funding for resilient infrastructure projects in the future as the Climate Bond Initiative promulgates adaptation taxonomy [48–50]. Most of the green bonds for resilient infrastructure projects have been related to water, specifically for water or wastewater utility projects.

Another type of revenue bond also aimed at a socially responsible investor is an environmental impact or social impact bond. The investor receives a higher return when a sustainability objective has been achieved. Thus, the issuer informs bond investors about the project's impact. For example, Washington, D.C., promulgated an environmental impact bond that accomplished measurable nonmonetary goals for a resilient infrastructure project aimed at capturing and preventing a certain amount of runoff from going into its sewer system [51,52]. The "pay for performance" concept rewarded investors for the stormwater benefits in addition to the fee revenue from the project.

Although some utility and city infrastructure project sponsors may worry about increased transaction costs for green or environmental bonds, anecdotal evidence suggests market demand for "green" labeled investment products may increase as investors seek investment portfolio diversity and aim to achieve social responsibility commitments. This increased demand can improve terms of the bond issuance, reducing the cost of capital. The Massachusetts Bay Transportation Authority (MBTA) case study provides an example of this scenario. In addition, these sustainably focused bonds can attract new investors, providing issuers with a new community of financiers.

Tax Increment Finance bonds

Tax increment finance (TIF) bonds capture land value and are considered revenue bonds securing the future anticipated increase in tax revenues generated by a development project as bond payback. This land value capture supports borrowing more money in bond issuance for infrastructure related to the area, such as sewer and water upgrades, parks, and roads [53]. Receipts from these greater amounts of debt can be used for projects that might not otherwise be financeable. Typically, a government designates a TIF area (or "district") while negotiating redevelopment terms, putting in place a tax increment based on the proposed redevelopment program, market feasibility, and estimated property value increases. Among the advantages of TIF bonding is that it allows governments to finance improvements without raising taxes for all tax payers or dipping into capital reserves. There are criticisms of certain TIF deals that offer financial incentives to encourage companies to relocate or that capture revenue from areas that would have appreciated in value regardless of the TIF designation.

Practitioners should have a general understanding of the various types of investments that fund resilient infrastructure. Of course, many of these bond types are used for private and noninfrastructure projects, but corporate bonds are not the subject of this chapter.

Case Study: Washington, D.C., Urban Heat Island

Washington, D.C., leveraged various finance sources to support resilient infrastructure projects. The District Department of Energy and Environment established a green building fund paid for with development-assessed fees. The fund provided grants for technical work needed for urban heat island mitigation projects such as green roofs and green alleys. Government funds also were leveraged by a local university to produce tools to assist with the initiative, and local building codes were altered to reflect the International Green Construction Code of the International Code Council that requires a certain ratio of green space. By requiring that ratio, resilience was integrated into the design process, allowing for existing project funds to be used to support resilience projects [54].

Case Study: MBTA Government Station

Boston's Government Center transit station underwent a renovation to improve handicap accessibility and climate resilience. The project was funded through various sources:

- Registered Retirement Income Fund loan for $220 million.
- Transportation Infrastructure Finance and Innovation loan for $162 million.
- MBTA revenue bonds for $99 million.
- Federal loan for $11.0 million [55].

The MBTA revenue bonds funding the project are landmark bonds in the United States, representing the first tax-exempt sustainability bonds. MBTA issued a combined $300 million in subordinated sales tax bonds and bond anticipated notes, $99 million of which were dedicated to the Government Center station project. Comparing this issuance to a traditional bond issued at the same time, the sustainability bond had more subscribers [56].

Insurance-Linked Securities

In addition to general obligation and revenue bonds, a relatively new category of bond is linked to insurance. These insurance-linked securities (ILS) include both catastrophe and resilience bonds. ILS are risk financing mechanisms that help asset owners transfer catastrophe and extreme weather risk from their balance sheets. In these transactions, in addition to the investor and issuer, there is a bond sponsor—generally an insurance company connected to an investment bank—that provides loans to the issuer. Some ILS require a special purpose vehicle (SPV), a company whose sole activity is to carry out the project by subcontracting construction and operations, serving as a third party for swaps of dollars between public and private investors [57].

SPVs are used to create projects beyond those funded by ILS as they protect municipal credit while stewarding resilience projects. They are a company with no previous business or record and, therefore, can achieve a favorable credit rating while remaining "off balance sheet" for the sponsors and the government issuers. The SPV has no assets other than the project.

Catastrophe bonds

In the case of a catastrophe bond, these ILS are designated for specific perils that might impact infrastructure, such as hurricanes, earthquakes, and windstorms. Each bond has a measurable trigger, a parameter such as wind or earthquake intensity at a specific location. If a catastrophe triggers this specified parameter during the life of the bond, the issuer gets the invested money to use for response and recovery, paying insurance claims and emergency relief (and investors lose their principal and any future dividends). Like insurance policies, but in the form of debt, these risk-linked securities transfer risks from an asset owner to investors when a pre-defined loss event occurs. An example in the US is the New York Metropolitan Transit Authority' $125 Million catastrophe bond [57]. One reason the investor would agree to the catastrophe bond is that the bond is not tied to any other financial mechanism (such as stock, bond, or treasury) and, thus, is used to diversify portfolios. If no catastrophe triggers during the life of the bond, investors get their money back plus interest.

Catastrophe bonds do not reduce physical risks and only pay out to repair infrastructure of any type after disasters strike. However, there is nothing that bars the recipient of a payout from using funds to pay to rebuild better or use the bond proceeds for things such as buyouts or other resilience measures.

Resilience bonds

On the other hand, resilience bonds fund projects that reduce physical risk, mitigating the impact of disasters by creating more resilient infrastructure. Resilience bonds capture future insurance premium savings from a reduced future risk. The future savings are capitalized and loaned to the issuer to invest in the resilient infrastructure projects that create that future risk reduction. Stated another way, as a funding source for resilient infrastructure projects, resilience bonds create the financial structure to collect funds today to fund improvements that lead to future benefits to insured properties. They do so by calculating the future reduced insurance costs and pooling this projected reduced cost, creating finance to loan to infrastructure projects aiming to protect properties [58]. Other types of resilience bonds combine a debt instrument (a bond) with a risk transfer element (insurance), with bond proceeds dedicated to an ex-ante resilience intervention and insurance available in the event of an extreme event during the bond's life. This definition places a particular emphasis on exposure to natural hazards and the potential impact that such hazards may have on a projects' finances.

In another use of the term, outside of the reinforcing of existing insurable assets, resilience bonds have sometimes entered the market as a way to fund seemingly "unbankable" resilient projects that do not have direct revenue generation potential but provide a public good, such as sea walls [59]. In this instance, the source of repayment might be a municipality or a nongovernmental organization.

Parametric insurance policies may complement these ILS, which are particularly considered when there is a concern about the potential impact that natural hazards may have on a projects' finances. Parametric policies differ from indemnity insurance, with payouts based on pre-defined index triggers such as wind speed, storm surge inundation depth, water flow rates, temperature or mortality. They can eliminate the costs and delays of investigating and adjusting loss following an event, minimizing the impact of shock and stress events on public entities, and they provide funds untied to the recovery of a specific asset.

ENABLERS OF RESILIENT INFRASTRUCTURE FINANCE

There are trillions of dollars in assets under management, but only a small portion of these assets are funding government infrastructure projects. This section explores several means for increasing private investment in resilient infrastructure, including via public–private partnerships, leveraged through state revolving loans or other government incentive programs, secured through a green bank and stewarded through a regional resilience collaboration.

Public–Private Partnerships

By general definition, a public–private partnership (PPP) is an agreement between a public agency and private sector entity to use the specific assets and skills of each to deliver a service to the general public that protects the public interest and generates private return on investment. Typically, PPPs are considered because they can attract greater net investment, enable management efficiencies, and transfer risks to the private from the public realm.

The level of engagement from the public and private sector varies from project to project. Fig. 6.2 shows the spectrum of engagement for PPPs [60]. A private entity generally may have ownership of the project with divestiture from the private sector so that all fees and assets come to the private entity.

At least four keys exist to a successful PPP for resilient infrastructure:

1. A clear revenue source to repay the initial investment.
2. A detailed understanding by all parties of the infrastructure's current and future risks—including given future climate change scenarios—and benefits—encompassing secondary or collateral

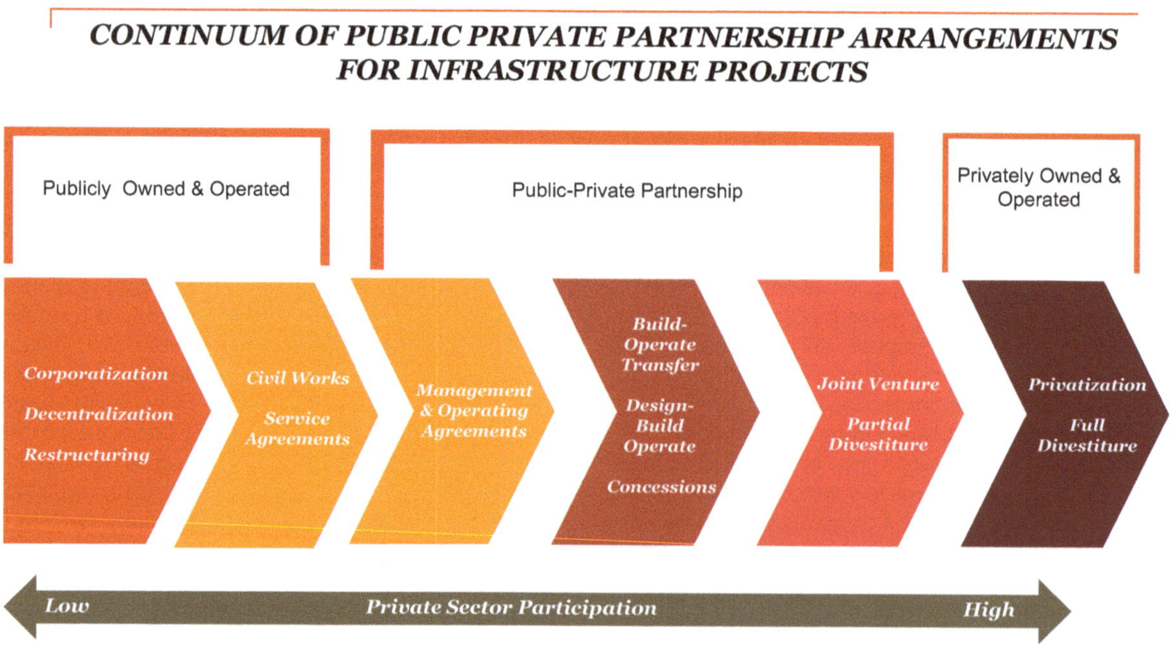

FIG. 6.2 Public–private partnership spectrum of engagement. (Source: Joyce Coffee.)

benefits from diverse project objectives. This includes allocation of revenue risk and operating cost risk as well as the distribution of future cash flows.

3. Appropriate staffing for PPP deal-making and ongoing operations, including contractual and technical expertise familiar with both traditional and innovative infrastructure finance.
4. A desire for innovation in contractual approach, which may be particularly important in the United States where fewer PPPs exist and which is particularly necessary where insurance-linked securities, green bonds, or other less common financing is used [61].

State Revolving Loan Funds

State Revolving Loan Funds provide low-interest loans to municipalities from states for investments in water and sanitation infrastructure, including sewage treatment, stormwater management facilities, and drinking water treatment [62]. In a project that blends public and private funds, these loans can make an infrastructure project investment more attractive to private investors by establishing a low-interest government source of funds as a foundation for the project, thus spreading the project's risk holders to include the State and increasing assurance of expected returns to private investors.

Although a very traditional means of investment, state revolving loans that could be used for water-related resilient infrastructure are not used as frequently as they could be, research indicates [63].

Property-Assessed Resilience

Property-assessed clean energy (PACE), or increasingly called property-assessed capital expenditures to reflect its expanded scope, assists building owners with financing energy improvements, providing them with funding and allowing them to pay the money back over time [64]. Uniquely, the repayment of these funds is assessed as part of the property tax bill or mortgage rather than directly with the property owner as with a consumer or home equity loan.

Similar programs at the state and city level assist property owners to improve resilience. One example is the NYC Build It Back program funded by the federal CDBG-DR program after Hurricane Sandy. The program has provided 12,500 properties with funds through reimbursement checks, construction starts, and acquisitions to help rebuild homes above the base flood elevation level [65,66]. The main difference between PACE and NYC Build It Back is that the funds are a grant, not a loan and, thus, payback is not required [67].

Many Florida PACE programs already finance hurricane protection improvements [64].

Green banks

Green banks differ from the conventional view of a bank. In the United States, green banks are legislated at the state or local level. They have the authority to partner with the private sector to use market development tools to accelerate green projects through a blend of public and private finance. Their intent is to increase the number of green projects by stimulating private sector investment. In practice, they serve as a go-between for the green marketplace and investors.

In 2018, green banks included the Connecticut Green Bank, Hawaii GEMS, IBank California CLEEN, Rhode Island Infrastructure Bank, and New York Green Bank. Montgomery County, Maryland, the first nonstate-based green bank [67a] and the District of Colombia Green Bank launched in 2018.

Green banks use a range of tools with the goal of increasing private lending activity and/or improving the terms of private financing such as [62]:

- Loan loss reserve, an expense set aside as an allowance for uncollected loans and loan payments due to issuer defaults and lower-than-estimated payments.
- Loan guarantees, for example, in combination with revenue bonds, a promise to assume the debt obligation if a loan defaults.
- Securitization (a/k/a warehousing) pools various projects to make a larger issuance to attract investors who otherwise shy away from the transaction cost and potential revenue ratio associated with smaller projects.

Green banks are legislated to focus primarily on greenhouse gas reduction, energy generation and distribution, and water security and stability. However, they may in the future help to increase funding significantly in other areas of resilience.

Regional Resilience Collaborations

Regional resilience collaborations are an emerging tool for infrastructure resilience finance that, potentially in partnership with green banks, can increase blended private and public finance of resilient infrastructure in the regions they target [68]. One such collaboration is the emerging Regional Resilience Trust Fund, part of the Regional Plan Association for New York, New Jersey and Connecticut [69]. Another example is the Urban Development Investment Funds within the Resilience Brokers Programme, an international network that provides integrated tools and a platform for collaboration on finance and decision making for resilience [70].

Practitioners should partner with finance experts to further enable resilience infrastructure finance.

RESILIENT INFRASTRUCTURE FINANCE CHALLENGES AND SOLUTIONS

To increase resilience infrastructure financing, several challenges including and beyond project bankability need to be anticipated and solved for. This section pairs challenges that many resilient infrastructure projects face with solutions that practitioners can consider.

Project Scale Resilience Risk and Impact Measurement Is Immature

Both successful finance and successful infrastructure rely on strong quantification. Yet, several elements of resilience measurement make resilience infrastructure quantification more difficult. First, actual and predicted hazard data are imperfect at the scale of many of the proposed resilient infrastructure projects, and second, no unique measure of resilience exists.

More generally, the financial services industry associates climate change and resilience with sustainability, particularly the United Nation's 2030 Sustainable Development Goals (SDGs) [3]. Many corporations have begun to use the SDGs as the framework for their corporate social responsibility reporting. Goal 11 specifically focuses on sustainable and resilient cities and goal 13 addresses climate change mitigation and adaptation. However, resilience intersects with many other SDGs, touching such areas as access to basic services, energy security, urban issues, and poverty.

More specifically for a particular infrastructure project, understanding future risk in the location the project operates and serves is key. However, finding location-specific risk information is not always straightforward. In fact, the location of future risk that official tools predict can differ from where impacts have been observed after extreme events. For example, as previously noted, flood maps developed by FEMA may not overlap with where the greatest flooding risks exist. In addition, FEMA flood maps are only based on past flood disasters for model inputs and current topography/elevation/bathymetry. They do not look forward and consider sea-level rise, potential for extreme weather events, etc. In reality, they can indicate building is okay in a place that is becoming less safe as precipitation patterns and sea levels change.

Furthermore, a unique measure of resilience—something as elemental as climate change mitigation's metric ton of greenhouse gas emissions—does not exist for resilience. Thus, a single resilience index does not exist that allows managers to easily compare the resilience of various infrastructure projects.

Solution: Data

Fortunately, the market already responds to the need for project-scale actual and predicted hazard data and, in addition, site-specific risk data are increasingly available. Since the TCFD released its mid-2017 guidelines [71] calling on financial institutions to examine their climate change risk, innovators are working to meet this market need. Community and parcel-scale risk analytics are becoming more available and robust [72].

With respect to measurement standards, the Global Adaptation and Resilience Investment work group has identified six ways to measure resilience. They are government indices and rankings; insurance risk ratings; corporate use data; risk screening tools; scorecards; and engineering due diligence and design analysis [73].

Several useful resources for resilient infrastructure measurement are available, including The Standard for Sustainable and Resilient Infrastructure and the Partnership for Resilience and Preparedness [72,74], USGBC's ReLi [75], and Envision [76], all of which offer standardization of resilience indicators, risk metrics, and performance for infrastructure projects. The Alliance for National and Community Resilience is also in the process of identifying resilience benchmarks for important community functions [77].

In the absence of a specific measures, comparison with other projects can prove helpful. The Private Participation in Infrastructure database [78], CIDA Good Practice Laboratory [79], ICLEI Solutions Gateway [80], and The Atlas [81] are potential sources for identifying comparable projects.

In sum, project deal makers should seek out the latest information while not letting the perfect be the enemy of the good. In many cases, enough is known about future hazards to act now to avoid their impact while improving quality of life.

Investment and Climate Impact Horizons Are Mismatched

Traditional infrastructure projects attract a certain type of investor: "patient capital" such as institutional investors like pension and sovereign wealth funds that invest for years rather than quarters. As infrastructure can take years to build and then lasts decades, it is not necessarily suited to the quarterly returns many market investors demand. Resilience projects add additional complexity to this longer-term time frame, as they also address future anticipated climate risk, which will be increasing during the decades of the infrastructure project's service. Some projects, for instance, need to anticipate periods when less freshwater dilutes their stormwater paired with more stormwater coming at once. Others need to anticipate growing king tides and sea-level rise eroding roadways and flooding streets and buildings. Generally, the market's quarterly timeframe is at odds with the decade-long climate change timeframe [82].

Solution: Collateral benefits that provide benefits now and in the future

The measurements described earlier will help investors to understand the value of investing in longer-term infrastructure with these future climate risk scenarios in mind. This will help them to avoid future project costs and failures borne not only by investors but also by owners, tenants, and taxpayers. At the same time, when analyzing the credit worthiness of a 30-year bond, for example, credit rating analysts look at values for their rating 30 years out, even as they review the strength of bond repayment every 5 to 7 years. So, the increasing attention that credit rating agencies pay to the physical risks from climate change may increase the overall market's interest in examining longer-term risk.

In addition to calculating the value of avoided losses, the market should consider it key that resilient infrastructure projects calculate collateral benefits from the multiple issues they address that pay back more immediately. Some of these collateral benefits will be social, with qualitative data associated with them, while others will be financial and can be quantified with dollars.

Climate Change Impacts Exacerbate Discrepancies in Vulnerability and Wealth

Climate change is expected to also severely impact the middle class, which comprises a majority of the US population. The Union Bank of Switzerland projects that the middle class will need to increase its spend on housing to secure and maintain housing [83]. Furthermore, climate gentrification will force low- and middle-class populations into higher risk areas, increasing vulnerability of housing and potentially forcing the middle class to increase spend on housing even more [84]. Marginalizing these populations will only result in greater inequality [85]. As wealth shifts from changes in middle class spending and growing social inequality, markets may become less stable. Poor communities are expected to get poorer, and Northern regions will likely experience an increase in wealth [86]. However, this model is based solely upon GDP. Social vulnerability to environmental hazards also will play a critical role in the distribution of economic losses. Key books on the topic include *The Disaster Profiteers: How Natural Disasters Make the Rich Richer and the Poor Even Poorer* [87] and *Extreme Cities: The Peril and Promise of Urban Life in the Age of Climate Change* [88].

Fig. 6.3 shows the social vulnerability index of US counties to environmental hazards from 2010 to 2014 based on 29 socioeconomic variables that contribute to reduction in a community's ability to prepare for, respond to, and recover from hazards.

Solution 1: Equate a lack of resilience with a decrease in growth in the middle-class market

A Metropolitan Planning Council/Urban Institute study put numbers on what many practitioners know: inequity is bad for all communities, not just ones with limited resources. In Chicago, inequity means that thousands of young people do not obtain the education they need to fulfill their potential every year. Hundreds of lives are lost by violence annually, and there are billions of dollars in lost wages each year [90]. A recent study post-Hurricane Harvey in Houston illustrates the differential finance harms of disaster response [91].

Solution 2: Visualize the risks and solutions using maps

One potentially powerful tool for practitioners is mapping climate risks with poverty or other equity-related data at the neighborhood scale. An archival Chicago project did this when considering ways to improve the city's marathon, resulting in trees planted along the route in areas both impoverished and with an elevated urban heat island effect. Similar map analysis was used to set priorities for on-street stormwater retention system repair.

Information Ownership and Power Are Mismatched

Resilience champions in government are becoming more common, especially with the prominence of 100 Resilient Cities, a program pioneered by the Rockefeller Foundation that supports a Chief Resilience Officer (CRO) and development of a city resilience strategy in participating cities. For instance, in Pittsburgh and Oakland, CROs are helping to create and finance green infrastructure projects for managing storm water runoff, reducing urban heat island effect, creating multibenefit street improvements for community experience and improving habitat and air quality [92]. Still, in most instances, no government focal point exists for resilient infrastructure to make the connections between

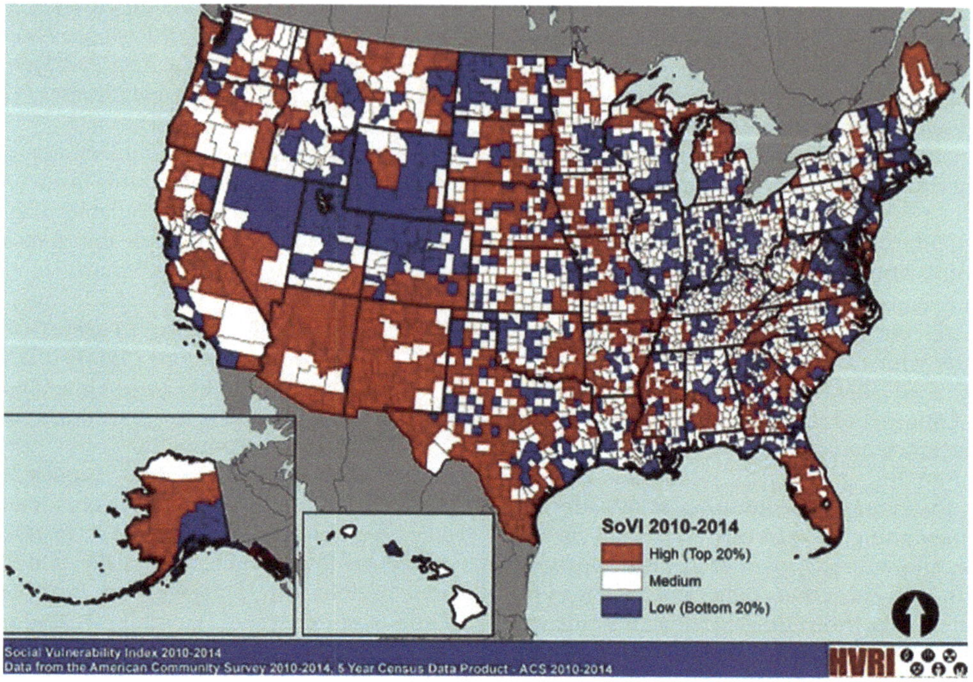

FIG. 6.3 Social vulnerability to environmental hazards from 2010 to 2014. SoVI: Social Vulnerability Index for the United States 2010–2014. (Source: University of South Carolina Hazards & Vulnerability Research Institute, https://artsandsciences.sc.edu/geog/hvri/sovi%C2%AE-0.)

infrastructure success and the served locality's current and future risk management, health, and operations.

This is also true beyond government. A 2015 survey by Business for Social Responsibility, Notre Dame Global Adaptation Index, and FourTwentySeven found that many executive leaders rely on sustainability leadership, which typically is an unfunded satellite function indirectly related to primary revenue generation and executive decision making [93].

Solution: Cross-sector collaboration and establishing a focal point

Resilience project success demands cross-sectoral skills (technical, political, and financial) and collaboration and coordination from seemingly unrelated fields or departments. Some cities experience their first-ever all-cabinet working meetings when creating and putting their resilience plans in place. Cities with or without a strong history of sustainability or resilience leadership can take advantage of cross-disciplinary teams that have worked together around other issues, such as security, risk management or public health. Indeed, the effort to communicate resilience concepts at the earliest stages of project development across a municipality's professionals is good practice for making the resilience infrastructure case to investors.

In terms of resilience finance, perhaps the most crucial benefit of this interdisciplinary work is illuminating the sources of revenue for a project..

Cost–Benefit Analysis Do Not Include Future Risk

Most cost–benefit analysis frameworks do not adequately calculate infrastructure resilience. They lack three crucial elements: counting avoided losses from risk mitigation, accounting for project collateral benefits, and evaluating projects, say for a 10 to 30-year timeframe. Typically, basic project cost–benefit analyses evaluate direct financial benefits (e.g., project revenues, decreased operational costs) and direct byproducts (e.g., labor days, taxes from business transaction revenue). Resilience-oriented cost–benefit analyses must incorporate impacts that are avoided in the future in addition to current benefits, such as outdoor community amenities, job creation for project maintenance, changes in property values, changes in public health, value of land-based amenities, and positive and negative impacts on lower income or minority

FIG. 6.4 Potential sources of taxes and fees. (Source: Joyce Coffee.)

populations [94]. (It is notable that TCFD's guidelines do assert that cost–benefit analyses for projects should incorporate avoided losses, although they do not offer guidance on best practices to incorporate this avoided loss into traditional cost benefit analysis.)

Solution: Use the latest ratios and methods and compare traditional to resilient

To start, resilience infrastructure projects could use the Multihazard Mitigation Council/National Institute for Building Science's benefit–cost ratio methodology to identify the ratio for their sector. This would show that when resilience is incorporated into a project, a positive payback results for every dollar invested [95]. As a simplified assessment of avoided losses, these ratios can call to mind the general benefits of investing in resilient infrastructure projects. Building upon this and continuing to dig into cost avoidance with actuaries and others on the project team will help evolve how cost–benefit analyses applies to individual projects. Suggested methodologies include those contained in the University of Massachusetts–Boston's Sustainable Solutions Lab's Recommendations for Resilience Finance [94].

Cost curves should be developed that compare potential resilience infrastructure projects, or even resilience measures within an infrastructure project, in terms of their cost/benefit and their avoided future loss.

In general, practitioners should consider the different approaches to infrastructure when working to attain funding for their resilient projects. Table 6.1 distinguishes traditional, sustainable, and resilient infrastructure.

Resilient Infrastructure Projects May Be Too Small to Generate Financier Interest

Finally, resilience infrastructure projects are sometimes smaller scale than traditional infrastructure. For instance, resilient infrastructure might call for distributed generation through decentralized power facilities or might rely upon green infrastructure solutions throughout a watershed. Generally, more fragmented projects can increase financial transaction costs and investors tend to focus on large investments [96].

Solution: Warehouse resilient infrastructure projects

Financing platforms such as green banks can warehouse or group projects together to entice investors looking for larger projects. On the other hand, resilience infrastructure proponents can gain scale in their projects by presenting a resilience portfolio rather than a solitary project to the market. Finally, regional projects that gain scale and thus cost and benefit can make a project more attractive.

TABLE 6.1
Comparing Traditional, Sustainable, and Resilient Infrastructure, in Terms of Specific Elements of Revenue, Finance, Design, Construction, and Operation.

		Traditional Infrastructure	Sustainable Infrastructure	Resilient Infrastructure
REVENUE GENERATION	Values captured	Captures economic costs and values.	Capture economic, environmental, and social costs and values.	
				Captures resilience values (e.g., lower insurance premiums, operational and maintenance efficiencies).
	Sources	Generally, a single source of revenue.	When developed as a multipurpose system, draws from different types of revenue streams.	
FINANCE	Types of investors	Large and experienced investors.	Attracts investors interested in investments with additional social or environmental impact.	
	Risks	Increasing risk of stranded assets due to physical climate impacts.	Accrues savings from increased productivity and resource use efficiency.	Better positioned to withstand shocks and stressors. Accrues savings from hazard events preparedness (reduced human and material losses, less disruption).
	Insurance	Increased insurance costs (due to hazard vulnerability).	New technologies can be harder to insure.	
	Finance instruments	Benefits from proven performance of traditional financing vehicles and instruments.	Benefits from new forms of financial instruments such as pay for success bonds and pooled investment vehicles.	
DESIGN	Political risk	Often exposed to risks related to community and/ or political opposition.	Inclusion of environmental and social considerations may diminish community and/ or political opposition risks.	Envisions future stability and success based on current and future risks.
	Team structure	Typically, one infrastructure department.	Team of infrastructure experts with a resilience focal point.	
CONSTRUCTION	Timeframes	Large scale centralized infrastructure requires long construction timeframes.	Some decentralized systems can be implemented faster or gradually (at lower cost) and cause less disruption.	
	Resources	Resource intensive.	Multipurpose can foster resource efficiency.	
OPERATION	Performance	Siloed nature can create operational deficiencies.	Generates operational efficiencies and less impact on the environment.	
	Flexibility	Highly structured.	Can adapt to existing urban structures more easily.	

Note That Sustainable Infrastructure and Resilient Infrastructure Share Some but Not All Characteristics.
Credit: Joyce Coffee.

Investor Requirements for Resilient Infrastructure Projects Vary

Investors have various expectations for the money invested in projects. First, almost all investors seek diversity in the project types, finance types, and regions in their investment portfolio. Traditionally, investors have desired returns commensurate with market indices such as the S&P 500. However, social impact investors—a growing investor category—may prioritize development impacts such as measurable changes in economic, social, or environmental targets, policy modifications, or the leveraging of their funds to inspire other related investments [7]. These investors may be content with no loss on their funds rather than market-indexed returns. Other investors might be willing to see returns lower than market indices, or they might be willing to wait a long time—beyond 5 years—to see returns. The funds that this category of investors invests in are known as "patient capital," and this type of investment is associated traditionally with sovereign wealth funds, insurance investors, and pension funds.

In addition, investors engage at various infrastructure project stages with governments, project owners, impact investors, and philanthropists providing debt capital during planning, design, and construction; commercial banks providing funds during construction and operation; and institutional investors—especially those with "patient capital" such as pension funds, insurance companies, and sovereign wealth funds—providing funds for operations [97].

At a minimum, investor requirements for resilient infrastructure include measurable local economic, social, or environmental improvements; climate change risks mitigated; policy mandates implemented; and no financial losses anticipated. Then, depending on investor type, the investor expects (a). returns that are not fully commercial, (b) returns but in a longer time frame than the majority of the market, or (c) returns commensurate with the market.

CONCLUSION

Resilient infrastructure finance depends on practitioners understanding the flows of funds in and out of the financial markets, the types of traditional and innovative financial mechanisms, enablers of resilient infrastructure finance, and trends in the finance industry related to physical climate risks. A great opportunity exists for practitioners to help create momentum in the financial services industry toward resilient infrastructure, especially as finance sector trends like credit rating agency physical risk integration, new financial tools, and finance industry guidelines about avoiding risk could create more supply of resilient infrastructure finance capital. Although there are many barriers to financing resilient infrastructure projects, solutions are continuing to emerge to ease the process in securing funding for these projects.

Practitioners are encouraged to consider bankability from the first moment of project creation, defining financial feasibility by calculating revenue potential and social impacts and describing projects in terms of maintaining creditworthiness by protecting them from risks that would impact payback of any debt.

Here are eight keys for practitioners helping to create a successful resilient infrastructure deal for financial markets.

1 Strategic objectives, priorities, and programs. Make clear the intent of resilient infrastructure in options analysis, procurement documents, engineering and feasibility studies, and cost–benefit assessments.
2 Stakeholder impact on priorities. Develop project priorities and program in collaboration with affected stakeholders, including community leaders who represent lower resourced populations and corporate decision makers.
3 Governance, vehicles, and focal point. Examine existing governance related to the project and ensure that a focal point interacting with financiers is well versed not only in the project but also in related jurisdiction laws and project design and intention.
4 Mechanisms for contracting by project stage. Do a dry run of the project through permitting departments to anticipate and solve for challenges and delays.
5 Overview of legislative, regulatory, and licensing elements. Examine existing laws in all jurisdictions that relate to the project and design to comply while working with regulatory bodies to understand key features of resilient infrastructure.
6 Clear, comparative, and analyzed data. Gather all project-related data, including on collateral benefits, and monetize these data where possible. Put plans in place to collect and monitor data throughout the project life.
7 Budget strategy. Ensure that budgets include both avoided loss and accrued benefits from all project elements throughout the project life. Focus on bankability from project conception to fruition.
8 Don't overlook new, future "utopias." Given current impacts and projections of future risk, look for opportunities in regions less prone to risk including America's heartland, where there are relatively ample supplies of freshwater, less wildfire risk, and no coastal sea-level rise.

ACKNOWLEDGEMENTS

Thanks to Sarah Dobie, Research Associate in the Rochester Institute of Technology Collaboratory for Resiliency and Recovery and the Arizona State University/NationalScience Foundation Urban Resilience to Extremes Sustainability Research Network for her collaboration and to Rob Moore, director of the Water and Climate Team at the National Resources Defense Council, Josh Sawislak, international expert on climate and disaster resilience and sustainable infrastructure and John Macomber, impact investor and Harvard Business School finance faculty for their expert review and critique of this chapter.

REFERENCES

[1] United Nations Framework Convention on Climate Change, Paris Agreement, 2015. https://unfccc.int/sites/default/files/english_paris_agreement.pdf.

[2] United Nations Office for Disaster Risk Reduction, Sendai Framework for Disaster Reduction, 2015. https://www.unisdr.org/we/coordinate/sendai-framework.

[3] United Nations Sustainable Development Goals Knowledge Platform. Sustainable Development Goals. 2019. https://sustainabledevelopment.un.org/topics/sustainabledevelopmentgoals.

[4] American Society of Civil Engineer (ASCE), 2017 Infrastructure Report Card, Investment, 2018. https://www.infrastructurereportcard.org/solutions/investment.

[5] American Society of Civil Engineer (ASCE), Failure to Act: Closing the Infrastructure Investment Gap for America's Economic Future, 2016. https://www.infrastructurereportcard.org/wp-content/uploads/2016/10/ASCE-Failure-to-Act-2016-FINAL.pdf.

[6] I. Simpson, L. Lopez, U.S. Infrastructure Gets D-Plus Grade in Civil Engineers' Report Card, Again, 2017. https://www.reuters.com/article/us-usa-infrastructure-reportcard/u-s-infrastructure-gets-d-plus-grade-in-civil-engineers-report-card-again-idUSKBN16G21I.

[7] Bloomberg New Energy Finance, Chatham House The Royal Institution of International Affairs, Frankfurt School, Finance Guide for Policymakers: Renewable Energy, Green Infrastructure, 2016. https://www.bbhub.io/bnef/sites/4/2016/08/Finance-Guide-for-Policymakers-RE-GreenInfra-August-2016.pdf.

[8] McKinsey Global Institute, Bridging Global Infrastructure Gaps, 2016. https://www.un.org/pga/71/wp-content/uploads/sites/40/2017/06/Bridging-Global-Infrastructure-Gaps-Full-report-June-2016.pdf.

[9] J. Alexander, R. Abadie, M. Tanda. Investor Ready Cities. 2019 https://www.pwc.com/gx/en/psrc/publications/assets/pwc-investor-ready-cities-v1.pdf.

[10] D. Wiener, N. Didillon, Financing Sustainable and Resilient Infrastructure by Creating a New Asset Class for Institutional Investors, 2016. http://www.gib-foundation.org/content/uploads/2014/03/Financing_Sustainable_and_Resilient_Infrastructure_GIB.pdf.

[11] J. Koh, E. Mazzacurati, S. Swann, Bridging the Adaptation Gap: Approaches to Measurement of Physical Climate Risks and Examples of Investment in Climate Adaptation and Resilience, 2016. http://427mt.com/wp-content/uploads/2016/11/GARI-2016-Bridging-the-Adaptation-Gap.pdf.

[12] S. Marshall, K. McCormick, Returns on Resilience. The Business Case, Urban Land Institute, Washington, DC, 2015.

[13] F. Banaji, J. Corfee-Morlot, I. Gençsü, J. Jaeger, H. Mountford, J. Rydge, The Sustainable Infrastructure Imperative. Financing for Better Growth and Development, The New Climate Economy, London and Washington, DC, 2016. http://newclimateeconomy.report/2016/wp- content/uploads/sites/4/2014/08/NCE_2016Report.pdf.

[14] C. Reinhard, Infrastructure-oriented Companies Have Outperformed in the New Millennium with Less Risk, 2018. https://www.privatebank.citibank.com/home/fresh-insight/the-outlook-for-investing-in-us-infrastructure.html.

[15] S. Moser, J. Coffee, A. Seville. Rising to the Challenge Together. https://kresge.org/content/rising-challenge-together.

[16] Interestco.nz. Credit Ratings Explained, 2019. https://www.interest.co.nz/credit-ratings-explained.

[17] Moody's Investors Service, Environmental Risks: Evaluating the Impact of Climate Change on US State and Local Issuers, 2017. http://www.southeastfloridaclimatecompact.org/wp-content/uploads/2017/12/Evaluating-the-impact-of-climate-change-on-US-state-and-local-issuers-11-28-17.pdf.

[18] ThinkProgress, Climate Change Poses 'nightmare Scenario' for Florida Coast, Bloomberg warns, 2017. https://thinkprogress.org/bloomberg-coastal-real-estate-638716394641/.

[19] S&P Global, Aftermath: Assessing the Impact on Local Government Credit Quality, 2017. https://www.spglobal.com/our-insights/In-A-Storms-Aftermath-Assessing-The-Impact-On-Local-Government-Credit-Quality.html.

[20] S&P Global, How Our U.S. Local Government Criteria Weather Climate Risk, 2018. https://www.spratings.com/documents/20184/4918240/HowOurUSLocalGovernmentCriteriaWeatherClimateRisk.pdf.

[21] J. Kanne, D. Malek-Madani, S. Bendix, Climate Change and Commercial Real Estate: How Resilient Is Your Portfolio?, 2017. http://natadvisors.com/wp-content/uploads/2017/10/274174-6929.ccincre-highres.pdf.

[22] Government Accountability Office, 2017 Hurricanes and Wildfires: Initial Observations on the Federal Response and Key Recovery Challenges, 2018. https://www.gao.gov/products/GAO-18-472.

[23] Federal Emergency Management Agency, FEMA Expands its Reinsurance Program by Transferring $500 Million in Flood Risk to Capital Markets, 2018. https://www.fema.gov/news-release/2018/07/31/fema-expands-its-reinsurance-program-transferring-500-million-flood-risk.

[24] The New York Times Magazine, In Nature's Casino, 2007. https://www.nytimes.com/2007/08/26/magazine/26neworleans-t.html.
[25] J. Macomber, Personal Communication, September 2018.
[26] W. Cobleigh. Connecticut's Model for Property Assessed Financing for Affordable Flood Resilience. https://www.floods.org/Files/Conf2015_ppts/C4_Cobleigh.pdf.
[27] Climate Central, Surging Seas Sea Level Rise Analysis, 2018. http://sealevel.climatecentral.org/maps.
[28] Zillow, Climate Change and Housing: Will a Rising Tide Sink All Homes?, 2017. https://www.zillow.com/research/climate-change-underwater-homes-12890/.
[29] Oasis Loss Modeling Framework. Our Modeling Platform. 2019 https://oasislmf.org/our-modelling-platform.
[30] Acclimatise, FourTwentySeven Climate Solutions, Climate Finance Advisors, Lenders' Guide for Considering Climate Risk in Infrastructure Investments, 2018. http://www.acclimatise.uk.com/wp-content/uploads/2018/01/Lenders_Guide_for_Considering_Climate_Risk_in_Infrastructure_Investments.pdf.
[31] World Economic Forum, The Global Risks Report, 2017. http://www3.weforum.org/docs/GRR17_Report_web.pdf.
[32] World Economic Forum, The Global Risks Report, 2018. https://www.weforum.org/reports/the-global-risks-report-2018.
[33] Forum Pour L'Investissement Responsible, Article 173-VI: Understanding the French Regulation on Climate Reporting, 2016. https://www.frenchsif.org/isr-esg/wp-content/uploads/Understanding_article173-French_SIF_Handbook.pdf.
[34] Conservation Law Foundation, Climate Adaptation and Liability: A Legal Primer and Workshop Summary Report, 2018. https://www.clf.org/wp-content/uploads/2018/01/GRC_CLF_Report_R8.pdf.
[35] P. Brandley, Presentation to Municipal Analysts Group of New York, May 11, 2018.
[36] United Kingdom Climate Impacts Programme (UKCIP), Adaptation Reporting Power, 2014. https://ukcip.ouce.ox.ac.uk/learning/arp.
[37] C40 Cities, How FSCI Is Accelerating Sustainable Investments, 2017. http://financingsustainablecities.org/.
[38] United Nations Conference on Climate Change, Biennial Assessment and Overview of Climate Finance Flows, 2016. https://unfccc.int/news/biennial-assessment-and-overview-of-climate-finance-flows.
[39] Nature Climate Change, Millions Projected to Be at Risk from Sea-Level Rise in the Continental United States, 2016. https://www.nature.com/articles/nclimate2961.
[40] Nature Climate Change, Migration Induced by Sea-Level Rise Could Reshape the US Population Landscape, 2017. https://www.nature.com/articles/nclimate3271.
[41] E&E News Extreme Weather, Cuts May Undermine Disaster Recovery, 2018. https://www.eenews.net/stories/1060075241.
[42] B. Benson, AB2800 Climate-Safe Infrastructure Webinar: Financing the Future III, July 5, 2018.
[43] Environmental Protection Agency. Learn About the WIFIA Program. 2019. https://www.epa.gov/wifia/learn-about-wifia-program.
[44] Blackrock, Deconstructing Infrastructure Debt, November 2017. https://www.blackrock.com/institutions/en-gb/insights/investment-actions/deconstructing-infrastructure-debt.
[45] Center for Innovation at Arizona State University, Smart Cities Financing Guide: Expert Analysis of 28 Municipal Finance Tools for City Leaders Investing in the Future, 2014. https://urbaninnovation.asu.edu/sites/default/files/smartcitiescouncil_-_financing_guide-3_31_14.pdf.
[46] M. Ruibal, K. Van Der Leeden, Increasing the Benefits from Resilience. Using the Approach of 'value Realization' to Bringing the Greatest Possible Gains in Resilience, 2016. http://nebula.wsimg.com/4dac2bcef2cc59b18d6efca3c17ef070?AccessKeyId=7BDE3FD796DC1A85B624.
[47] Climate Bonds Initiative. Bonds and climate change state of the market report 2017. https://www.climatebonds.net/2017/10/climate-bonds-state-market-report-2017-green-bonds-climate-aligned-universe-now-stands-895bn.
[48] A. Whiley. Climate Bonds State of the Market Report 2017: Green Bonds & Climate-Aligned Universe Now Stands at $895bn: Launch Today at Climate Week NYC Event. https://www.climatebonds.net/2017/10/climate-bonds-state-market-report-2017-green-bonds-climate-aligned-universe-now-stands-895bn.
[49] IISD, Adaptation Finance Update Green Bonds on the Rise but Will Adaptation Projects Benefit?, 2018. http://sdg.iisd.org/news/adaptation-finance-update-green-bonds-on-the-rise-but-will-adaptation-projects-benefit.
[50] Climate Bonds Initiative, Taxonomy. 2019. https://www.climatebonds.net/standards/taxonomy.
[51] District of Columbia Water and Sewer Authority, DC Water's Green Bond Report, 2017. https://www.dcwater.com/sites/default/files/finance/FY%202017%20Green%20Bond%20Report_revised.pdf.
[52] Kennedy Street Project, Kennedy Street Revitalization Project, 2016. http://kennedystreetproject.com.
[53] National Association of Realtors, Tax Increment Financing (TIF), 2002. https://www.nar.realtor/smart_growth.nsf/docfiles/TIFreport.pdf/$FILE/TIFreport.pdf.
[54] C40 Cities. New Perspectives on Climate Finance for Cities, Washington DC. 2019. https://www.c40.org/other/washington-dc-case-study.
[55] P. Burton, MBTA Plans to Issue First Tax-Exempt Sustainability Bond, 2017. https://www.bondbuyer.com/news/mbta-plans-to-issue-first-tax-exempt-sustainability-bond.
[56] The Bond Buyer. Video deal of the year 2017 - Northeast: Massachusetts Bay Transportation Authority. https://www.bondbuyer.com/video/deal-of-the-year-2017-northeast-massachusetts-bay-transportation-authority.
[57] Reuters, New York's Transit Authority to Sell $125 Mln 'catastrophe' Bond, 2013. https://www.reuters.com/

article/mta-bond-idUSL6N0FL1NG20130715?feedType =RSS&feedName=bondsNews.
[58] Re:focus Partners, A Guide for Public-Sector Resilience Bond Sponsorship, 2017. http://www.refocuspartners.com/wp-content/uploads/pdf/RE.bound-Program-Report-September-2017.pdf.
[59] The Brookings Institution, Financing Infrastructure through Resilience Bonds, 2015. https://www.brookings.edu/blog/the-avenue/2015/12/16/financing-infrastructure-through-resilience-bonds.
[60] J. Delmon, Understanding Options for Public-Private Partnerships in Infrastructure, 2010. https://openknowledge.worldbank.org/bitstream/handle/10986/19947/WPS5173.pdf?sequence=1&isAllowed=y.
[61] The Brookings Institution, Modernizing Infrastructure Policies to Advance Public-Private Partnerships, 2018. https://www.brookings.edu/blog/the-avenue/2018/05/22/modernizing-infrastructure-policies-to-advance-public-private-partnerships.
[62] National Resource Defense Council, Coalition for Green Capital, Climate Finance Advisors, Green & Resilience Banks: How the Green Investment Bank Model Can Play a Role in Scaling up Climate Finance in Emerging Markets, 2016. https://www.nrdc.org/sites/default/files/green-investment-bank-model-emerging-markets-report.pdf.
[63] Re:focus Partners, RE.invest: A Roadmap for Resilience, Investing in Resilience, Reinvesting a Communities, 2015. http://www.refocuspartners.com/reinvest/.
[64] United States Department of Energy, Office of Energy Efficiency & Renewable Energy. Property-Assessed Clean Energy (PACE) Financing Resources. 2019 https://www.energy.gov/eere/solarpoweringamerica/downloads/property-assessed-clean-energy-pace-financing-resources.
[65] Georgetown Climate Center Adaptation Clearinghouse. New York City Build it Back program. http://www.nyc.gov/html/recovery/html/home/home.shtml.
[66] The City of New York, Build it Back Stronger and Safer: Welcome to NYC Housing Recovery, 2018. http://www.adaptationclearinghouse.org/resources/new-york-city-build-it-back-program.html.
[67] Mayor's Office of Housing Recovery Operations, Build it Back Progress Update, 2016. http://www.nyc.gov/html/recovery/downloads/pdf/build-it-back-update-10-20-16-final.pdf.
[67a] Coalition for Green Capital, Nation's First Local Green Bank Designated in Montgomery County, MD, 2016. http://coalitionforgreencapital.com/2016/08/03/montgomery-county-green-bank.
[68] J.M. Keenan, Regional Resilience Trust Funds: An Exploratory Analysis for the New York Metropolitan Region, President and Fellows of Harvard College, Cambridge, 2017.
[69] Natural Resource Defense Council. Go back to the well: states and the federal government are neglecting a key funding source for water infrastructure. http://protectcleanwater.org/wp-content/uploads/2017/09/SRFs-Go-Back-to-the-Well-NRDC-FINAL.pdf.
[70] Resilience Brokers. Funding the Global Goals in City Regions. 2019 https://resiliencebrokers.org/funding.
[71] Task Force on Climate-related Financial Disclosures. About the Task Force. 2019. https://www.fsb-tcfd.org/about.
[72] Darob Malek-Madani Presentation to the National Association of Real Estate Investment Managers, January 18, 2018.
[73] Global Infrastructure Basel, SuRe – the Standard for Sustainable and Resilient Infrastructure, 2015. http://www.gib-foundation.org/content/uploads/2015/07/SuRe-Summary_June_2015.pdf.
[74] Perkins + Will. RELi Credit catalogue. http://c3livingdesign.org/?page_id=5110.
[75] Institute for Sustainable Infrastructure. Envision 3.0. https://sustainableinfrastructure.org/.
[76] Alliance for National & Community Resilience. http://www.resilientalliance.org.
[77] The World Bank. Private Participation in Infrastructure Database. 2019 http://ppi.worldbank.org.
[78] Cities Development Initiative for Asia (CDIA). City Case Studies. 2019 http://cdia.asia/what-we-do/col-good-practices/.
[79] Solutions Gateway. Case Studies. 2019. http://www.solutions-gateway.org/discover?area=cases.
[80] The Atlas. Navigate to your Future City. 2019. https://www.the-atlas.com/.
[81] Mott MacDonald, Climate Change and Business Survival: The Need for Innovation Delivering Climate Resilience, 2015. https://www.mottmac.com/download/file/127/6772/climate%20change%20and%20business%20survival.pdf.
[82] D.A. Koehler, B. Bertocci, J.J. Buonocore, P. Donovan, A. Gordon, H. Kunreuther, Climate Change: A Risk to the Global Middle Class, UBS, 2016. https://www.ubs.com/microsites/climatechange/en/home.html.
[83] O. Milman, Climate Change Set to Worsen Inequality in US if Greenhouse Gas Emissions Aren't Reduced, The Guardian, 2018. https://www.theguardian.com/environment/2017/jun/29/climate-change-income-inequality-environment.
[84] S.N. Islam, J. Winkel, Climate Change and Social Equality, United Nations Department of Economic and Social Affairs, October 2017. http://www.un.org/esa/desa/papers/2017/wp152_2017.pdf.
[85] B. Dennis, Climate Change in the U.S. Could Help the Rich and Hurt the Poor, 2017. https://www.washingtonpost.com/news/energy-environment/wp/2017/06/29/climate-change-in-the-u-s-could-help-the-rich-and-hurt-the-poor/?utm_term=.8e4c1da5e031.
[86] J.C. Mutter, The Disaster Profiteers: How Natural Disasters Make the Rich Richer and the Poor Even Poorer, MacMillan, 2015.
[87] A. Dawson, Extreme Cities: The Peril and Promise of Urban Life in the Age of Climate Change, Verso Books, 2017.
[88] S.L. Cutter, The Perilous Nature of Food Supplies: Natural Hazards, Social Vulnerability, and Disaster

Resilience, 2017. https://www.researchgate.net/figure/Spatial-patterns-of-damages-social-vulnerability-and-community-resilience-A-total_fig2_312428695.

[89] University of South Carolina. SoVI: Social Vulnerability Index for the United States – 2010-2014. https://artsandsciences.sc.edu/geog/hvri/sovi%C2%AE-0.

[90] Metropolitan Planning Council. The cost of segregation. http://www.metroplanning.org/costofsegregation/default.aspx.

[91] Texas Housers, A Costly and Unequal Burden, an Equity Analysis on the City of Houston's Revision to its Floodplain Management Code, 2018 (Chapter 19), https://texashousers.net/2018/08/30/new-report-details-how-houston-floodplain-ordinance-can-worsen-inequality-in-face-of-future-floods.

[92] Georgetown Climate Change Center Adaptation Clearinghouse. 100 Resilient Cities Resilience Advisory Council. 2019. http://www.adaptationclearinghouse.org/networks/100rc-resilience-advisory-council/resilience-strategies.html.

[93] A. Seville, C. Gannon, E. Mazzacurati, J. Coffee, C. Erickson, Corporate Adaptation Survey, 2015. https://news.nd.edu/assets/164209/2015_corporate_adaptation_survey.pdf.

[94] University of Massachusetts Boston Sustainable Solutions Lab. Recommendations on Financing Climate Resilience. https://www.umb.edu/news/detail/sustainable_solutions_lab_recommendations_on_financing_climate_resilience.

[95] Multihazard Mitigation Council of the National Institute for Building Sciences, Natural Hazard Mitigation Saves: 2017 Interim Report, 2017. https://www.nibs.org/page/mitigationsaves.

[96] McKinsey & Company, How Impact Investing Can Reach the Mainstream, 2016. https://www.mckinsey.com/business-functions/sustainability-and-resource-productivity/our-insights/how-impact-investing-can-reach-the-mainstream.

[97] World Economic Forum, Integrating Natural Disaster Risk into the Financial System Private Workshop, 2014. http://www3.weforum.org/docs/WEF_ENV_NaturalDisasterRiskFinancialSystem_WorkshopSummary_2014.pdf.

CHAPTER 7

Addressing Climate Risk in Financial Decision Making

NATALIE AMBROSIO, BSC • YOON HUI KIM, PHD, MPHIL • STACY SWANN, BA, MBA, MTS • ZIYUE WANG, MEM, BA

INTRODUCTION

The integrity of our communities and the social, economic, and environmental systems that underpin them are directly affected by the resilience (or lack thereof) of a community's water, power, transportation, and municipal (e.g., schools, hospitals) infrastructure. Developing climate-resilient infrastructure to support thriving communities necessitates leveraging an understanding of physical climate impacts on infrastructure to finance adaptation actions that make existing and new infrastructure resilient to future climate impacts. Financing infrastructure often occurs as a "blend" of public and private investments, with the private sector helping to finance, develop, and build the projects. Catalyzing private investment for climate-resilient infrastructure requires understanding how investors assess, quantify, value, and price risks embedded within projects. This in turn calls for unpacking how physical climate risks may affect a project's financial stability and how those impacts will shape investment returns through direct impacts on an asset as well as indirect impacts on the community within which an asset is situated.

CLIMATE RISKS AND OPPORTUNITIES: WHY THEY MATTER FOR INFRASTRUCTURE LENDING

Climate change poses both acute and chronic physical risks to infrastructure, but also presents investment opportunities. Weather- and climate-related losses in the United States reached a record high of $306 billion in 2017, with 16 events each resulting in costs greater than $1 billion [1]. From 1980 to 2016, climate-driven extreme weather events caused EUR436 billion in losses in Europe with average annual losses doubling over the last 30 years [2]. The failure to adapt to acute and chronic climate risks can diminish performance and reliability (e.g., reduced water availability for cooling of power plants or data centers will limit operations), increase operating costs, decrease generated revenues, and/or increase expenses to repair damaged assets. Increasingly frequent and intense climate-driven acute risks, including hurricanes, extreme precipitation, coastal storm surge, floods, and wildfires, may disrupt infrastructure service delivery and cause damage that results in catastrophic failure and reduces the lifespan of infrastructure assets.

Chronic risks caused by gradual climatic shifts, such as increasing temperatures, water stress, or sea-level rise, can also pose substantial risks to infrastructure. For instance, sea-level rise may exacerbate chronic flooding that is beginning to affect coastal real estate and critical regional transportation arteries, with implications for asset value. The flood-prone Miami-Dade area in Florida has already experienced a loss of $465 million in the real-estate market between 2005 and 2016 due to sea-level rise [3]; in the New York City tri-state area, from 2005 to 2017 coastal housing value declined by $6.7 billion due to sea-level rise [4].

Climate change will also present opportunities for infrastructure investors, including through new markets. For example, higher temperatures may enhance the attractiveness of existing or new tourist destinations that feature sports or other recreational infrastructure that is adapted to warmer temperatures (e.g., facilities with a retractable roof), offering increased revenue generation potential. Changing climate conditions also provide project developers and financers with opportunities to invest in resilience features that enable infrastructure to endure extreme events with minimal damage and to support the resilience of the local communities in which the asset is embedded. For instance, electricity grids with distributed energy resources and

Optimizing Community Infrastructure. https://doi.org/10.1016/B978-0-12-816240-8.00007-0

energy storage systems can restore service faster after outages. Telecommunication networks based on fiber-optic cables, which are waterproof unlike traditional copper wires, can save operating costs by reducing network faults, while also improving network reliability. As urbanization and population growth continue, global infrastructure investment demand is estimated at $6.3 trillion per year, highlighting the enormous potential to build resilience into existing or planned infrastructure investment opportunities [5].

This chapter will look at the ways in which climate change can affect investment returns for infrastructure investors. It will then explore how these investors can address the exposure and vulnerability of infrastructure to anticipated physical climate impacts to mitigate their own financial risk, while also financing projects that foster community resilience.

THE INFRASTRUCTURE FINANCE LANDSCAPE

Sources of Infrastructure Finance

The elements of financing sustainable and climate-resilient infrastructure vary depending on the market, project, policy context, and regulatory certainty. In most developed-country contexts, funding infrastructure investment involves ex ante public planning, particularly for larger projects such as airports, road infrastructure, and water infrastructure. Government concessions and tenders for procurement are also frequently necessary in the early stages of the funding process. Large-scale public infrastructure projects are often part of a municipal, state, or federal government planning process that sets in motion a well-defined series of activities that enable private sector actors—including developers, construction companies and (in some cases) operators—to bid on the project. On the other hand, in some sectors such as energy and telecommunications, project developers drive the development of infrastructure projects at the outset, planning and designing the project prior to engaging government actors for concessions and permits. Capital-to-finance infrastructure projects usually involve a combination of both private debt and equity, but in many cases are also attracted from capital markets through bond issuances.

In many emerging economies legal and regulatory environments are weaker or uncertain and there are other more substantial barriers to investment and doing business than in developed markets. In these regions investor risk perceptions often hinder private investment in infrastructure projects. High perceived project risk results in a heavy reliance on multilateral, bilateral and domestic development finance institutions (DFIs) and development banks, which can help to mitigate risks, particularly market, currency, and regulatory risks, or transfer them off of the project investors' balance sheets. These banks often have strong relationships with country governments, which help to manage default risks. International, regional, and domestic capital markets can also provide additional routes for raising capital, although in some cases support from development banks is critical for successfully mobilizing capital from these markets.

Generally, public and private capitals are the two primary sources of infrastructure funding. Public capital can take two forms: budgetary or financial. Budgetary funding refers to direct appropriations from public budgets, whether at the national, state, or local level, and is often derived from tax revenue and/or general obligation bonds issued by government entities.

DFIs, including national or subnational infrastructure banks, are the other common type of public funding source. These institutions provide a variety of different instruments to fund infrastructure, including grants (usually for project preparation or capacity building), debt to the project, equity, and risk mitigation coverage through guarantees, insurance, or export credits. In some cases where the infrastructure investment is able to provide carbon credits, carbon markets can be used to provide an additional source of finance for these projects.

Many DFIs and infrastructure banks invest with the private sector and use financing procedures similar to those used by commercial banks. Private capital for infrastructure is derived from a range of sources and channeled through a number of different actors, including commercial banks, infrastructure banks and, in some cases, infrastructure funds and asset managers (Fig. 7.1).

Infrastructure Life Cycle

When it comes to financing, large infrastructure projects are generally characterized by three distinct phases of funding requirements:

- **Project preparation:** This phase involves putting together a project for financing, including technological designs, engineering and the necessary permitting, easements, rights of way and concessions (particularly for projects that require government permits). This phase of a project requires significant time and effort, and developers will often seek preparation funding, and/or public entities will underwrite and finance these activities. For large and complex infrastructure projects, particularly those that are financed through a project–finance

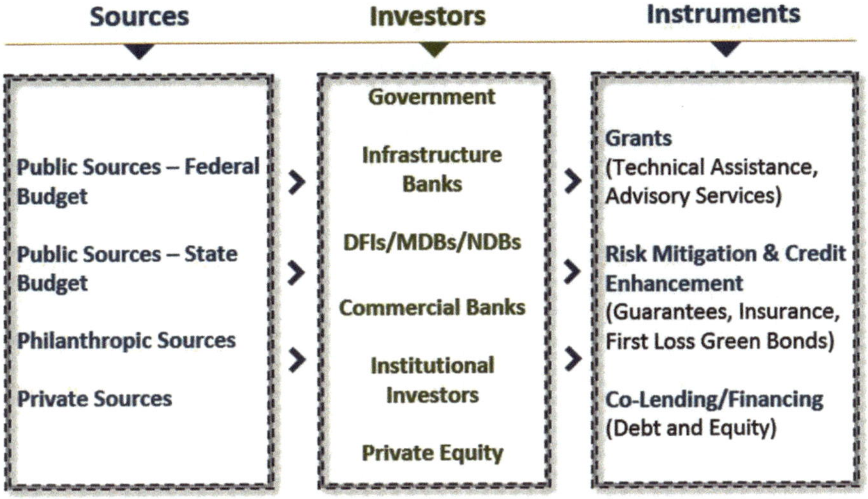

FIG. 7.1 Sources of infrastructure finance. (Credit: Climate Finance Advisors.)

approach, significant effort is required to prepare the legal, engineering, procurement, and construction agreements. Project preparation also includes engaging funders and financers on debt funding agreements, equity contribution agreements, security documents, and potentially intercreditor agreements.

- **Construction and development:** This phase encompasses all the work of building the infrastructure and occurs after the project is financed. The construction phase is capital-intensive and incurs significant risk due to high levels of uncertainty. Although this phase often presents significant risks such as construction delays, design deficiencies, or environmental impacts, these risks are closely tracked (often quarterly) by all lenders and equity investors with a stake in the development of the infrastructure project. The successful delivery of infrastructure largely depends on effective project management and a professional construction team and can be affected by various physical climate hazards such as hurricanes or floods.
- **Operation:** This phase occurs after the development of the project is complete and operations have commenced. Investors often consider risks that may arise during the first few years of operations. During the operational phase, debt costs can be reduced through refinancing, during which initial investors release some capital and a wider range of investors are brought on. This allows investors who are less willing to invest in the preparation and development of a project-to-finance infrastructure once it is operational.

Each phase of project development has risks, some of which occur across phases and some of which are unique to a particular phase. In general, the earlier the stage of infrastructure investment, the greater the risk to the investor. Risks shared by all stages of a project include regulatory, legal, environmental, credit or counterparty, interest rate, and inflation risks. The preparation phase can also pose significant risks related to design and obtaining sufficient financing, while at the operational phase risks may relate to the supply of inputs, markets, or labor.

Climate change-related risks can manifest at all stages of infrastructure development. Although most infrastructure projects undergo an environmental impact assessment, climate risks are not yet consistently integrated into infrastructure planning. Fig. 7.2 illustrates how climate-driven physical and transition risks can be assessed alongside other regularly examined risks. Rather than operating in isolation, climate risks will interact with and potentially exacerbate risks that infrastructure lenders or developers are familiar with, such as regulatory risk and credit risk.

Because of high perceived and actual risks of early-stage infrastructure investment, project developers often leverage their own equity and rely on public funding to cover predevelopment costs, while structuring deals to finance the later construction process. DFIs and multilateral development banks (MDBs) with development missions most commonly fund this stage of a project as they are less deterred by the uncertainty of a project's feasibility (see below). However, some commercial banks with extensive in-house infrastructure project lending experience will also fund the preparation phase.

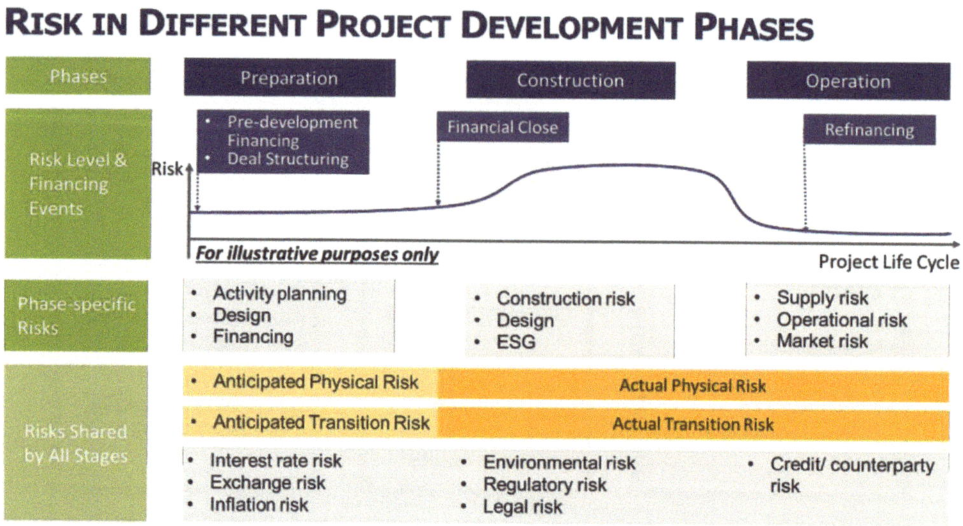

FIG. 7.2 Risk in different project development phases. (Credit: Climate Finance Advisors.)

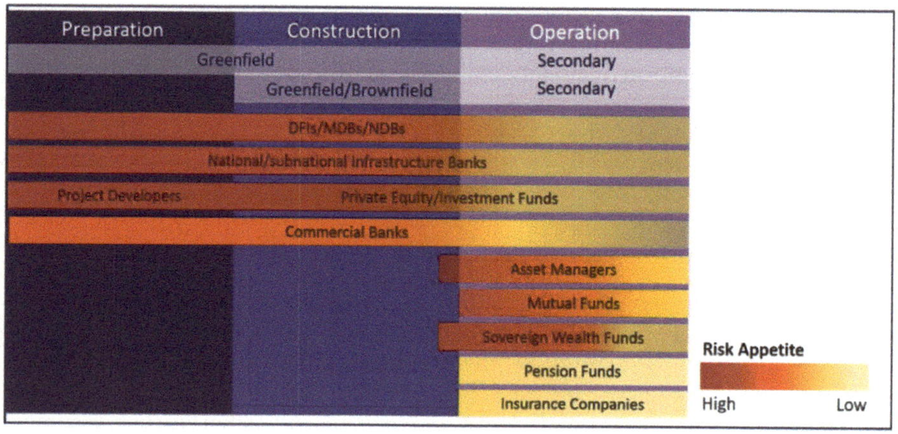

FIG. 7.3 Investment cycle. (Credit: Climate Finance Advisors.)

At the end of the preparation period, before initiating the construction process, the project developer must obtain additional capital from private equity, infrastructure investment funds, asset managers, mutual funds, and sovereign wealth funds.

Refinancing and regular project revenue streams help to attract investors with low risk appetite, particularly for projects at the operation stage (or nearly operation-ready). Although the operation stage presents the lowest risk and is widely pursued by risk-averse investors, climate change can exacerbate risks stemming from supply shortages, market demand changes, and daily operations leading to cash flow reductions and, in severe cases, negative cash flow. Fig. 7.3 illustrates where infrastructure investors, both banks and institutional investors, tend to invest along the project cycle.

Depending on which stage of a project investors are involved in, infrastructure investments can be categorized as greenfield, brownfield, and secondary. Greenfield refers to projects being constructed on an undeveloped site, which tend to have high risk exposure and consequently high returns. Brownfield projects typically involve an existing infrastructure asset that requires a significant retrofit or expansion. These types of projects usually have high potential for material growth and involve less risk than greenfield projects, as investors can avoid risks associated with permitting for undeveloped sites. Secondary infrastructure investment happens at the operational stage where the risks are lower, while returns are more moderate but steady.

Infrastructure banks (including commercial banks) and institutional investors are the two primary categories of private investors that invest in infrastructure. To understand how these institutions can help catalyze sustainable, climate-resilient infrastructure, and how they can integrate climate risk into their investment decisions, it is important to understand how these investors allocate their funding and at what point in the project life cycle this occurs.

Infrastructure Banks

Infrastructure banks, including DFIs, MDBs, national development banks (NDBs), national and subnational infrastructure banks, and commercial banks, are traditional mainstream infrastructure investors that largely use debt to finance infrastructure projects. However, infrastructure banks that are oriented to support public efforts, such as DFIs and NDBs, also provide grants for project preparation. Infrastructure banks fund both the project preparation phase and the development and construction of projects.

In many parts of the world, infrastructure, particularly in the development and construction phases, is financed by development banks. The capital that DFIs provide is critical and often takes the form of grants and technical assistance for project preparation as well as risk mitigation instruments such as guarantees, export credits, insurance, and debt for project financing. This funding plays a key role in mitigating risks perceived by private investors and can serve to "crowd-in" private capital. DFI's investment selection criteria also differentiate them from commercial banks.

Infrastructure banks' openness to funding the development and construction phases of projects highlights a risk appetite that is higher than other types of institutional investors who are more focused on operating and revenue-generating assets. The role of their investments in the infrastructure development cycle makes infrastructure banks vulnerable to physical climate risks that occur during the construction period and disrupt project construction or increase building costs, which may reduce revenue or increase operational costs in the long term. Moreover, physical risks can stretch funding capacity that is already limited, curbing infrastructure banks' ability to fund future projects.

Institutional Investors

Institutional investors include insurance companies, pension funds, infrastructure funds, sovereign wealth funds, mutual funds, and asset managers. As of May 2017, pension funds, insurance companies, and sovereign wealth funds together represented 77% of global infrastructure investment, with pension funds and insurance companies accounting for 55% and 20%, respectively [6].

Although coupling institutional investors' capital with the increasing need for infrastructure in developing countries sounds like a logical match, it is not happening naturally. Despite both the large and rapidly growing demand for infrastructure investment in emerging markets and institutional investors' outsize role as the global financier of infrastructure, such investors devoted only 0.67% of all private sector infrastructure investment worldwide to so-called emerging and "frontier" markets [7]. The low appetite from institutional investors is partially a result of a lack of well-structured projects and of risk and return profiles that are not aligned with their strategies [8]. Moreover, infrastructure projects in emerging markets tend to be greenfield projects, which expose investors to high construction risks.

Most institutional investors do not invest directly in project development or construction, and few support any type of project preparation for infrastructure projects. These investors largely invest in operating assets. The long time horizon of infrastructure investment and the ability of established infrastructure to generate steady cash flows over decades of operation match the long duration of these investors' liabilities and asset allocation strategies. Infrastructure investment at the operating stage is highly attractive to institutional investors because of the potential to generate attractive yields that exceed the returns of other fixed-income products. This stage of an infrastructure asset is also less sensitive to capital market volatility, construction challenges, and other risks that are more prevalent during the preparation, development, and construction phases.

Although operating infrastructure is considered an ideal asset for risk-averse institutional investors, exposure to physical risks across all phases of the infrastructure project cycle can lead to financial vulnerability, underperformance, and in some cases substantial financial losses, if not managed properly.

TRANSLATING PHYSICAL CLIMATE RISKS INTO INVESTMENT LIFE CYCLES

Physical climate-related risks and opportunities can have numerous financial impacts on infrastructure investors, affecting revenues, expenditures, assets, liabilities, and capital and financing. Although each of these risks may not undermine a project's financial viability, climate-related impacts can interact with other risks during an asset's operating lifetime, with implications for the investment's financial viability over loan tenors or investment horizons, weakening the case for new investments in the future.

- **Revenues**: Climate-driven risk events can disrupt business operations and affect income derived from business activities. For example, flooding due to hurricanes and extreme rainfall may limit travel and reduce the daily number of vehicles on toll roads, affecting their revenues. For infrastructure assets that depend on goods and services, such as municipal infrastructure including hospitals and recreation facilities, supply chain risks may be particularly threatening to operations and lead to decreased revenues.
- **Costs/Expenditures**: Restoring infrastructure to operating condition following damage from extreme weather events may increase unplanned maintenance expenditures. Although expenses to adapt infrastructure to climate change impacts may save money in the long term, these efforts may require unplanned operational and capital expenditures, particularly if a reactive (ex post) rather than proactive (ex ante) approach is taken. Although insurance will play an increasingly important role in asset protection, increasing premiums and stricter exclusionary clauses may make it increasingly difficult to obtain insurance at reasonable rates. Insurance premiums and uninsured risk event-related costs are both likely to increase for infrastructure assets exposed to climate hazards.
- **Assets**: Physical climate-related impacts may also negatively affect the value of tangible as well as intangible assets, such as reputation. Extreme weather events or temperature variability causing disruption to operations, service performance, and delivery of infrastructure services can lead to a decrease in overall asset value, especially on land and leasing contracts. Damage from extreme weather events can shorten asset life and increase depreciation rates, decreasing asset value. The increasing frequency of storms, for example, may undermine the brand value of infrastructure operators who experience frequent disruptions and cannot provide consistent service.
- **Liabilities:** Impacts related to physical climate risks may affect current and contingent liabilities of infrastructure projects. In response to physical climate risks, the evolution of regulations, technologies, and markets may increase capital expenses and the costs of supplies and production, which could lead to increased current liabilities. Changes to laws, case law, and regulations related to a company's preparedness for climate change may also increase the incidence or probability of contingent liabilities. Noncompliance with laws and regulations can result in contractual, civil, or criminal liabilities for the project owner, which may adversely affect cash flow or market capitalization due to increased costs and loss of reputation.
- **Capital and financing:** Physical climate-related risks can also affect long-term debt and equity capital. Increased capital and operational expenditures in response to more frequent and severe extreme events or chronic stresses, alongside lower cash flows and revenues may lead to greater debt. At the same time, the ability to raise debt, refinance debt, and attain adequate tenders may be affected by the altered operations and revenues. In the case of equity investments, lower cash flow may decrease project valuation, in turn lowering the attractiveness of the asset in capital raising or feasibility of timely exit by equity investors. Profitability ratios estimating future return on equity may decline if decreases in interim payments received (dividends) and the company's long-term market value are attributable, in part, to enduring climate change-related impacts.

MANAGING PHYSICAL CLIMATE-RELATED RISKS

Infrastructure investors can take concrete steps to manage physical climate risks. The foundation for these actions is assessment and quantification of potential impacts to understand where and how physical climate risks are likely to affect infrastructure investments. Understanding where risks are concentrated, the primary risk drivers, the potential consequences for investments, and asset-level resilience then provides a basis for identifying and implementing strategies that help to mitigate physical climate risks and build resilience to their impacts.

Risk assessment entails overlaying climate, economic, financial, and other relevant data to evaluate the exposure and vulnerability of infrastructure investments to a set of climate hazards and the potential value at risk. To understand their exposure to physical climate risks at the portfolio level, investors can conduct a climate risk screening that provides a view on where risk is concentrated and what the primary risk drivers are across their portfolio. Climate risk assessments that seek to inform potential investments or provide a more detailed understanding of asset-level risk exposure can explore how characteristics such as infrastructure type, age, and location interact with climate hazards and contribute to vulnerability. These more detailed analyses set the stage for the identification of risk mitigation or resilience-building strategies.

Understanding an infrastructure asset's exposure to physical climate risks is only a piece of the puzzle, as an asset's vulnerability or resilience to these risks determines if its exposure will lead to material financial impacts. Many factors influence an asset's resilience to climate impacts, including property management, financial stability, physical adaptations, and resilience of the surrounding community, as well as the resilience of other revenue-generating customers and clients. Institutional investors and infrastructure banks can leverage their understanding of risk exposure and infrastructure resilience to strategically construct portfolios, promote asset-level resilience-building, and support community-wide adaptation. Investors can also transform climate risks into investment opportunities focused on resilience.

Strategically Assessing Physical Climate Risks

Developing a new paradigm for infrastructure design and investment

Climate change necessitates that infrastructure design and investment consider the future, rather than base decisions on the past, as has been standard practice to date. The past is no longer a reliable indicator of what the future will bring, as storms increase in severity, extreme rainfall events become more frequent, sea levels rise, and heat waves become more prevalent. Thus, although actions such as upgrading building codes are important, simply building to a new standard—for instance, to withstand a 1-in-500-year flood event rather than a 1-in-100—may not be sufficient to address the impacts of changing climate risks.

Designing infrastructure for an uncertain climate future not only requires measures to mitigate the impacts of specific physical climate risks, but also to incorporate principles such as flexibility, modularity, and redundancy to enhance infrastructure resilience. The uncertainty and the variable nature of climate impacts demand that infrastructure is made to withstand fluctuating conditions. For instance, infrastructure may be built to serve dual purposes: in Kuala Lumpur, the Stormwater Management and Road Tunnel transitions from a tunnel to stormwater management asset to help the city endure flash floods [9]. Microgrids can serve as redundant and modular infrastructure systems that help critical municipal infrastructure, such as hospitals, to continue to function in the face of disasters that render grid-scale utilities inoperable.

Institutional investors and infrastructure banks must evaluate how an infrastructure project has assessed physical climate risks and incorporated relevant risk mitigation and/or resilience building measures into project design. Furthermore, investors need to determine whether the measures that have been implemented are sufficient to ensure longer-term investment viability and profitability. Investors must be equipped with the right questions and be prepared to conduct their own due diligence to properly assess whether and to what extent climate risks have been identified and mitigated.

Understanding risks "beyond the fence"

Infrastructure assets are typically part of broader infrastructure systems, which in turn are embedded in the communities and economies they serve [10]. Due to this interconnectedness, understanding how physical climate risks are likely to affect local communities and economies, and how local actors are responding to these risks, enables investors to understand the full extent of physical climate risk an investment faces. For example, an inland airport in a coastal city may not be directly affected by flood risks. However, if major arteries connecting the airport to key economic centers in the city and region are inundated, the airport's operations will be affected by the surrounding community's ability to manage stormwater, with implications for staff and travelers' ability to reach the airport. This, in turn, can affect operations and revenue generation. Thus, understanding the surrounding community's climate risk exposure and ability to manage climate risks is essential for assessing asset-level risk and resilience.

Evaluating risks that extend beyond a specific asset can be difficult for investors due to the "black box" nature of government and the many interacting components of resilience. Frameworks, such as the United Nations International Strategy for Disaster Reduction (UNISDR) scorecard discussed later, can help investors engage with the wider community and gain actionable information by asking relevant questions.

Assessing climate risk in infrastructure

To identify potential risk hotspots and inform appropriate next steps for managing climate risks, it is critical for institutional investors and infrastructure banks to identify material risks, understand the degree to which their investments are exposed and vulnerable to these risks, and make meaningful comparisons across a portfolio. Research on the materiality of climate hazards for different sectors, physical climate risk assessment tools, and good practices for the disclosure of climate risks can help guide risk assessments and inform the determination of risk management priorities.

An initiative by the European Bank for Reconstruction and Development (EBRD) and the Global Centre of Excellence on Climate Adaptation [11,12] convened industry leaders from asset management firms, insurers, banks, and corporations to compile actionable recommendations for physical climate risk and opportunity disclosure. The resulting paper includes a sensitivity matrix by industry (see Table 7.1), providing an indication of which hazards present material risks to the industries considered.

For example, industrials including transport infrastructure are highly susceptible to adverse impacts from storms, extreme rainfall and floods, and sea-level rise, and face medium sensitivity to extreme heat. For instance, extreme heat can disrupt operations at airports by preventing airplanes from achieving adequate lift and taking off efficiently in the hotter, less dense air. At Sky Harbor International Airport in Phoenix, Arizona, several days of temperatures exceeding 120°F (49°C) caused service disruptions that decreased airport revenues and resulted in additional costs to operate airport facilities to accommodate customers from delayed flights [13].

Understanding how material physical climate risks will manifest through both direct impacts on infrastructure assets and indirect impacts on economies, people, and ecosystems that are situated beyond an asset's boundaries is critical for managing financial risk. Infrastructure assets often have lifetimes of 20 to 50 years, and they can extend to 100 years. To assess direct physical climate risks in the near-term—from 5 to 20 years—investors can use climate projections. Recent changes in infrastructure performance due to extreme weather events or climate-driven stresses over the past 5 to 10 years can provide an indication to investors of the potential financial impacts of climate change on the asset [13]. Recent impacts to look for include lower revenues, greater operational expenditures, and higher capital expenditures during or after extreme events.

To explore longer-term direct risks as well as indirect risks, investors can employ scenario analysis, which accounts for the uncertainty of long-term climate and socioeconomic conditions due to the changing landscape of global climate policy and other macroeconomic factors. In addition, increasingly sophisticated climate risk assessment tools for financial institutions, including those applicable to infrastructure projects, are now being developed by the private sector. An Institute for Climate Economics (I4CE) ClimINVEST research paper released in December 2018 explores climate risk assessment approaches for financial institutions developed by leading climate risk and resilience firms, examining a range of methodologies and dimensions of risk exposure [14]. The I4CE analysis identifies a range of risks including systemic and idiosyncratic, various counterparty risks, and those driven by different time horizons, among others that can be explored to assess a project's vulnerability.

Investors can also leverage tools that have been created by DFIs to screen public sector finance projects for physical climate risks and assess how these risks might affect their investments. The World Bank has developed a set of interactive tools to support climate risk screening at the project level for coastal flood protection, energy, water, roads, agriculture, and health sectors, as well as general development [15]. It enables project developers and reviewers to (1) determine whether an infrastructure or development project is subject to climate hazards such as extreme temperature, extreme rainfall and flooding, sea-level rise and storm surge, strong winds, and earthquakes, (2) assess the historical and potential projected impacts of each hazard on the project's physical (e.g., physical work related to infrastructure) and nonphysical (e.g., emergency and project management protocols) components, and (3) evaluate the extent to which the infrastructure system and its development context are likely to mitigate or exacerbate physical climate risks. This information provides the basis for users' rating of a project's physical climate risk. The tool also provides high-level recommendations for next steps, largely focused on conducting more detailed climate risk assessments.

Credit rating agencies are also exploring ways to include forward-looking climate risks as an element of the environmental, social, and governance criteria they examine when determining ratings for corporations, including utilities, and for municipalities. Rating agencies already incorporate past occurrences of extreme events into their ratings. For instance, in November 2018, PG&E Corp.'s credit rating was downgraded by Moody's to the second-lowest level of investment grade, due to $10 billion of exposure to California's 2017 wildfires and uncertainty regarding its liability in the 2018 California Camp Fire [16]. Credit agencies' frameworks for integrating climate-related risks into their corporate and municipal ratings help to illustrate how climate hazards may translate to credit risk through lower revenues, higher costs, impaired assets, higher liabilities, and increased debt [17]. By focusing on the hazards known to pose material risks to the asset's sector, leveraging available risk assessment tools, and seeking informative disclosure

TABLE 7.1
Climate Sensitivity Matrix, by Industry

GICS Sector	GICS Industry Group	Storms and Cyclones	Extreme Rainfall and Flood	Extreme Heat	Rainfall Variability	Temperature Variability	Water Stress	Sea Level Rise	Other climate Hazards
Energy	Energy	High	High	High	Medium	High	Medium	High	Ice melt, permafrost melt
Healthcare	Healthcare equipment and services	High	High	High	Low	Low	Medium	High	Wildfires, humidity, degraded air quality
Industrials	Transport	High	High	Medium	Low	Low	Low	High	Permafrost melt, ice melt
Information Technology	Technology hardware and equipment	High	High	High	Low	High	Medium	High	
Real estate	Real estate	High	High	Low	Low	Low	Low	High	
Telecommunication services	Telecommunication services	High	High	Low	Low	Low	Low	High	
Utilities	Utilities	High	High	High	High	High	High	High	Wildfires

Credit: Four Twenty Seven, from E. Mazzacurati, J. Firth, S. Venturini, N. Ambrosio, F. Freitas, L. Ross, C. Foubet, R. Hamaker-Taylor: Advancing TCFD guidance on physical climate risks and opportunities. European Bank for Reconstruction and Development and the Global Center of Excellence on Climate Adaptation, 2018 [11].

from infrastructure asset managers and owners, investors can gain insights to inform management of risk exposure across their portfolio and to screen potential investments.

Assessing climate resilience in infrastructure

Understanding an asset's or portfolio's exposure to climate risk is an essential first step in effective risk management. However, whether an asset will incur material financial damage from climate risk depends on the asset's resilience. Climate resilience is affected by many factors ranging from property management, on- or off-site fortifications against hazards such as floods and storms, diversified energy sources, and site access for employees and customers. Approaches to evaluating infrastructure resilience range from industry self-reporting to frameworks for understanding systemic resilience and approaches for quantifying resilience benefits, several of which are explored later. Although no approach provides a complete picture on its own, investors can leverage the ideas and indicators from these and other relevant frameworks to assess the resilience of potential investments and strive to build resilience in their portfolios.

GRESB [18] assesses the environmental, social, and governance (ESG) elements of real estate and infrastructure, and in 2018 the group released a Resilience Module. It strives to respond to the increasing demand by investors for transparent data and to improve knowledge sharing around strategies to build resilience in real estate and infrastructure [19]. The 10 indicators in this module are divided into four sections including leadership and team; resilience assessment; management objectives and strategies; and implementation and improvement. Questions focus on which members of the team are responsible for resilience, stakeholder engagement around resilience, vulnerability assessment, implemented resilience strategies, specific actions to develop resilience, and concrete responses taken during extreme events occurring within the reporting period [19]. Although the GRESB reporting results can provide insight into the resilience of specific assets, the framework also provides a lens for investors to tailor to their own knowledge-gathering efforts.

As noted earlier, UNISDR created a scorecard for city governments to assess their disaster resilience, structured around 10 "essentials." Increasing infrastructure resilience is 1 of the 10 essentials, and the publicly available assessment includes nine preliminary questions for city leaders to assess the resilience of their public infrastructure. The assessment strives to capture whether there would be significant loss of service across a city in the case of an extreme event and if areas surrounding energy and transport infrastructure, for example, would remain safe in the case of failure. Although the target audience is the public sector and the scope is municipal level, investors can utilize these indicators to vet potential infrastructure project investments for their resilience and to understand where an asset is situated in the larger infrastructure system. Questions cover protective infrastructure, water and sanitation, energy, transport, communications, education, first responders, and healthcare.

To support the quantification of the benefits derived from projects that explicitly incorporate climate resilience considerations, the EBRD [20] has developed a set of resilience results metrics. The metrics are two-tiered, accounting for both physical and financial elements of projects. Under this approach, physical outcomes (e.g., reductions in energy consumed) are translated into financial impacts (e.g., dollars saved) across the life of an infrastructure asset to determine the climate resilience benefits.

As regulatory and market pressures propel increasing physical climate risk disclosure, investors will have more access to self-reported data on resilience, such as that provided by the GRESB reporting module. Investors can also leverage publicly accessible resilience frameworks, such as the UNISDR scorecard and the EBRD metrics, to guide their own screening and engagement processes and inform the quantification of benefits. By building on these approaches, investors can learn to ask pertinent questions related to management, planning, assessment, and action and to leverage analyses that support more climate-resilient investment decisions.

Translating Physical Climate Risks into Opportunities

Investing in resilient infrastructure

Having a thorough picture of an asset's exposure and resilience to physical climate risks lays the groundwork for physical or financial climate risk management strategies that actively build resilience. Physical measures include design adjustments that mitigate the impacts of acute or chronic climate risks and vary in complexity and cost. Low-cost measures, such as better insulation and windows to protect against the impacts of extreme temperatures or building critical elements at higher elevations in flood-prone areas, can be promoted as new standards for infrastructure construction. More complex and costly investments should be based on an understanding of which risks are most relevant for a given area. For example, building with fireproof materials can be expensive but important in fire-prone areas, while water recycling and efficiency measures can be critical for water management in water stressed regions [21].

It is important to consider opportunities for both existing and new infrastructure projects outside of traditional hard infrastructure, leveraging the benefits of gray infrastructure, green infrastructure, and nature-based design. For example, to build resilience to urban flooding, bioswales, retention ponds, green roofs, and permeable paving can be integrated into the development of new urban infrastructure or built as standalone resilience projects. Replenishing wetlands and investing in living beaches can build resilience to coastal storms and sea-level rise, supporting an ecosystem that can serve as a natural flood barrier while giving the water somewhere to go. A study by the Nature Conservancy in Howard Beach, Queens in New York, examined the impact of using different types of defenses for coastal storms, including nature-based, gray infrastructure, and hybrid defenses. It found that hybrid defenses leveraging both nature-based and gray infrastructure could lead to up to $244 million of avoided losses in the current 1-in-100-year storm event, highlighting the importance of considering all types of potential infrastructure adaptation [22].

Investing in resilient infrastructure also calls for reducing the potential financial impacts that physical climate risks pose to an investment and effectively managing the remaining risks. This calls for pricing risks accurately, assigning them to the party best situated to manage them based on an understanding of parties' capacity to shoulder risks, and more explicitly defining risks that qualify as events for which a contractor is financially responsible versus those that are unforeseeable. Insurance requirements may also be expanded to include different or higher coverage of physical climate risks and ensure that infrastructure providers have adequate coverage to protect them from relevant risks [23].

Obtaining resilience dividends

The Rockefeller Foundation describes the resilience dividend as the net social, economic, and environmental benefits that are derived in the near- and long-term from forward-looking investments that focus on better service delivery, more strategic resource use and response to multiple challenges [24]. In the face of a changing climate, resilient infrastructure may sustain less damage and destruction, resulting in fewer operational disruptions. This can help to limit outlays for repairs and result in longer asset life and more sustainable revenues. Benefits accrue to investors and asset owners as well as to the communities, whose physical well-being and economic well-being rely on these infrastructure assets, enabling both to thrive.

Leveraging engagement to build shared resilience

Engagement is a powerful tool that can benefit both investors and the infrastructure assets in which they invest by fostering long-term sustainability to increase competitiveness, protecting the value chain, improving reputation, and capitalizing on opportunities innovatively to address climate risks [25]. Investors are in a unique position to engage with actors involved in prospective and current infrastructure investments to understand their exposure to climate risks. They can also leverage their position as financers or lenders to promote climate resilience strategies in processes specific to asset development and management as well as within the broader community or region.

Infrastructure investors can foster resilience by engaging with asset managers and communities around planning for the duration of an asset's life cycle. Engagement with asset developers and managers can help investors to evaluate whether they have assessed an asset's exposure and vulnerability to climate risks and adequately accounted for changing climate conditions through design adjustments and other measures that enhance climate resilience. Questions may include the following: Has asset location been informed by an understanding of projected climate conditions? What types of measures have been incorporated into design to mitigate the impacts of relevant acute and chronic climate hazards? How have principles such as flexibility, modularity, and redundancy been accounted for in asset design?

Engagement with communities helps investors to understand how critical regional systems (e.g., energy networks) are likely to experience damage, changes, or shifts in customer behavior. It also enables investors to assess how well positioned a community is to manage changing climate risks, which is an important component of asset risk. Questions in flood-prone areas, for example, may include the following: How has existing stormwater infrastructure endured past floods? What types of adjustments are being made to promote this infrastructure's ability to manage the impacts of more frequent and intense rainfall events? In addition to providing greater transparency, these conversations can yield insights into how investors can foster climate-resilient practices within their own assets to advance shared climate resilience priorities.

Staying ahead of a shifting regulatory landscape

Voluntary and regulatory frameworks are emerging that impact or evaluate infrastructure projects, encompassing both climate risk disclosure as well as risk assessment and management. The Task Force on Climate-related Financial Disclosures recommended that investors and corporations disclose the actual and potential material impacts of climate-related risks and opportunities on their businesses, strategy, and financial planning, including both transition and physical climate risk [26]. Regulations for climate risk disclosure are emerging in different geographies and investors have the opportunity to differentiate themselves by staying on the leading edge of climate risk disclosure.

Europe is leading the way in establishing regulation around financial risk disclosure, which will both improve investors' understanding of potential risks and provide increased incentive for investors to share their own meaningful disclosures. France's landmark law on Energy Transition and Green Growth and its Article 173 passed in August 2015 requires that publicly traded companies, banks and credit providers, asset managers, and institutional investors disclose physical climate risks [27]. The European Commission's Action Plan: Financing Sustainable Growth, released in early 2018, establishes a regulatory framework to support the goals of the Paris Agreement and sets a 2-year timeline for implementation [28]. This framework is likely to drive change in financial markets globally and set standards on reporting, disclosures, and infrastructure resilience. Later in 2018, the UK Parliamentary Environment Audit Committee recommended that Britain requires large companies, pension funds, and other large investors to disclose climate risks based on scenario analysis by 2022 [29]. Proactively incorporating physical climate risks into investments can help infrastructure investors anticipate the changing landscape of regulatory requirements.

CASE STUDIES

Port of Durban—Lessons Learned from Past Losses

Setting the scene: South Africa's port system and the Port of Durban

The Port of Durban is one of eight commercial ports in South Africa, managed by the state-owned Transnet National Ports Authority (TNPA) [30]. Located on South Africa's East coast, the Port of Durban is the largest handler of containers in the country, handling 65% of containerized cargo for South Africa and 60% of all the nation's imports and exports. The Port contributed 54% to TNPA's overall revenue or R4.9 billion during 2017 [31]. It includes five business terminals operated by the Transnet Port Terminals, including the Durban container terminal, Pier 1 container terminal, multipurpose container terminal, Durban car terminal, and Maydon Wharf terminal [32]. Transnet National Port Authority has 6200 employees at Durban Port and is expected to reach over 9000 by 2018/19. Over 30,000 people work for the port indirectly, making it an important regional employer [33].

The TNPA is one of four operating divisions within Transnet SOC Ltd., a state-owned enterprise owned by the South African government and operating as a corporation. The other four operating divisions are Transnet Freight Rail, Transnet Rail Engineering, Transnet Port Terminals, and Transnet Pipelines. The TNPA serves as the landlord of South Africa's ports, while the Transnet Port Terminals manage port and cargo terminal operations [34].

The Port of Durban is a marine port with facilities along the coast that provide equipment for ships to dock and transfer passengers and products. Its infrastructure includes piers, basins, stacking or storage areas, warehouses, and equipment, such as cranes. This infrastructure requires large amounts of capital investment and is vulnerable to acute and chronic climate hazards [35].

Climate change impacts on revenues

Revenues brought in by marine ports can be affected by acute climate hazards as well as chronic stresses. Acute hazards, such as extreme heat events and extreme rainfall, can compromise asphalt integrity or flood roads, hindering access to ports and potentially slowing business. Similarly, these events can also damage port infrastructure, which may disrupt operations and decrease revenues.

For example, as the climate changes, South Africa's eastern coast is expected to have more severe extreme rainfall events and increasing storm surge, along with rising seas [36]. Key infrastructure elements exposed to climate hazards at the Port of Durban are the entrance navigation channel, berths, and equipment used in shipping operations. According to a 2011 presentation by the eThekwini Municipality and TNPA, changing wave direction and energy, along with increased sedimentation are expected to have the greatest impact on the entrance channel and berths [36].

These factors, along with more severe storms, are expected to affect daily shipping operations. Changes in rainfall and wave direction may lead to increased sedimentation in ports and affect sediment transport rates, which can impact channel efficiency. Both extreme rainfall events and enduring drought may affect South

Africa's commodity supply and demand, which may in turn impact port revenues. For example, Transnet cites drought as a leading cause of reduced grain export, although increased demand for grain imports managed to partially outweigh this impact [37].

In October 2017, a supercell storm caused widespread destruction at the Port of Durban, blowing a container ship across the harbor mouth and initiating a 2-day closure at Durban Container Terminal and Maydon Wharf. Damaged ship-to-shore cranes, derailed gantries, a straddle carrier being blown into the bay and a workshop made inoperable by flooding were some of the damages incurred [38]. Disrupted revenue was estimated at $3.38 million, and congestion and delays persisted at the port for 12 days, with port and stack occupancy up to 90% [38].

Climate change impacts on costs

Acute and chronic climate hazards also increase operating costs at marine ports such as the Port of Durban. Damage caused to asphalt or port infrastructure may lead to increased maintenance and repair costs. Similarly, fines may be incurred if operating standards are violated due to failure to properly repair ports after storms. If chronic temperature rise leads to increased port traffic due to ice melt, ports will likely see an increase in daily operating costs alongside an increase in business. Chronic flooding from sea-level rise may necessitate that port owners incur the costs of raising the ports.

Transnet's 2017 Annual Report says that climate change-induced weather patterns including storm surges, rising seas and inland flooding may affect port and rail infrastructure [39]. These risks represent the potential for increased operating costs for Transnet. For example, while revenue grew in 2018 due to increased revenue from automotive and break-bulk sectors, Transnet's 2018 Port Terminal report states that operating expenses grew by 11.8% in 2018 compared to 2017, partially driven by repair expenses at Durban after this storm [40]. October's storms are also mentioned as a driver behind the declines in average moves per gross crane hours, which missed targets across terminals and likely led to increased operating costs. Generally deteriorating weather conditions are also mentioned as a reason for the decrease in crane efficiency.

Climate change impacts on assets and liabilities

Asset value is likely to decrease if ports are repeatedly damaged or disrupted by extreme weather events. A port's reputation may also be at risk if it incurs frequent operational disruptions that lead to negative ratings by vessel operators and clients. For example, during the October 2017 storm, shipping line MSC issued a General Rate Increase of $100 per container, and the Transnet Port Terminals received negative feedback from other shipping lines. Similar delays were already experienced at the port during adverse weather conditions in July and August, likely exacerbating the impact of this event and further threatening the port's reputation [38].

In contrast, ports with resilient infrastructure that has been adapted to climate change impacts may see an increase in asset value. If Transnet leverages lessons learned to raise capital for adaptation projects, the Port of Durban could see an increased reputation and higher value over time.

Climate change impacts on capital and financing

Recurring damage, disrupted operations, and higher costs from increasingly frequent and severe rainfall and flooding events could affect Transnet's debt, capital, and reserves. This in turn can affect its ability to repay loans or attract new capital investment. Furthermore, public assets such as Trasnet's are particularly affected by the macroeconomic characteristics of the surrounding community, as climate factors interact with other social, economic and political factors. For example, enduring drought can decrease farmers' yields, leading to increased migration to cities and exacerbated social inequity and income gaps, which can strain national economies.

Investors have expressed concern over policy uncertainty affecting Transnet, which is coupled with strained socioeconomic conditions in South Africa. Transnet's inextricable connection to the national government means that national challenges exacerbated by climate change will also threaten Transnet's financial viability and ability to raise capital. Although institutional investors, infrastructure banks, and development agencies funding specific infrastructure projects will want to ask questions of the climate impacts for those projects, an initial screening will indicate that a specific asset's risk is also related to the greater enterprise to which it belongs and socioeconomic context in which it sits.

Risk management

Transnet's 2017 Port Terminals report cites the top five risks for the Port Terminals, including inability to attract and sustain additional volumes, capital affordability risk, deterioration in the macroeconomic environment, operational efficiency risk, and shortage of water and energy. Each of these risks can be exacerbated by climate change impacts on commodities, trade routes, and port infrastructure. However, the listed risk mitigation measures do not explicitly mention climate adaptation except those addressing water shortages, which include updating the Transnet Port Terminals' energy plan, obtaining standby generators, installing water meters, repairing leaks, refurbishing the water recycling plant in Richards Bay, building a desalination plant, and exploring options for loading woodchips with no water [37].

The disruption caused by the October 2017 storm initiated intensive reviews of the port's emergency planning and operational leadership training. The port was also slated to receive 29 additional haulers and 23 additional straddles because this event highlighted that the old straddles break down and create a weak link in a system that must work uniformly. Transnet Port Terminals also plan to continue improving fleet utilization and replacement efforts, fostering flexible human capital to meet demand [38]. Acting upon lessons-learned during past extreme weather events can help to build resilience and mitigate future financial loss.

Transnet's integrated report cites climate, energy, and water-related risks as components of its risk management, and members of its team participate in the Industry Task Team on Climate Change, indicating that the enterprise is aware of climate risk. Tracking progress made by these efforts would help investors to monitor the risk to their particular investment over time. The South Africa Department of Public Enterprise released a Climate Change Policy Framework for State-Owned Companies in 2011, stating that they must begin integrating climate change in planning and operations, which indicates that the Transnet Port Terminals should be including climate risk in their planning [41]. However, it is unclear if physical risk is prioritized alongside mitigation efforts. In addition, in 2011, the presentation by the eThekWini Municipality and the TNPA shared the expected climate impacts on South Africa's ports, reporting that risk assessments at the ports and design standards for new ports were being implemented [36]. These are valuable first steps to building resilience to climate change. However, the results of these efforts are not easily accessible and potential lenders would want to thoroughly understand progress made to date.

Implications for investors

Although some infrastructure projects are specific to ports, or even the Port of Durban, loans are taken out by Transnet, rather than its operating divisions. For example, in May 2018, BRICS Bank loaned Transnet R2.7b ($200m) for container terminal work at Durban [42]. Thus, investors funding specific infrastructure projects would benefit from a thorough understanding of the ways in which climate risks may manifest for the larger entity taking out the loan. Transnet borrowed R10.8 billion in 2017 with no government guarantees, according to its 2017 financial report. This funding came from commercial paper, TN25, TN30, and TN40 bonds, bank loans, development agencies, and export credit agencies [43]. In June 2017, Moody's downgraded Transnet's senior unsecured rating to the lowest investment grade level, Baa3 and issued a negative outlook for Transnet's other ratings [44]. Around the same time, Standard and Poor's (S&P) lowered Transnet's foreign currency rating from BBB- to BB+, maintaining standalone credit at bbb [45]. These downgrades were due to the national government's poor credit outlook and Transnet's close connection to the government. It is important to evaluate climate risks with this macroeconomic context in mind. As credit ratings agencies increasingly integrate forward-looking climate risks into their ratings, climate impacts may also play a more direct role in ratings of highly exposed entities, which can help investors view climate-driven factors alongside other considerations that influence an entity's ability to repay debt.

Identifying mention of climate risk and resilience in Transnet's annual reports and thinking about how identified risks may affect an investor's priority projects, such as port improvements, is a valuable step in the initial investment screening process. First, Transnet's participation in climate initiatives indicates that it is aware that climate risks are important to address. However, understanding the depth and breadth of Transnet's approach to climate risks is essential for identifying how risks might translate to material impacts for investors. For example, if Transnet is beginning to budget in extreme weather-driven revenue losses and increases in operating costs, it is more likely to maintain sufficient capital reserves.

Transnet's integrated report cites climate adaptation as an external variable affecting its businesses, alongside social inequality, slow economic growth, unemployment, volatile commodity prices, energy and water shortages, and South Africa's creditworthiness, which are all impacted by climate change [39]. Understanding how Transnet is addressing these risks and how they will be applicable to a project's timeline and asset's life cycle is essential. When evaluating the Transnet Port Terminal's general risk mitigation and development strategies, investors can add a resilience perspective by asking how these strategies support continued operations in the long term and how they may be affected by climate impacts.

For example, despite the widespread impacts during the October 2017 storms, the Port of Durban did recover to surpass expected volumes by the end of the year. Transnet's Port Terminals' plans for 2019 include efforts to improve operations and supply chain efficiency to improve capacity in the bulk sectors, provide additional manganese transport channels, and improve service delivery in the Durban Container precinct. The ports' ability to achieve these goals will be affected by climate change and approaching this strategically provides an opportunity to integrate climate adaptation into plans and budgets.

A potential lender for a loan slated to support these projects may want to ask questions around planned investments to improve capacity. Is new equipment going to improve operations during extreme rain events or is it likely to incur costly damage and exacerbate congestion? Are their opportunities to invest in improved drainage or other explicit adaptation? How will channel improvements be affected by increased sedimentation? The port plans to partner with the private sector to increase maize and wheat volumes at the Port of Durban. How will climate change affect this project over the course of several decades? Are there opportunities for leveraging this public—private partnership to adapt agricultural practices to support both farmers and the TNPA? Potential investors can support resilience building when pushing infrastructure managers to consider all elements of climate risk and how it interacts with project priorities.

San Diego Airport—Embracing Opportunities in Climate Resilience

Setting the Scene: San Diego International Airport

The San Diego International Airport (SAN) is on the San Diego Bay in Southern California. SAN is run by the San Diego County Regional Airport Authority, which is governed by a nine-member board along with three ex officio members. The Authority itself employees about 400 aviation professionals [46].

The San Diego Airport relies on an array of infrastructure including airside infrastructure such as runways, airfields, taxiways, gates, and airbridges; landside infrastructure such as terminals, parking facilities, baggage claim facilities; security and customs infrastructure; and additional airport facilities such as airplane maintenance buildings. Each of these is important for efficient airport operations and revenue generation and can be affected by both acute climate hazards and chronic stresses.

Climate change impacts on revenues

SAN's 2017 Financial Report acknowledges that its 661 acres of waterfront property are sensitive and require significant water and energy that must be carefully managed [47]. Chronic climate stresses such as increasing average temperatures and water stress can lead to recurring revenue losses unless an airport invests in adaptation. Revenues from associated services such as retail stores, rental car operations and land rentals from hotels may be affected by increasing temperatures and water stress, as consumer behavior shifts. San Diego is threatened by enduring water stress and faces longer-term risks from sea-level rise. If the greater San Diego area experiences regular flooding in the coming decades, SAN revenues could be impacted by its employees' and travelers' inability to access its facilities.

Climate change impacts on costs

Chronic water stress may lead to increased water restrictions or use fees for airports or require costly upgrades in water-saving infrastructure. Similarly, in the long term, chronic flooding from sea-level rise may necessitate costly updates to drainage systems, runway surfaces, and connecting infrastructure.

Airports in arid climates use large amounts of water for cooling and irrigation and pavement power washing. Water-saving adaptation measures may increase costs in the short term but will likely reduce operating costs over time. SAN's 2017 sustainability report cites recent investment in building resilience to water stress, including the stormwater collection and reuse system that were commissioned with the new parking plaza to be used in the central utility plant's cooling towers. It also cited installation plans for wireless sensors at connection tanks used for air condition units in some passenger boarding bridges to monitor volume, rate, and condensation generation [48]. SAN also planned to install a battery energy storage system to reduce peak energy demand and costs, which will also build resilience to heat stress.

Although these investments can be costly, SAN has been able to raise funds to invest in adaptation, which can alleviate some of the financial burden. For example, the airport authority has received grants from the Federal Aviation Administration (FAA) to integrate sustainability into long-term planning, with climate resilience explicitly listed in the budget as a component of this goal [49]. By investing in resilience now, SAN is likely to be saving money over the duration of its facilities' life cycles.

Climate change impacts on assets and liabilities

Airport facilities that incur repeated damage and operational disruptions from extreme weather events and chronic stresses will likely see a decline in asset value, while those that invest in proactive and resilient infrastructure may see increasing asset value. An airport's reputation and traveler and airline preferences are likely to shift if it becomes known for recurring delays and operational disruptions or heavy resource use.

Water stress can be an important contributor to reputation risk, particularly for large, publicly visible entities that are known to use a lot of water. However, by investing in water-saving technology and participating in regional initiatives to build climate resilience, SAN is likely improving its reputation.

Climate change impacts on capital and finance

SAN's Capital Improvement Program (CIP) provides improvements and additional assets on a 5-year basis. In 2017, the CIP projects extended through 2022 and included $424.3 million for airside projects, $422 million for landside and ancillary projects, $140.3 million for terminal projects, and $45.6 million for administrative projects [47]. All of these projects can be impacted by climate hazards. Funding for CIP projects comes from airport revenues and facility charges, the FAA's Airport Improvement Program, Transportation Security Administration grants, airport revenue bonds, special facility bonds, and short-term borrowing with commercial paper or revolving lines of credit [47].

Obtaining loans and selling bonds can be made more difficult if an airport experiences recurring operational disruptions, costly infrastructure damage, or consumer backlash based on an airport's water use, for example, which can all weaken an airport's financial standing. On the other hand, by demonstrating the resilience benefits of prospective projects, SAN has the opportunity to attract additional funding for climate adaptation.

SAN has favorable credit ratings, with Moody's rating its general airport revenue bonds senior debt as A1 and subordinate debt as A2 in 2018 and S&P providing A+ and A ratings for those bonds in 2018. Bonds contribute largely to SAN's infrastructure projects. In 2017, the Airport Authority issued $291,210,000 series A and B Subordinate Airport Revenue Bonds to finance capital improvements such as the Parking Plaza and Federal Inspection Station facility [47]. Retail investors purchased $4 million and 80 institutional investors bought $287 million. Bonds contributed the bulk of funding to SAN's 2013 terminal 2 improvement projects, along with Authority cash and the Passenger Facility Charge, paid by airlines [50]. Each of these projects integrated water-saving technology, addressing SAN's most prominent climate risk and potentially increasing investor confidence.

Risk management

In November 2017, SAN signed the "Airports Sustainability Declaration," supporting climate action in line with the United Nations Sustainable Development Goals. In 2015, the airport authority began developing a Water Stewardship Plan to address the growing risks of drought and stormwater management. The plan outlines the airport's goal of having a closed-loop water system [51].

SAN's new $316 million Rental Car Center was opened in January 2016 and received the Leadership in Energy and Environmental Design (LEED) Gold certification by the US Green Building Council. In addition to fostering energy efficiency, which can help maintain operations during extreme heat events, the facility includes water collection and reuse in its car wash bays, reclaiming 85% to 90% of water used by the facility. Its landscaping includes drought tolerant plants to further reduce water use [47]. Although potential lenders to this project may have flagged San Diego's water stress and the water and energy intensity of rental car facilities as a risk to the asset's long-term financial sustainability, SAN's plans for integrating resilience to water stress may have addressed lenders' concerns.

The San Diego Airport Authority participates in the San Diego Regional Climate Collaborative, which is a knowledge and resource sharing network to prepare for regional climate change impacts [52]. In 2010, the San Diego Foundation and the Tijuana National Estuarine Research Reserve joined a project with ICLEI (Local Governments for Sustainability) to assess vulnerability to sea-level rise and create an adaptation strategy in the San Diego Bay [53]. The San Diego County Regional Airport authority served on the project steering committee and aviation services was 1 of the 13 analyzed sectors. The project's risk assessment found that SAN faces severe inundation risks by 2100 with regular flooding of the main access road and flooding of the airfield during extreme events. By 2050, localized flooding and drainage backups may occur even though inundation from the coast is unlikely.

The project suggested that SAN leverage sea level rise flood scenarios and consider different building site options, as well as diversifying airport access beyond a few main roads that are likely to become more frequently inundated. Participating in this regional initiative helped SAN identify its vulnerabilities as part of the broader region, define itself as a collaborative stakeholder in resilience efforts, and communicate more clearly on what does and does not yet pose a risk. It would also be valuable to the airport to consider the surrounding community's vulnerability to sea-level rise and how this may affect local and regional infrastructure and the ability of travelers and employees to reach the airport.

Implications for investors

Upon initial investment screening, an investor is likely to notice SAN's clear recognition of climate risk. During the due diligence and in-depth review of risk mitigation strategies, investors will want to ask questions about how SAN is implementing climate risk planning in practice. Examining the results of SAN's risk assessments to date provides an understanding of the risks the airport is focused on. For example, the results of the sea-level rise project showed that the airport has both minimal short-term risk from sea-level rise and more severe long-term risks from sea-level rise. This knowledge can inform investors' understanding of the risks of potential infrastructure projects over the lending period and the asset's life cycle. It can also guide questions to SAN on its long-term resilience strategy.

Understanding the climate resilience of past airport projects provides an indication of SAN's infrastructure planning. SAN's Green Build Project, a terminal 2 renovation, earned the Institute for Sustainable Infrastructure's Envision rating system's platinum award. The rating considers quality of life, leadership, natural world, resource allocation, and climate and risk in the planning and construction of infrastructure projects. In addition to building on a former landfill to preserve greenfields, while eliminating the permitting risks of greenfields, the project included intensive stormwater management practices, including monitoring for spills, leaks, and water quality [54].

Potential funders of infrastructure improvement projects at SAN will likely find mention of environmental and sustainability concerns at a quick glance. SAN's participation in climate change initiatives and the sustainability accolades of its past projects provide positive indication that climate change is on the radar. It also suggests that SAN's infrastructure projects could provide opportunities for investors to invest in climate resilience.

However, understanding exactly how climate hazards are integrated into infrastructure planning is important to understand the long-term resilience of potential projects. Potential investors may ask if the airport has considered other climate hazards that may become more severe over the course of the century, including sea-level rise and heat stress. Is the Airport Authority using current climate change projections to assess their risk and developing infrastructure with these risks in mind? How might traveler needs and preferences shift and place new demands on airport facilities?

In summer 2018, SAN released a draft Environmental Impact Report for public review for a $3 billion airport development plan to replace its terminal 1 [55]. The project's expected timeline is for state and federal environmental review in 2018 and 2019, with construction beginning in 2020 and the terminal replacement opening in 2023. Phase 1 project elements include 30 gates with more seating area, shops, restaurants and a security checkpoint as well as a new on-airport entry road, new parking, divided arriving and departing curb-fronts, and taxiway improvements [55]. Funding is not yet secured, though user fees are expected to make up the primary funds. Potential lenders may want to understand how the new entry-road is prepared for incremental increases in flooding from gradual sea-level rise. How does it connect to the roads of the surrounding area? Do the plans include updated drainage systems or will it be cost-effective to make improvements throughout the asset's life cycle if risks develop? Although the project was criticized for its lack of a public transit connection and contribution to traffic, the Draft Environmental Impact Report reviewed climate change among other environmental considerations and the project includes LEED certification and stormwater capture and reuse [56].

SAN largely presents its climate change considerations as sustainability and stewardship practices. Although these are important values and can lead to efficient risk mitigation along with sustainability cobenefits, it is important that the airport assess and quantify its own financial risk to climate change. The airport is a local entity that relies on the region's residents being healthy enough to travel and having the public infrastructure necessary to reach the airport. SAN is an essential component of the region's economy, fostering the cultural and economic value of the region's tourism, and it can leverage this role to help build community resilience and foster climate change awareness across the region. Infrastructure investors fund the development of assets that often rely on a much larger community than that in which they sit. They can help prompt infrastructure managers to consider the broader picture of risk and resilience in which they function and go further from building onsite resilience to building integrated community resilience.

CONCLUSION

Climate change has become a driver of risk and performance in financial markets, and the imperative to integrate physical climate risk into investment decisions will only grow over time. Local and regional communities and economies rely on infrastructure assets and thus, resilient infrastructure is critical to resilient economies at large. Investors can understand and manage climate risks by focusing on how specific assets are materially affected by climate hazards, identifying components of resilient infrastructure, and investing in projects that consider asset-level resilience while fostering community resilience.

Climate risks translate into financial risks for infrastructure through impacts on costs, revenues, assets, liabilities, and financing. The impacts of different hazards on a specific asset will depend on the asset's particular vulnerabilities as well as its adaptive capacity, which are driven by its location, resource needs and operating conditions. Similarly, the impacts of climate hazards on the community surrounding an infrastructure asset will also determine how these risks affect the asset itself, as its operations and revenues depend on employees, customers, and often goods, from the broader community.

Investors should also consider how risks may manifest differently across a project's life cycle. For example, projects in the construction stage have risks distinct from operating assets, which lead to different climate risk implications. The risk/return profile of infrastructure projects at the development and construction stages deter most institutional investors even absent climate-related risks, which is a challenge to consider when seeking such sources of capital for infrastructure projects. Development and infrastructure banks can be powerful partners in assessing and addressing climate risk in infrastructure, as they already play a critical role in the development and construction stages of infrastructure projects, particularly in the developing world.

The development of regulation on physical climate risk disclosure will enable investors to obtain an increasingly detailed and nuanced view on the risk and resilience of a particular asset. In the meantime, investors can leverage data on physical climate risk, alongside existing frameworks for evaluating infrastructure resilience, to gain an understanding of the risk drivers for particular assets and across a portfolio. When evaluating climate risks of a project, investors need to consider their investment time frame and varying risk exposures of different project activities.

Stakeholder engagement is a powerful tool for gaining clearer insight on a project manager's understanding of climate risk and resilience considerations in project design. ESG safeguards, which are employed robustly by many development finance institutions as well as published thought-leadership on climate risks for investors, can serve as useful toolkits for such engagement. Investors can combine their own knowledge of asset-level risk exposure, with information gained from asking direct questions, to understand both how an asset may be materially impacted by climate risks and to encourage asset managers to integrate resilience-building strategies. Particularly for institutional investors who primarily invest in operating assets, asking questions will not only help them understand the embedded climate risks, but may also encourage project developers, financiers, and managers to integrate built-in resilience features. Moreover, investors can also leverage engagement to push for continued uptake of clearer climate-related financial disclosure across the infrastructure and real asset sectors.

By assessing the risk and resilience of infrastructure assets, investors can understand risk drivers across their portfolio, screen potential investments for climate-related risks, and also capitalize on resilience opportunities. Investing directly in adaptation projects is a way to minimize investment risk while building resilience in the broader community. By encouraging project managers to consider resilience elements in design and operations and examine an asset's relationship with the climate vulnerability and resilience of the broader community, investors can promote the resilience of infrastructure projects and of the economies they underpin and rely upon.

REFERENCES

[1] National Oceanic and Atmospheric Administration (NOAA) website, Assessing the U.S, Climate in 2017, 2017. Available from: https://www.ncei.noaa.gov/news/national-climate-201712.

[2] European Environment Agency. 2018. "Economic Losses from Climate-Related Extremes." European Environment Agency. February 2018. https://www.eea.europa.eu/data-and-maps/indicators/direct-losses-from-weather-disasters-3/assessment-1. Cited in Hubert, Romain, Julie Evain and Morgane Nicol, 2018. "Getting started on Physical climate risk analysis in finance - Available approaches and the way forward: ClimINVEST Research Project Work Package 1". Institute for Climate Economics (I4CE), December 2018. https://www.i4ce.org/wp-core/wp-content/uploads/2018/12/I4CE-ClimINVEST_2018_Getting-started-on-physical-climate-risk-analysis.pdf.

[3] S.A. McAlpine, J.R. Porter, Estimating Recent Local Impacts of Sea-Level Rise on Current Real-Estate Losses: A Housing Market Case Study in Miami-Dade, Florida, 2018. Available from: https://doi.org/10.1007/s11113-018-9473-5.

[4] First Street Foundation, As the Seas Have Been Rising, Tri-state Home Values Have Been Sinking, 2018. Available from: https://assets.floodiq.com/2018/08/17ae78f7df2f7fd3176e3f63aac94e20-As-the-seas-have-been-rising-Tri-State-home-values-have-been-sinking.pdf.

[5] OECD, Technical Note on Estimates of Infrastructure Investment Needs, 2017. Available from: https://www.oecd.org/env/cc/g20-climate/Technical-note-estimates-of-infrastructure-investment-needs.pdf.

[6] Cliffwater, Investments in Infrastructure, 2017. Available from: https://www.cliffwater.com/Research/DownloadFile?path=docs%2FInvestments%20in%20Infrastructure.pdf&title=Investments+in+Infrastructure.

[7] World Bank Group, Contribution of Institutional Investors, 2017. Available from: http://ppi.worldbank.org/~/media/GIAWB/PPI/Documents/Global-Notes/PPI_InstitutionalInvestors_Update_2017.pdf.

[8] A. Joshi-Ghani, D. Saha, Behold the White Knights! New Research on Institutional Investor Participation in Financing EMDE Infrastructure, 2018. Available from: http://blogs.worldbank.org/ppps/behold-white-knights-new-research-institutional-investor-participation-financing-emde-infrastructure.

[9] Road Traffic Technology, SMART (Stormwater Management and Road Tunnel), Kuala Lumpur. Available from: https://www.roadtraffic-technology.com/projects/smart/.

[10] N. Ambrosio, Y.H. Kim, Assessing Local Adaptive Capacity to Understand Corporate and Financial Climate Risks, Four Twenty Seven, 2019. Available from: http://427mt.com/wp-content/uploads/2019/01/Assessing-Local-Adaptive-Capacity_427_1.15.19.pdf.

[11] E. Mazzacurati, J. Firth, S. Venturini, N. Ambrosio, F. Freitas, L. Ross, C. Foubet, R. Hamaker-Taylor, Advancing TCFD Guidance on Physical Climate Risks and Opportunities, European Bank for Reconstruction and Development and the Global Centre of Excellence on Climate Adaptation, 2018. Available from: http://427mt.com/wp-content/uploads/2018/05/EBRD-GCECA_final_report.pdf.

[12] The GCECA is now the Global Center on Adaptation.

[13] J. Hedding, Planes at Phoenix Sky Harbor Airport Grounded when it Gets Too Hot, Trip Savvy, 2017. Available from: https://www.tripsavvy.com/planes-at-phoenix-skyharborairport-2682513.

[14] R. Hubert, J. Evain, M. Nicol, Getting Started on Physical Climate Risk Analysis in Finance - Available Approaches and the Way Forward: ClimINVEST Research Project Work Package 1, Institute for Climate Economics (I4CE), 2018. https://www.i4ce.org/wp-core/wp-content/uploads/2018/12/I4CE-ClimINVEST_2018_Getting-started-on-physical-climate-risk-analysis.pdf.

[15] World Bank, Climate & Disaster Risk Screening Tools. Available from: https://climatescreeningtools.worldbank.org/.

[16] A. McNeely, PG&E Credit Cut to Brink of Junk by Moody's on Wildfire Risk, 2018. Available from: https://www.bloomberg.com/news/articles/2018-11-15/pg-e-credit-cut-to-brink-of-junk-by-moody-s-on-wildfire-risk.

[17] K. Okuji, M. Wertz, K. Kurtz, L. Jones, Evaluating the Impact of Climate Change on US State and Local Issuers, Moody's Investor Service, November 28 , 2017.

[18] GRESB, About. Available from:https://gresb.com/about/.

[19] GRESB, 2018 Resilience Reference Guide, April 1, 2018. Available from: http://gresb-public.s3.amazonaws.com/2018/Assessments-and-Reference-Guides/2018-GRESB-Resilience-Reference-Guide.pdf.

[20] C. Davies, Emerging Work by Financing Institutions on Climate Resilience Results Metrics, 2017. Available from: https://unfccc.int/sites/default/files/session_9_davies_ebrd.pdf.

[21] International Finance Corporation, A Member of the World Bank Group, New Ways for Cities to Tackle Climate Change, 2016. Available from: https://www.ifc.org/wps/wcm/connect/f0d2db33-13c6-4ca3-993b-dba6a4731aac/Note-12-EMCompass-New-Ways-For-Cities-to-Tackle-Climate-Change.pdf?MOD=AJPERES.

[22] The Nature Conservancy, Envisioning a Great Green City. Available from: https://www.nature.org/content/dam/tnc/nature/en/documents/Green_City_FINAL_PDF.pdf.

[23] L. Vallejo, M. Mullan, Climate-resilient Infrastructure: Getting the Policies Right, OECD Environment Working Papers, No. 121, OECD Publishing, Paris, 2017, https://doi.org/10.1787/02f74d61-en. Available from:.

[24] J. Rodin, Valuing the Resillience Dividend, 2017. Available from: https://www.rockefellerfoundation.org/blog/valuing-resilience-dividend/.

[25] B. Frey, A. gardaz, L. Karbassi, M. golberd, F. Luboyera, R. Fischer, G. Scott, J. Morrison, S. Woodward, G. Charles, H. Coleman, N. Shufro, L. Dougherty-Choux, E. Metzger, P. Terpstra, B. Dickson, E. Shiffer, S. Bee, A. Olhoff, C. Schaer, K. Levick, S. Reuvers, L. Hermans, C. Gannon, E. Mazzacurati, A. Seville, J. Hayward, J. Coffee, The Business Case for Responsible Corporate Adaptation: Strengthening Private Sector and Community Resilience. A Caring for Climate Report, United Nations Global Compact, 2015. Available from: http://427mt.com/wp-content/uploads/2015/12/Caring-For-Climate-Business-Case-Responsible-Corporate-Adaptation-2015-1.pdf.

[26] TCFD, Final Report: Recommendations of the Task Force on Climate-Related Financial Disclosures, 2017. Available from: https://www.fsb-tcfd.org/publications/final-recommendations-report/.

[27] L. Chatain, Art. 173: Lessons Learned from Climate Risk Disclosures in France, 2018. Available from: http://427mt.com/2018/03/21/art-173-lessons-learned-climate-risk-disclosures-france/.

[28] N. Ambrosio, EU Moves Towards Regulation for Climate Risk Disclosure, 2018. Available from: http://427mt.com/2018/03/15/eu-moves-towards-regulation-for-financial-risk-disclosure/.

[29] UK Parliament, Climate Risk Reporting. Available from: https://publications.parliament.uk/pa/cm201719/cmselect/cmenvaud/1063/106306.htm#_idTextAnchor040.

[30] South Africa Maritime Safety Authority, Ports Services – Industry Profile. Available from: https://www.samsa.org.za/sites/samsa.org.za/files/port_services_industry_profile_2_0_0.pdf.

[31] Transnet, Transnet 2017 Annual Report, 2017. Available from: https://www.transnet.net/InvestorRelations/AR2017/OD%20Reports/8335_Transnet%202017_NPA%20HR.pdf.

[32] KwaZulu-Natal Department of Transport. Available from: http://www.kzntransport.gov.za/public_trans/freight_databank/kzn/ports/Durban/index_xml.html.

[33] Exportgov, South Africa- Port and Maritime Infrastructure, 2018. Available from: https://www.export.gov/article?id=South-Africa-port-infrastructure.

[34] Department: Public Enterprises, Republic of South Africa, Transnet. Available from: http://www.dpe.gov.za/soc/transnet/Pages/default.aspx.

[35] Acclimatise, Climate Finance Advisors, Lenders' Guide for Considering Climate Risk in Infrastructure Investments, Four Twenty Seven, 2018. Available from: http://427mt.com/wp-content/uploads/2018/01/Investment-Guide_17012018.pdf.

[36] A. Mather, South African Coasts and Ports: What Will Climate Change Bring, 2011. Available from: https://unctad.org/sections/wcmu/docs/AHM2011_2_10_Mather_en.pdf.

[37] Transnet, Port Terminals 2017, 2017. Available from: https://www.transnet.net/InvestorRelations/AR2017/OD%20Reports/8335_Transnet%202017_Port%20terminals%20HR.pdf.

[38] B. Whitehouse, Heavy Weather Interferes with Port of Durban Efforts to Relieve Congestion, 2017. Available from: http://www.sashippingnews.com/2017/10/16/heavy-weather-interferes-port-durban-efforts-relieve-congestion/.

[39] Transnet, Transnet Integrated Report 2017, 2017. Available from: http://pmg-assets.s3-website-eu-west-1.amazonaws.com/Transnet_IR_2017.pdf.

[40] Transnet, Port Terminals 2018, 2018. Available from: https://www.transnet.net/InvestorRelations/AR2018/TPT.pdf.

[41] Transnet, National Climate Change Response Dialogue, 2014. Available from: https://www.environment.gov.za/sites/default/files/docs/transnet_scontributiontowardssaefforts_cuttingdownemission.pdf.

[42] A. Speckman, Brics Loans to Eskom and Transnet in the Spotlight Ahead of Summit, 2018. Available from: https://www.businesslive.co.za/bd/companies/2018-07-23-brics-loans-to-eskom-and-transnet-in-the-spotlight-ahead-of-summit/.

[43] Transnet, Annual Financial Statement 2017, 2017. Available from: https://www.transnet.net/InvestorRelations/AR2017/Transnet%20AFS%202017.pdf.

[44] Moody's Investors Service, Rating Action: Moody's Takes Action on Five South African Corporates Following Sovereign Downgrades and Outlook Change, 2017. Available from: https://www.redefine.co.za/view-file/sa_corporateaction2_13june2017.pdf.

[45] M. Van Wyngaardt, Transnet Says Well-Positioned to Withstand Downgrade, 2017. Available from: http://www.engineeringnews.co.za/print-version/transnet-says-well-positioned-to-withstand-downgrade-2017-04-07.

[46] San Diego International Airport, About the Airport Authority. Available from: https://www.san.org/Airport-Authority/About-the-Authority.

[47] S. Brickner, K. Kiefer, Kathryn, Comprehensive Annual Financial Report 2017 & 2016, San Deigo County Regional Airport Authority, 2017. Available from: https://san.org/DesktopModules/Bring2mind/DMX/Download.aspx?EntryId=10992&Command=Core_Download&language=en-US&PortalId=0&TabId=197.

[48] San Diego International Airport, Effectively Managing Energy and Water Use. Available from: http://sustain.san.org/operational/.

[49] San Diego County Regional Airpot Authority, FY 2019 Adopted Budget & FY 2020 Approved Conceptual Budget, 2018. Available from: https://san.org/DesktopModules/Bring2mind/DMX/Download.aspx?EntryId=12241&Command=Core_Download&language=en-US&PortalId=0&TabId=197.

[50] San Diego International Airport, Passenger Facility Charge Project Description and Financial Plan, 2018. Available from: http://www.san.org/Portals/0/Documents/PFC/2018-05-10_PFC-13-Public-Notice-Detail-v2.pdf.

[51] San Diego International Airport, Seeking New Ways to Conserve Water and Adapt to 'New Normal,', 2015. Available from: http://2015-sustain.san.org/operational/.

[52] The San Diego Foundation, Climate. Available from: https://www.sdfoundation.org/programs/programs-and-funds/climate/.

[53] Airport Cooperative Research Program, Airport Climate Adaptation and Resilience A Synthesis of Airport Practice, 2012. Available from: http://www.trb.org/Publications/Blurbs/167238.aspx.

[54] San Diego International Airport, Green Build Project at San Diego International Airport Earns Envision Sustainable Infrastructure Platinum Award, 2016. Available from: https://www.san.org/News/Recent-Press-Releases/ArtMID/951/ArticleID/139.

[55] San Diego County Regional Airport Authority, Terminal 1 Replacement: The Airport Development Plan, 2018. Available from: http://san.org/Portals/0/Documents/Environmental/2018-Draft/Primary_Components_Map-02.pdf.

[56] San Diego County Regional Airport Authority, Terminal 1 Replacement: The Airport Development Plan Fact Sheet, 2018. Available from: http://san.org/Portals/0/Documents/Environmental/2018-Draft/General_Fact_Sheet-02.pdf.

Introduction

A 2008 flood along the Cedar River brought significant devastation to downtown Cedar Rapids, Iowa. Flooding covered 10 square miles after the river crested at a historic high of 31 ft. More than 5000 residential structures were flooded, 1300 of which were beyond repair. More than 18,000 residents and 9000 employees were evacuated. At the time, the Federal Emergency Management Agency (FEMA) called the event the fifth-worst national disaster in US history. Community services were impacted, as municipal facilities became inaccessible. Communication systems failed including 911 dispatch. The wastewater facility flooded and lost power. Potable water systems became incapacitated [1].

Following the flooding, the community came together to identify solutions so this devastation would not happen again. They explored multiple options from building a wall along the river to bringing the floodplain closer to its natural state (see Fig. 1). The ultimate solution was a combination of natural and engineering-based solutions. A detachable storm wall system was deployed along the waterfront. About 1500 houses in the floodplain were bought out, and a new stormwater system was built at street level. The community also looked to implement a strategy that benefited the entire community—a new 220-acre greenway that acted as a levee while providing new community amenities [2].

The flood mitigation strategies were put to the test in 2016 as the Cedar River would reach its second highest level on record—only dwarfed by the 2008 floods. Although many of the flood protection projects were not yet complete, the city fared well [3].

Cedar Rapids is not alone in seeking natural solutions to enhance resilience. Communities across the United States are looking to nature to mitigate everything from localized stormwater management to community-wide flood risk. Although nature-based solutions are generally considered to manage water, other hazards such as extreme heat and urban heat islands can be mitigated through shading trees and green areas in lieu of dark surfaces. Green roofs have also been shown to reduce temperatures in the surrounding area.

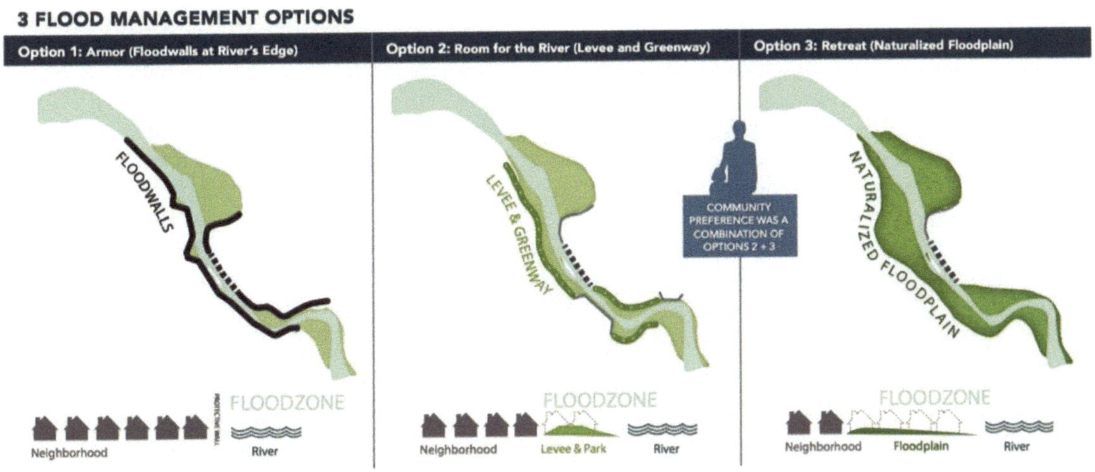

FIG. 1 Flood management options for Cedar Rapids. (Source: Courtesy of, Sasaki [2].)

In addition to using natural solutions to deliver ecosystem services, nature is assisting resilience in other ways. Engineers and architects are looking to nature to inform the design of engineered systems. This approach, called biomimicry, examines how nature is resilient [4]. According to Beynus [4], biomimicry is based on the following principles and strategies from nature:

1. Nature runs on sunlight.
2. Nature only uses the energy it needs.
3. Nature fits form to function.
4. Nature recycles everything.
5. Nature rewards cooperation.
6. Nature banks on diversity.
7. Nature demands local expertise.
8. Nature curbs excess from within.
9. Nature taps the power of limits.

It is therefore not hard to see how natural systems provide a context for framing community resilience. Diversity better prepares communities to absorb shocks and stresses, renewable energy sources support self-sufficiency, resilience decision-making is local, and waste is unnecessary.

Biomimicry has been used to help structures resist the forces of nature. Designers mimicked the structure of a honeycomb to allow a building in New Songdo City, South Korea, to resist buckling in high winds [5]. Design firm HOK and the Biomimicry Group examined various biomes to help unlock new design principles that more closely match the strategies deployed by nature. For example, the die off of the American Chestnut in the northeastern forests of the United States and the rise of tree species that perform similar functions in the ecosystem reflect the principle that redundant functional groups create resilience. A similar approach can be applied in the context of multimodal transportation networks. A system offering overlapping, redundant modalities creates resilience. Furthermore, the intersections of these modalities provide opportunities for social interactions, community amenities, and commerce—all leading to enhanced resilience [6].

Nature also contributes to the social and economic resilience of communities. Research is emerging finding that building occupants with views of nature and access to natural light are more productive and generally more satisfied with their job [7–10]. Students in schools have exhibited similar benefits, achieving higher grades and increased satisfaction (and their teachers are happier as well) [11,12].

Although this book does not go into depth about the need to mitigate greenhouse gases as a resilience-enhancing (or at least risk-reducing) strategy, it would be shortsighted to not mention the role nature can play in capturing and storing greenhouse gases such as carbon dioxide. Retaining and enhancing these natural capacities will be an essential part of any climate mitigation strategy. Currently, US land absorbs and stores (mostly in forests) about 17% of the annual US fossil fuel–based emissions [13]. Some emerging research suggests that planting 1.2 trillion trees globally (above the 3 trillion that currently exist) would significantly reduce greenhouse gases in the atmosphere [14]. Clearly, to limit the impacts of these greenhouse gas emissions, a strategy based on reducing the supply of emissions and increasing demand in nature is necessary.

Nature is clearly entrenched in the resilience dialogue. As Jacob and York point out in Chapter 8, the concept of resilience is drawn from the ability of ecosystems to adapt to stresses and emerge improved. Using nature's capabilities to achieve community resilience can provide multiple cobenefits. Engineered solutions alone will not achieve the level of resilience communities' desire. Setting performance expectations (amount of stormwater absorbed or diverted in a set period of time following an event of a certain magnitude) will allow communities to determine the best combination of natural and engineered approaches. Communities should also be sure to capture the cobenefits each strategy provides.

We should also heed nature's warnings on where to build. Medlock and Schwab in Chapter 9 discuss how land-use decisions lead to vulnerabilities and offer insight into how and why such decisions are made.

NATURE-BASED SOLUTIONS

The elegant sophistication of nature is often hard to fully comprehend. One species' waste is another's food source. The water cycle purifies water and delivers it as life-giving rain. Plants turn our respirated carbon dioxide into oxygen for people and other animals to breathe. The food chain establishes a cycle whereby each species is food for the next until the cycle begins anew with decomposition. Whole ecosystems go through stages of succession following a disturbance to ultimately face another disturbance that triggers another succession pathway. Without these approaches, humans would not long survive on this planet.

Harnessing the lessons learned from these elegant systems is just beginning. The circular and sharing economy and cradle-to-grave processes (or cradle-to-cradle) attempt to institutionalize these approaches

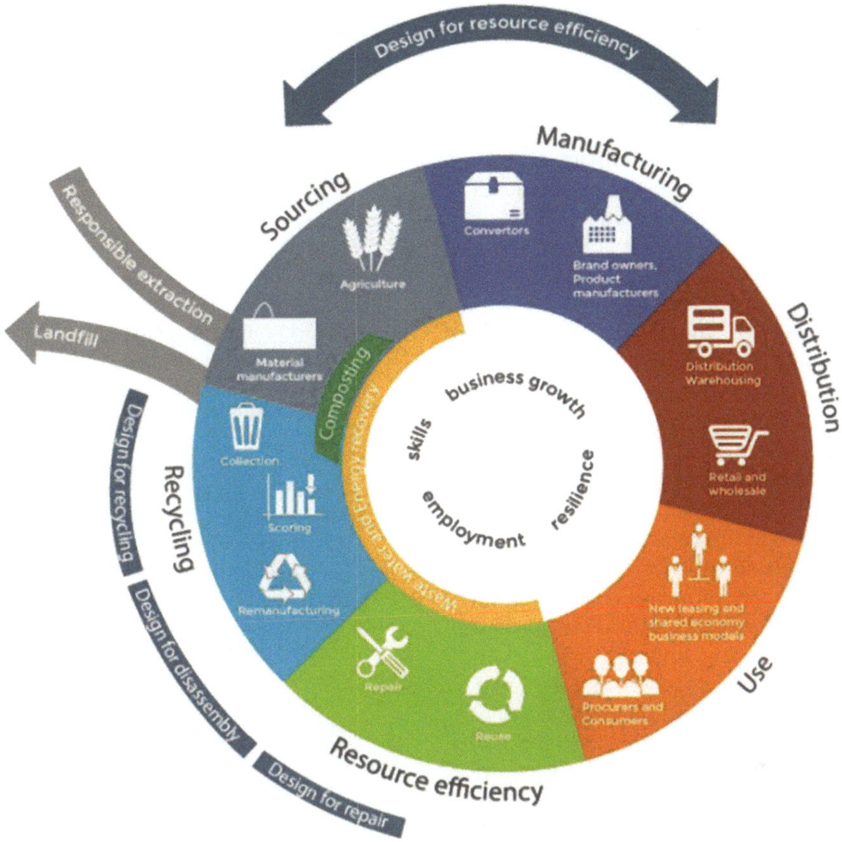

FIG. 2 The circular economy. (Source: From, Anthesis Group.)

in how products are designed and used (Fig. 2) [15,16]. The farm-to-table and snoot to tail movement aims for efficiency and the reduction in food waste. Such efforts contribute to overall community resilience. The efficient use of raw materials benefits human health through reduced pollutants. Keeping products local keeps money in the local economy and reduces the likelihood of disruptions in supply chains. However, solely relying on local resources is also not very resilient. Few communities or companies today can claim to be self-sufficient. Natural systems operate at multiple scales with inputs and outputs entering the systems from various sources. There are certainly lessons to be learned from nature.

Jacob and York examine the broad set of resilience strategies provided by nature. Their approach looks at multiple scales, from large, undisturbed natural areas far from urban areas to the small-scale bioretention areas along city streets. In addition to the stormwater management and water quality benefits provided by wetlands and other forms of green infrastructure, they capture the benefits provided by increased population density including walkability, neighborhood cohesion, and community diversity.

Case studies from New York and Houston demonstrate the economic value that nature-based solutions provide and the cobenefits that can be captured. Regardless of the benefits a solution provides, the examples demonstrate the importance of community involvement throughout the process of securing property and restoring the natural system.

LAND-USE POLICIES

In Chapter 9, Medlock and Schwab explore the intersection of infrastructure, land use, and hazard mitigation planning and how a coordinated approach leads to reduced risk and better outcomes for communities. As they state, "The connection between land use and infrastructure is often overlooked, but community development plans and regulations drive decision-making for nearly every infrastructure investment from regional and citywide scales down to the neighborhood, block, and lot."

Many communities are looking to reverse or mitigate poor land-use decisions of the past. The buyout or elevation of properties located in the flood plain is an effective strategy for reducing the impact of hazard events. The National Institute of Building Sciences (NIBS) Mitigation Saves Study identified a $7 benefit for every $1 invested by the Federal Emergency Management Agency (FEMA) in flood mitigation, which included buyouts and elevations [16]. Many of the properties that have been bought out have suffered repetitive losses under the National Flood Insurance Program (NFIP).

Of the roughly 1.5 million properties enrolled in NFIP, about 2% have been the subject of repetitive losses. However, they have accounted for about 30% of the claims since the initiation of the program [17].

How we build is just as important as where we build. Examination of the role of building codes in establishing criteria for the risks a building could face is captured in Chapter 12. To be resilient, communities must support effective decision-making. In many instances, individual property owners may not have the savvy to know the implications of their decisions. Zoning and building codes function at multiple levels, protecting individual building owners and homeowners from specific risks while working to protect a community's tax base and reduce the risk of costly recovery when a disaster strikes.

REFERENCES

[1] S. Fowler, Flood Management and Rebuilding Plans Help Iowa Town Recover, Government Technology, July 14, 2009. https://www.govtech.com/em/disaster/Flood-Management-and-Rebuilding.html.

[2] P. Sisson, How community-led design can help a city rebound after a storm, Curbed (August 31, 2017). https://www.curbed.com/2017/8/31/16234698/cedar-rapids-flood-houston-hurricane-harvey-rebuild.

[3] R. Smith, Lessons learned from floods of 2008 helped Cedar Rapids this time, Gazette (September 30, 2016). https://www.thegazette.com/subject/news/flood-2016/lessons-learned-from-floods-of-2008-helped-cedar-rapids-this-time-20160930.

[4] J. Benyus, Biomimicry: Innovation Inspired by Nature, 1997.

[5] S. Amandolare, Will biomimicry offer a way forward, post-Sandy? New York Times (January 4, 2013). http://green.blogs.nytimes.com/2013/01/04/will-biomimicry-offer-a-way-forward-post-sandy.

[6] Biomimicry Group, HOK Group, Genius of Biome, 2014. https://issuu.com/hoknetwork/docs/geniusofbiome.

[7] G.W. Evans, The built environment and mental health, Journal of Urban Health (December 2003).

[8] M. Boubekri, Daylighting, Architecture and Public Health: Building Design Strategies, Architectural Press, 2008.

[9] L. Heidari, M. Younger, G. Chandler, J. Hooch, P. Schramm, Integrating health into buildings of the future, Journal of Solar Energy Engineering 139 (1) (November 26, 2016).

[10] E. Largo-Wight, Cultivating healthy places and communities. Evidence-based nature contact recommendations, International Journal of Environment and Health (February 2011).

[11] J.A. Benfield, G.N. Rainbolt, P.A. Bell, G.H. Donovan, Classrooms with nature views: evidence of differing student perceptions and behaviors, Environment and Behavior 47 (2) (2015).

[12] P. Barrett, F. Davies, Y. Zhang, L. Barrett, The impact of classroom design on pupils' learning: final results of a holistic, multi-level analysis, Building and Environment 89 (July 2015).

[13] U.S. Global Change Research Program. Carbon Cycle, https://www.globalchange.gov/explore/carbon-cycle.

[14] (a) M. Tutton, The Most Cost Effective Way to Tackle Climate Change? Plant 1 Trillion Trees, CNN, April 17, 2019. https://amp.cnn.com/cnn/2019/04/17/world/trillion-trees-climate-change-intl-scn/index.html;
(b) R. Botsman, R. Rogers, What's Mine Is Yours: The Rise of Collaborative Consumption, 2010.

[15] W. McDonough, M. Braungart, Cradle to Cradle: Remaking the Way We Make Things, 2010.

[16] Multihazard Mitigation Council, Natural Hazard Mitigation Saves: 2018 Interim Report, K. Principal Investigator Porter, C. co-Principal Investigators Scawthorn, C. Huyck, Investigators: R. Eguchi, Z. Hu, A. Reeder, P. Schneider, Director, MMC, National Institute of Building Sciences, Washington, D.C. https://www.nibs.org/resource/resmgr/mmc/NIBS_MSv2-2018_Interim-Repor.pdf.

[17] R. Simon, One house, 22 floods: repeated claims drain federal insurance program, The Wall Street Journal (September 15, 2017). https://www.wsj.com/articles/one-house-22-floods-repeated-claims-drain-federal-insurance-program-1505467830.

CHAPTER 8

Harnessing Green Infrastructure for Resilient, Natural Solutions

CHARRIS R.H. YORK, MS • JOHN S. JACOB, PHD

INTRODUCTION

Nature knows best. It knows best because it has been evolving and adapting from the beginning. This notion is the basis of the movement to build green or natural solutions into infrastructure. The greater the proportion of green infrastructure versus gray infrastructure, the more resilient the entire system will be. Nature is infrastructure.

Resilience is a term originally defined for ecological systems and refers to the ability to bounce back from disturbances. More significantly, natural ecosystems are not only able to bounce back from disturbances, but their very composition and structure may also be dependent on disturbance. In coastal areas, for example, wind and storm surge associated with hurricanes will obliterate significant pieces of the natural landscape. The ecological succession that occurs in those "damaged" patches is key to the diversity of coastal prairies and woodlands and maintains their dynamic equilibrium in a shifting environment. Many forest ecosystems, for example, rely on the periodic introduction of fire to remain healthy.

No system can be more resilient than one where disturbance or catastrophe is part of the DNA.

If nature does know best, it follows that humans might know much less. Natural systems are not only more complex than we imagine them to be, they are more complex than we can imagine them to be. What we know about natural ecosystems is far outweighed by what we do not know. As such, we have much to learn from natural systems and therefore must be careful to preserve large-enough natural areas for the future discovery of their workings. We will have to draw from these preserves to identify new natural solutions.

Natural areas provide beneficial services that would otherwise be quite costly. For example, wetlands and soils can filter out most aquatic contaminants, ensuring a high-quality water supply. Native prairies can detain significant volumes of stormwater in their soils and in the bunch-grass plant cover, diminishing downstream peak flood heights. This chapter is about harnessing these and other services to provide natural solutions for our most troublesome infrastructure needs and doing so in a way that the resilience of the overall system improves. We focus on green infrastructure for water: supply, stormwater detention, and wastewater and stormwater purification.

Just because nature knows best does not mean *we* always know how to best apply natural solutions, nor does it mean that in every case a natural solution is appropriate. There will be many cases where a structural, gray solution will be the best solution. But the goal should be to use nature-based solutions wherever possible.

A NOTE ON TERMINOLOGY

The recognition of the ecosystem services provided by native prairies and forests began in earnest with the publication of the book *Green Infrastructure* in 2006 [1]. The concept had been around for many years, but the wide distribution of this book brought the concept to the forefront. Calling large-scale natural areas "green infrastructure" puts these areas on a par, at least conceptually, with gray infrastructure. Just as we recognize that we cannot exist very comfortably without the roads, sewers, and power lines provided by gray infrastructure, we might be even more destitute without green infrastructure, which is ultimately responsible for the air we breathe and the water we drink.

Over time, the term green infrastructure has expanded to include small-scale practices associated

Optimizing Community Infrastructure. https://doi.org/10.1016/B978-0-12-816240-8.00008-2

with low-impact development (LID). LID is essentially infiltration practices applied at the site scale (described below in the site-scale section). Both the site-scale practices and the large, natural prairie and forest tracts of land involve natural solutions, and so are both classified as green infrastructure. Most practitioners in this field recognize that in practice there really is no such thing as LID. All development is impactful no matter how many best management practices (BMPs) are put in place. Hence, the term green stormwater infrastructure (GSI) is emerging as the preferred term for micro- and midscale BMPs. BMPs simply refer to a practice such as pervious pavement or a stormwater wetland. We find the GSI term to be very descriptive of the kinds of natural practices used for stormwater management, while providing a genetic thread back to the ecosystem scale services.

In this chapter, we will explore the application of GSI BMPs at a variety of scales. As you will see, each application provides a benefit, but coordinated strategies across scales provide the greatest opportunity to harness nature's capacity.

MACROSCALE GREEN INFRASTRUCTURE

Macroscale green infrastructure (GI) refers to large, relatively undisturbed tracts of prairies, forests, and other ecosystems.

Large-scale GI provides the best return on investment compared with either mid- or site-scale applications of GI. There are ecologies of scale at the larger end that just cannot be replicated at smaller scales. A quarter-acre bioswale in the city is not part of a larger ecosystem the way a natural swale in a 10,000-acre native prairie might be. The small urban bioswale can provide some water cleansing and perhaps even a little runoff detention, but it is a largely isolated "natural" feature in an otherwise urban setting. The swale in the prairie is part of a much larger system and is integrated into that system through multiple pathways. Both swales may function similarly in terms of water throughput and cleansing, but the prairie swale contributes much more through interactions with nearby wetlands, mima mounds, upland flats, and other features.

Bigger is better suggests that the number one priority for natural systems mitigation should be the preservation and restoration of large tracts of land, large enough that they could be managed as ecosystems. Ecosystem properties, where the whole is greater than the sum of the parts, begin to emerge at the 1000-acre and above scale. Ecosystem services at this scale can be quite significant.

Ecosystem Services

Ecosystem services is the term used to describe the benefits that healthy natural areas can provide for humanity. These services range from the global (e.g., rainforests as global lungs) to the local (wetlands that detain stormwater). There is abundant literature about these services, far too complex for review here. These services are the heart of green infrastructure. Lost ecosystem services must almost always be replaced, sometimes at great expense, to avoid the negative impacts of that loss. For example, paving over wetlands and prairies results in a loss of stormwater detention, and most communities require some form of excavated detention volume to avoid additional flooding.

Ecosystem services (services from areas large enough to quality as ecosystems) benefit the larger community, whereas smaller-scale practices, such as rain gardens, provide benefits at the parcel scale. Both scales have contributions to make, but the difference in scale actually results in a difference in kind because of the greater diversity of services offered at the ecosystem or watershed scale.

Mitigation for lost ecosystem services is usually for a single service—as with the detention basins. Healthy ecosystems, however, simultaneously provide many services, more than we likely understand. The four broad categories of ecosystem services (with a few examples) are provided in Table 8.1.

Importantly, land managers should recognize that the many and varied functions of large-scale GI cannot be replaced by either mid- or small-scale practices placed in the city. Scale is paramount when it comes to ecosystem productivity. The loss of large tracts of natural lands cannot be mitigated by small-scale, distributed GSI practices, despite the many benefits that are provided by these smaller-scale BMPs.

The loss of large-scale natural areas can have a major impact. Houston, for example, after Hurricane Harvey, became much more interested in the state of the prairies to the west of the city that drain directly into the city. New York City has successfully incorporated large-scale watershed planning into its program for maintaining a sustainable and resilient water supply for the future (discussed below).

A plan to incorporate large-scale GI begins with the identification of natural areas with significant ecological services. For example, what are the source areas

TABLE 8.1
Four Types of Ecosystem Services.

REGULATING SERVICES	
Climate	Absorption of CO_2 and other greenhouse gasses. Feedback loops between oceans and terrestrial ecosystems. Flood water detention.
Waste treatment	Soil microorganisms naturally break down waste products, including organism remains and dead vegetation.
PROVISIONING SERVICES	
Pure water	Water is purified as it moves through ecosystems.
Food	Ecosystems provide "wild foods." Soils provide the basis for all agriculture.
CULTURAL SERVICES	
Recreation	Hiking, backpacking, rock climbing, skiing, bird watching, and other activities in green areas.
Inspiration	Aesthetic appreciation of natural beauty.
HABITAT SERVICES	
Habitat	Healthy ecosystems provide all that species need to thrive.

for drinking water? Are there "hot spots" of high-value ecosystems that are imminently vulnerable to loss by development? The advent and widespread use of geographic information systems makes this mapping exercise very accessible for most communities. Experts of course are needed to help identify significant parcels. Several decision-support tools are now available to help stakeholders and professionals prioritize natural areas in contributing watersheds (for example, the CHARM model [2]).

Once critical areas are identified and prioritized, the next step is to determine ways to preserve lands with existing ecosystem services as part of the green infrastructure of a city or town.

Tools for Incorporating Land and Ecosystems into a Community's GI Portfolio

Multiple tools are available for communities wanting to incorporate ecosystem services from surrounding hinterlands into their resilience portfolio. These tools range from outright purchase of land to instituting incentives that sustain the ecosystem services.

It is essential for policymakers to be aware that any meddling in local land use in the watershed hinterlands of a city or town will get a reaction, possible very negative, from the people who live and work in those hinterlands. Any successful project to control upstream land for beneficial uses by downstream users will require extensive stakeholder engagement. See the NYC case study below. Some specific tools include the following:

- Fee simple purchase: The simplest, but perhaps most expensive, way to control land. The purchase is simple, but local repercussions may not be so simple. Public control of land may reduce the tax value of land. Taxes that would support a smaller community may no longer be available.
- Conservation easements: The landowner gives up the development rights to a parcel of land, usually for a price significantly less than outright purchase. The landowner can still use the land, but only in ways consistent with the conservation easement. That use could include farming and ranching, for example.
- Payment for ecosystem services: Similar to conservation easements, but one step beyond (or backward depending on your perspective). Payment is given in return for managing the land in a certain way. Perhaps a payment for keeping a parcel forested on a steep slope where policies or regulations may not exist to keep the land forested. The major difference from conservation easements is that payment is for a specific time period, perhaps only a year or just 5 to 10 years.
- Zoning: Zoning is the classic regulatory framework for land use planning. Most often this tool is used for segregating specific urban land uses—such as residential versus commercial. Land can also be zoned to restrict uses in known hazardous areas, for example, floodplains or steep slopes that might erode or result in landslides. The takings issue (i.e., zoning being seen as taking away the property rights of the individual) looms large with this tool, but at least in the United States, this kind of zoning is usually upheld by the judicial system, if the use limitation is tied to well-defined issues of health and safety. In terms of ecosystem services or natural solutions, keeping development out of floodplains is the ultimate natural solution, versus, say, levees or other forms of engineering to reduce the size of the floodplain.
- Tax credits: A conservation easement on a tract of land can result in a very significant tax deduction for the landowner. The landowner can continue to use the property consistent with the terms of the

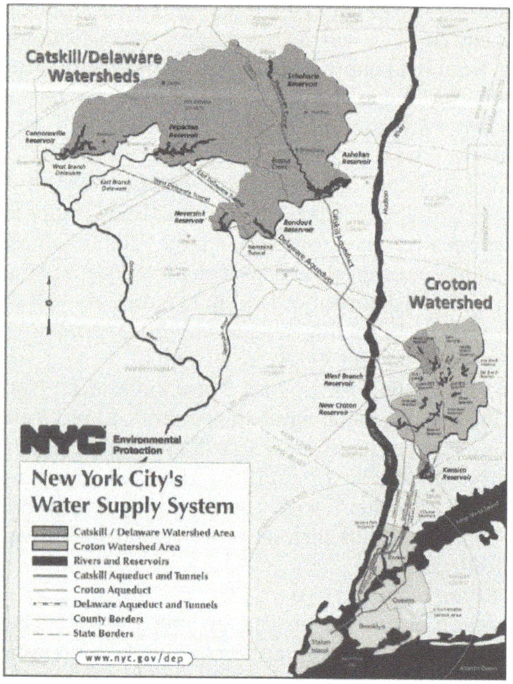

FIG. 8.1 Map of the NYC water supply system showing the green infrastructure used by NYC to ensure a safe and resilient water supply for the future. (*From*: New York City Department of Environmental Protection.)

easement but can offset the loss of development rights with the tax reduction.

- Transfer of development rights (TDR): TDR is based on zoning. Without a well-designed and defined comprehensive plan, TDR will not work. Even with zoning, it is difficult to put in place, but TDR is a very attractive tool because it is market based. The basic idea is to transfer development rights from sensitive areas to areas more suitable to development. The development rights are severed from the property just as in a conservation easement. Sending areas must be designated, as must receiving areas. For example, a base zoning of 1 house per acre could cover a TDR project area. A landowner in a sensitive sending area could develop his or her land, but at the maximum of 1 house per acre. A landowner in a receiving area could also develop at 1 house per acre, or he or she could purchase development rights from landowners in the sending area, to where he or she could develop at 2 units or more to the acre. Usually, there is a clearing house for these transactions. Thus, landowners in sensitive areas can get the money for development rights but retain their land for ranching, farming, or other allowed uses. Developments cannot be transferred from receiving areas to sending areas.

Case Study: New York City Source Water Protection

One of the best documented, and perhaps most famous, large-scale green infrastructure, projects is the long-term New York City Watershed Protection Plan. New York City (NYC) actively works with watershed stakeholders far removed from the city to ensure a safe drinking water source for more than 9 million people indefinitely into the future. From very early on, NYC recognized that most of its drinking water was going to come from watersheds east and west of the Hudson River, as far as 100 miles to the north as the crow flies. In the late 19th and early part of the 20th centuries, the City undertook massive public works engineering projects to secure and build water supply reservoirs west of the Hudson (WOH)—the Catskill/Delaware watersheds—and east of the Hudson (EOH) farther south and closer to NYC—the Croton watershed, as well as a massive network of large tunnels and aqueducts to get the water to the city (Fig. 8.1). The result was a long-term water supply of unparalleled quality. The other result was deep-seated, long-term resentment of watershed citizens owing to rather heavy-handed condemnation of lands through eminent domain.

The reservoir and conveyance system was working quite well when the Safe Water Treatment Rule (SWTR) was promulgated by the US Environmental Protection Agency (EPA) in 1989. The SWTR required public water supply systems using unfiltered surface water sources to either provide filtration or meet certain criteria. A water filtration plant capable of treating the NYC water supply would have cost more than $10 billion in today's dollars to build and maintain. Or, the city could establish certain watershed control criteria to minimize the potential for contamination. The watershed alternative would cost $1.5 billion—or just more than 10% of the cost of the treatment plant. The City chose the cheaper route and embarked on a journey of collaboration and discovery that would have implications for cities across the world.

There is abundant academic literature on the ins and outs of the NYC Watershed Protection Plan [3]. In addition, there are many reports held by the NYC Department of Environment Protection that detail the evolution of this plan. We only briefly review the many facets of this plan.

REBUILDING RELATIONSHIPS

NYC began its journey of watershed protection, while the scars from the use of eminent domain had yet to completely heal. Given the complexity of what was going to be required, the potential for much deeper wounds was significantly greater. NYC recognized this issue and worked hard to build win-win relationships with both government agencies and watershed stakeholders, with the main story in the WOH or Catskill/Delaware contributing watersheds. Too much development had taken place by this time to be able to get around a filtration plant for the Croton watershed. A Memorandum of Agreement (MOA) was eventually signed in 1997 by watershed stakeholders, nonprofits, and state and federal agencies.

ADAPTIVE COMANAGEMENT: EFFECTIVE PARTICIPATORY GOVERNANCE

The MOA involves a complex group of participants. One of the key aspects of the MOA was the establishment of at least two transjurisdictional, regional boards. The most prominent of these is the Catskill Watershed Corporation, which administers the watershed programs that constitute the heart of the Watershed Protection Plan. The Watershed Protection Partnership Council was established as an arbiter for conflict resolution. A robust system of structural, judicial, and popular safeguards were put in place to minimize opportunistic behaviors by actors in the governance framework.

THE NITTY-GRITTY: WATERSHED PRACTICES ON THE GROUND

A common misconception is that the main thrust of the Watershed Protection Plan was to purchase and preserve natural lands with significant ecosystem services available. Land acquisition turned out to be a much less prominent activity in the plan, although it is a very important one. Reversing some of the degradation in the WOH watersheds has been a central focus of the initial activities—activities that nonetheless maintain and enhance ecosystem services. Failing septic systems and community wastewater facilities contribute substantially to degraded water quality, and watershed monies have been appropriately targeted at remediating these systems. There is also an active stormwater program for upgrading stormwater management measures.

Very important to the overall plan, New York State established the New York City Watershed Rules and Regulations, which are managed by the NYC Department of Environmental Protection. These are regulations that govern activities far from NYC. One of the trade-offs was that there would be no more use of eminent domain. The rules and regulations have limited teeth, but what teeth they have can be leveraged for real environmental protection.

The NYC watershed regulations do not cover every aspect of land use in the WOH watershed, but they do cover things such as development near streams, wastewater treatment systems, and siting of junkyard or hazardous materials. Equally important, however, is that incentives are just as, if not more, important for the long-term watershed plan. But without the teeth of the rules and regulations, the incentives would not have as much of an impact as they do today.

Land acquisition is a key component of the plan. As of 2017, almost 40% of the land in the Catskill/Delaware watersheds is protected by fee simple acquisition or conservation easements, with some subwatersheds having in excess of 60% of lands similarly protected.

THE PLAN IS WORKING—EVEN IF IT DID NOT HAVE TO

There is little question that water quality of the WOH watersheds has improved. It is also likely that water quality from these watersheds would still be very good even if the plan had not been implemented. The most cogent critique of the watershed plan is that it really was not needed and that perhaps the requirement of the EPA that the city install an expensive filtration plant was not necessary after all. Although that may be a plausible argument, the fact is that the requirement for filtration was there, and the logical choice for the city was to invest in watershed protection. For some critics, the fact that the plan did not focus exclusively on land acquisition was an admission that this plan did not establish a market for ecosystem services.

But the plan was never about establishing markets for ecosystem services—it was about preserving and enhancing ecosystem services.

Whatever the shortcomings in its origin story, the NYC plan stands as a preeminent example for cross-boundary watershed management with the goal of increasing and maintaining ecosystem services.

MIDSCALE PRACTICES

Between the 3000-plus acre tracts of high-value natural areas and individual sites, there are many opportunities for midscale or community-level GSI practices that are not large enough to be considered wholly natural and yet have ecosystem characteristics not achievable at the site scale. Stormwater wetlands, for example, are larger than the small site-scale practices. As such, these kinds of wetlands cannot be considered as "distributed" BMPs. Stormwater wetlands collect runoff from much

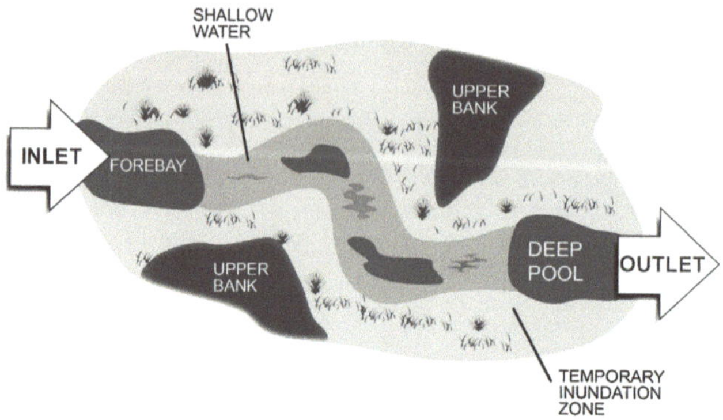

FIG. 8.2 Stormwater wetlands have a long flow path, which includes deep pools and areas that are temporarily flooded after a rainfall. This variety of depth zones allows for more water treatment opportunities. (*From*: J. Jacob, F. Jaber, M. Sipocz, C. York, Stormwater Wetlands for the Texas Gulf Coast, second ed., 2012.)

larger areas, at least at the neighborhood scale. They are thus large enough to be managed to some degree as ecosystems. Because of their size, these midscale solutions can offer more benefits than simply water quality improvement. The "ecosystem" benefits of these designed environments approach those provided by larger-scale ecosystems. For example, the diversity of flora and fauna provides educational and recreational amenities for communities.

Constructed Stormwater Wetlands

Constructed stormwater treatment wetlands are an engineered addition to detention and retention basins. These wetlands are designed as a drainage course with a primary channel and deep pools of permanent standing water. The rest of the system contains shallow slopes and ledges planted with native wetland vegetation designed to be inundated during storm events. The stormwater runoff enters the system through stormwater culverts and is treated through several biogeochemical pathways and then slowly released from the system over the course of 48 to 72 hours. Wetlands are not ponds, and they do not have to be saturated or wet all of the time.

These engineered systems differ from natural wetland systems. They are designed with a long, slow flow path to maximize the amount of wetland the stormwater encounters to clean the water in more ways. The components of the wetland remove pollutants in numerous ways: plants both remove sediments and uptake nutrients, deep pools create aerobic zones for nitrate removal, and in shallow water areas, photodegradation transforms and removes bacteria and pollutants. Finally, aesthetics cannot be overlooked in constructed wetland systems. Because these wetlands are intended to be landscape amenities, creating a pleasing design with color, texture, and interest in each season is essential. Special care should be taken to select flowering species, to layer plants of varying heights to create dimension, and to consider the look of the plants even during winter months (Fig. 8.2).

Constructed stormwater wetlands are not an eyesore like retention ponds, and communities frequently respond positively to the final product, asking for walking paths, benches, and other amenities to create a park-like atmosphere in this stormwater management practice. The Exploration Green project in Houston, Texas, is one example of this.

Case Study: Exploration Green Stormwater Wetland System

Exploration Green Nature Park is an award-winning flood reduction and stormwater cleansing system undergoing construction in southeast Houston, Texas. Only the excavation of the first phase was complete when Hurricane Harvey dumped over 40 in. of rain in the immediate area in 2017. Nearby low-lying houses that had previously flooded with 10- and 50-year storms did not flood at all, removing any lingering doubt held by the last few opponents.

The Clear Lake City area was developed beginning in the 1960s alongside the development of NASA's Johnson Space Center. A golf course built along major drainage ways of the original neighborhoods was a major amenity for Space Center staff. By the early 2000s, however, the golf course had lost its luster and was sold to a developer. When the developer closed the course in 2005 and announced that he was going to convert the green space into townhomes and apartments, local citizens sprang into action.

These citizens worked with the local Clear Lake City Water Authority (CLCWA) to explore solutions. The CLCWA is also the floodplain manager for the area, and the authority quickly recognized that the old golf course could be revamped for floodwater detention, while still providing the much-loved green space that the citizens wanted to maintain. Exploration Green thus emerged from two longstanding major needs identified by local citizens: flood protection and green space.

The CLCWA offered to buy the golf course, but the owner refused. The Authority was forced to condemn the property for flood control, initiating a years-long legal battle that the CLCWA eventually won. Citizens, however, immediately got to work exploring what additional uses could be made of this newly available green space, so that construction could begin quickly following the legal resolution.

Public meetings were held, some with several hundred neighbors in attendance. Committees emerged for each of the interests that community groups advocated for. These included trails and bike/running paths, native vegetation, athletic fields, and stormwater wetlands. The stormwater wetlands group researched constructed wetlands from around the country and toured detention basins in the region, which bore some resemblance to what might fit with the Clear Lake context. Periodic "all hands" meetings allowed sharing between committees and a chance for the larger public to see how the project was evolving. Once all the committees finished, the CLCWA contracted with a design firm to develop a master plan. The Texas Community Watershed Partners (TCWP), a part of the Texas A&M AgriLife Extension Service, provided the design specifications for the wetlands.

Dedicated citizen involvement turned out to be a key component of the success of this project. It helped of course that the project itself emerged from local needs. There were, however, a few dedicated critics who worked

Continued

FIG. 8.3 Exploration Green Natural Park Master Plan as developed by SWA Group Houston. Once a linear golf coursed along major drainage ways, the area was excavated enough to hold 16,000 acre-ft of water. The stormwater wetlands are shown as the hashed area adjacent to the lighter open water areas. Phase I is the southernmost section.

to impede the project. However, given the hundreds of citizens participating in committees and community meetings, the critics had little impact.

Also important to this project was the formation of the Exploration Green Conservancy to manage funding and construction of the park. The park area is 2000 acres, with an eventual total of about 40 acres of treatment wetlands.

Construction began in 2012 and is ongoing. The TCWP collects wetland plants from around the region and stages them in special inundated nurseries. Volunteers collect and stage the plants, and they also place the plants in the wetlands. Volunteers spend every Thursday on this kind of work. Most of the volunteers are members of the Texas Master Naturalist group.

The entire project is scheduled to be completed in 2022 (Figs. 8.3 and 8.4).

FIG. 8.4 Phase I wetland near the outlet of Phase I of the Exploration Green stormwater system.

GREEN STREETS/TREATMENT TRAINS

Green streets are a way to reimagine a traditional drive where multiple GSI practices work in sequence to slow down, spread out, and soak in runoff. Such approaches are more than just stormwater features; they also use sustainable elements and improve environmental and human health [5]. Envisioning a street as an ecosystem and a green corridor changes the paradigm of traditional roadway engineering. Instead of masses of concrete with little to no vegetation, green streets use a series of swales, rain gardens, street trees, and pervious pavement to create a lush oasis within a community. Including energy-efficient or solar-powered lighting, and recycled materials, adds benefits to these projects, making them "green" for more than just the plants.

Setting up GSI practices in sequence creates a treatment train. Water directed through multiple practices is filtered more thoroughly by various mechanisms, removing multiple pollutants. A treatment train might begin with a swale that directs water into a rain garden; after passing through the rain garden, the water travels through a series of pipes and ditches then into a constructed stormwater wetland, where additional treatment occurs. GSI BMPs can be configured in numerous combinations to create a treatment train that meets the stormwater treatment goals and fits into the available project space. Treatment trains also increase on-site retention volumes; each practice retains a certain volume of water, and each of these add together to create a larger overall retention system (Fig. 8.5).

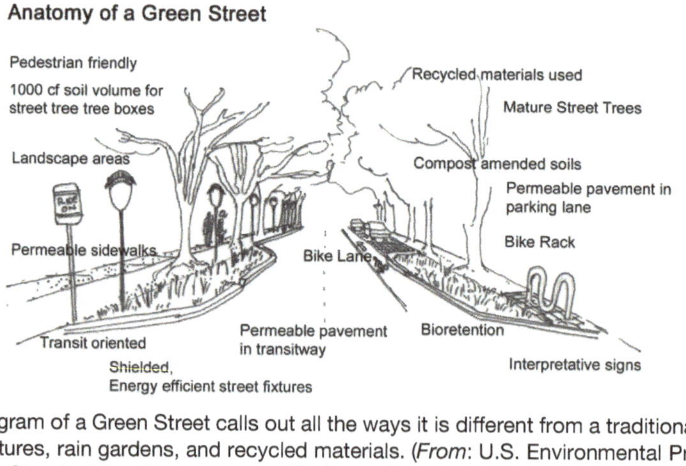

FIG. 8.5 This diagram of a Green Street calls out all the ways it is different from a traditional street including energy-efficient fixtures, rain gardens, and recycled materials. (*From*: U.S. Environmental Protection Agency, Learn About Green Streets. https://www.epa.gov/G3/learn-about-green-streets.)

CONSERVATION NEIGHBORHOOD DESIGN

Conservation design for neighborhoods is less about specific practices and more about reimaging how land is used in a new development. In true conservation subdivision design, half or more of the buildable land area is designated as undivided, permanent open space [7]. To achieve the same lot yield as traditional development, smaller, narrower lots are used for single family homes and other residential types such as townhomes are frequently included. Selecting the land for conservation should be the first step, not an afterthought. This allows the most appropriate tracts to be kept in their natural state and provide ecosystem services (see Macro-Scale Practices). Less pristine areas should be developed into parks and recreational areas that maintain open space and reduce impervious surfaces while still providing function and value to residents. When looked at as a whole, a conservation subdivision provides extreme value and access to amenities while reserving and harnessing the natural function of the land (Fig. 8.6).

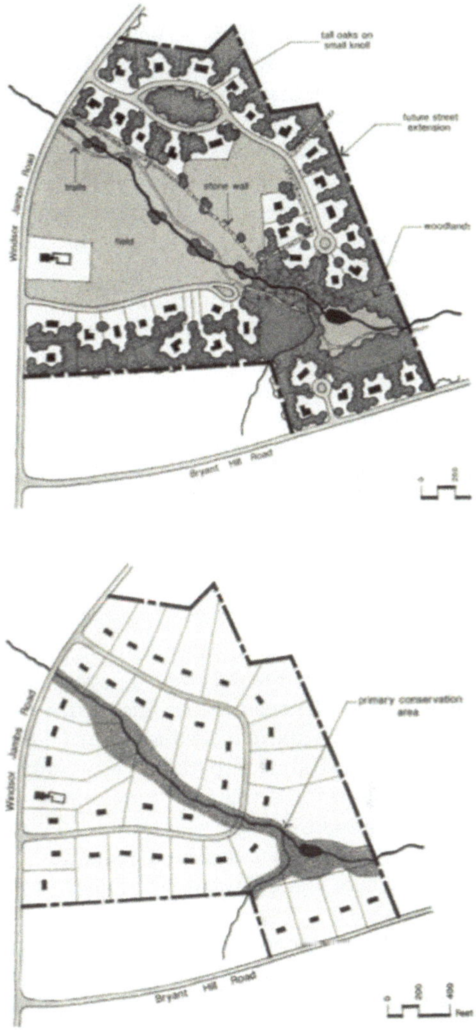

FIG. 8.6 Conventional (upper) and conservation (lower) subdivision design for a 96-acre site. Both designs have 124 lots, but the conservation design lots are $^1/_8$-acre versus ½-acre lots for the conservation design. Residents in the conservation subdivision have access to a much larger natural area. (*From*: Delaware Department of Natural Resources, Conservation Design for Stormwater Management, 1997. https://cfpub.epa.gov/watertrain/pdf/delawaremanual.pdf.)

SITE-SCALE PRACTICES

Small, site-scale green stormwater infrastructure practices have traditionally been called low-impact development (LID), a term coined by Larry Coffman in the 1990s for an alternative approach to stormwater management [9]. The idea was to capture and treat stormwater at the site level, where it falls, instead of shunting it directly into the storm sewer system. LID practices are designed to mimic and restore the natural hydrologic regime of a small area after development. Coffman envisioned these practices would minimize the impacts of development by reducing imperviousness and conserving natural ecosystems and drainage courses, and dispersing detention in small cells across the site instead of in one large detention pond.

The term LID has recently been replaced by Green Stormwater Infrastructure (GSI). This term expands the idea of LID beyond a single site and encompasses larger-scale practices such as constructed stormwater wetlands. Also, the "low impact" of LID practices refers only to stormwater and not the overall design or intent of the development pattern. The name LID is generally assumed by the layperson to mean that the overall site development has a lower impact, which is simply not the case. Leadership in Energy and Environmental Design (LEED) is a rating system developed by the US Green Building Council that better quantifies the entire development strategy. Use of the term Green Stormwater Infrastructure and now LEED is supplanting LID.

GSI uses the natural filtration of soil to infiltrate and clean stormwater. Each of these practices works to slow water down, spread it out, and soak it in. The flow from impervious surfaces is directed into a GSI BMP where this treatment happens. Numerous pollutants build up on roadways, parking lots, sidewalks, and roof tops. When it rains, typically the stormwater picks up both dissolved and particulate contaminants and moves them into the storm sewers that empty into local creeks and streams, resulting in impaired waterways. GSI BMPs are located to intercept stormwater runoff before it enters the storm sewer system. Practices such as rain gardens (bioretention cells), swales, and pervious pavement, all slow water and encourage it to soak in. Water that infiltrates into soil is cleansed by the naturally occurring microorganisms. Infiltration also reduces the amount of water moving into the storm sewer.

Site-scale GSI practices are appropriate for suburban and urban development patterns but are less effective in the hyperurban core due to the high percentage of impervious cover and lack of space for appropriately sized gardens and swales. In these dense urban areas, some practices can still be implemented including street trees, especially with a filter box design, green walls, and green roofs. The City of Chicago has embraced green roofs and now features them on more than 500 buildings including City Hall [10]. A green roof turns a flat impervious roof into a luscious garden stories above the ground. The roof must be engineered to bear the weight of soil, plants, and water held within the soil and must also be protected with a water shielding barrier. If all these criteria are met, the garden can take on any design or scheme to meet the desire of the owner.

Suburban development patterns, especially those featuring single family detached homes, offer many opportunities to install GSI BMPs. Traditional landscapes can be easily reimagined to include a grass swale, a flowering rain garden, and a hidden rain collection system used for irrigation.

Apartment complexes with expansive parking lots and large roofs can easily reap the benefits of GSI. Seldom used grassy areas can be reimagined as rain gardens that retain and filter parking lot runoff as it travels to the storm sewer. Big box stores and shopping centers are also logical locations for GSI practices.

The natural systems used in GSI have additional benefits including habitat for urban wildlife including song birds, humming birds, pollinators, butterflies, and migratory birds. Each small garden offers an oasis for these organisms, making their existence in a less than desirable location. Although one small area may not provide all the habitat requirements for a species or population, enough of these gardens situated in close proximity can become a habitat highway where the sum of the pieces is much more valuable than any one garden alone.

Design Considerations

GSI BMPs are intended to capture and treat small storm events, typically the 90th percentile storm, which encompasses a wide range of volumes depending on local rainfall patterns. The water retained is filtered through the BMP and then either piped into the storm drain system or infiltrated into the local soil structure. Either method reduces the peak stormflow volume leaving the site, thus reducing downstream flows. In soils with slow infiltration rates, filtration is typically the best option, allowing rain water to filter through the BMP and into the storm sewer in a short amount of time, instead of standing for days before soaking in. Whereas in high infiltration soils, the rain water passes quickly through the soil and recharges subsurface flows in a timely manner. Both methods provide water quality benefits.

During large rain events, for example, 100+ year storms, GSI provides little water quality benefit. Both gray and green infrastructure are only as good as the storm they are designed for and designing for the 100+ year storm event is almost always cost prohibitive. In large-scale flood events, GSI practices, like gray infrastructure, will be overwhelmed and will do little to reduce the impact of flooding. However, both the mid- and macroscale practices discussed in this chapter both offer real flood protection.

Pollutant Removal

How well does GSI actually improve water quality? This is an ongoing field of study across the United States, and as more practices are installed, the data on pollutant removal efficiencies continue to grow. The Center for Watershed Protection produced several versions of the National Pollutant Removal Performance Database to summarize available data [11]. Fig. 8.7 summarizes the percent removal efficiency for wetlands, rain gardens, infiltration practices (including pervious pavement), and open channels (including swales). Almost all the BMPs show a positive median removal efficiency for all of the pollutants tested.

There is still a great need for data on GSI practices, especially related to design criteria, as that has yet to be standardized across regions of the United States. Part of this is due to the variability in soil type, plant communities, and rain fall volumes. However, engineers have overcome these hurdles when designing gray infrastructure, and the same needs to be done for green infrastructure.

Types of Green Stormwater Infrastructure

	Definition	Considerations
RAIN GARDEN (BIORETENTION)		
 A rain garden at the Ghirardi Family Watersmart Park, League City, Texas. In the bottom right of the photo, you can see a curb cut; this is where the stormwater from the parking lot drains into the garden for filtration. This installation has a perforated under drain pipe that runs the length of the garden and transports water into the storm sewer system after it filters through the engineered soil mix.	A shallow, bowl-shaped depression that is filled with an engineered soil mix and planted with native and adaptive plants. These gardens are designed to accept runoff from parking lots, roof tops, and other impervious surface.	The exact design of a rain garden can vary depending on the size of the watershed for the garden. A small watershed (e.g., suburban residential driveway) might only need a small, shallow rain garden to capture and infiltrate the 90th percentile storm, whereas a garden that collects water from a large parking lot should be built much deeper to hold a greater volume of water.
SWALE		
 A vegetated swale at the Ghirardi Family Watersmart Park, League City, Texas. This swale was planted with native Texas plants, which slow water traveling through the swale, allowing sediments and pollutants to fall out of suspension.	A shallow drainage way with gently sloping sides that is typically vegetated but can be lined with rock.	These shallow, open ditches move water from one location to another while allowing for infiltration, evaporation, and evapotranspiration. Rocks and vegetation within the swale slow the velocity of the water, allowing solids to fall out of suspension. When planted with turf grass, swales can be easily mowed and maintained. Swales differ from rain gardens in that they do not typically contain engineered soil, nor do they pond and hold water.

Continued

Types of Green Stormwater Infrastructure—cont'd

	Definition	Considerations
RAINWATER HARVESTING		
 A highly decorated cistern at the Master Gardner Pavilion and teaching garden, Victoria, Texas. In the right of the photo, you can see the white pipe that transports water from the gutters of the building to the tank.	A system to collect and store runoff from an impervious surface for later, nonpotable uses.	Rainwater harvesting can take place at many scales, from an individual 40 gallon barrel, up to a 1500 gallon cistern, or a 25,000 gallon underground system. Watershed size, water needs, and budget should all be considered when determining the project scope.
PERVIOUS PAVEMENT		
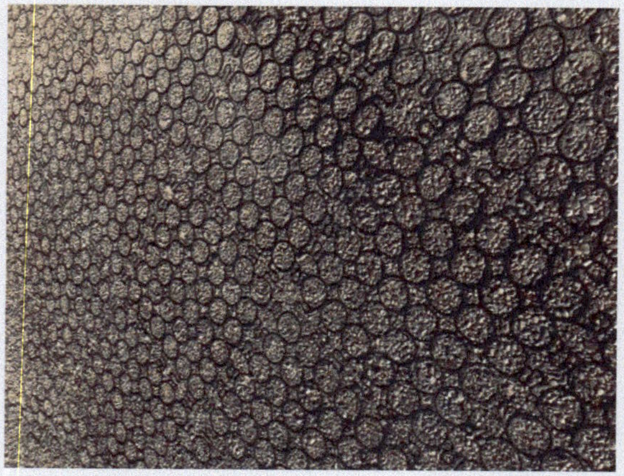 White Oak Music Hall, Houston, Texas. This parking lot is a plastic grid with gravel; the grid system gives the parking area more structure, prevents rutting, and prevents compaction of the rock, thus maintaining the permeability of the system.	An alternative to asphalt or concrete that allows stormwater to drain through the porous surface into a reservoir for temporary storage. It can take on many shapes including pavers, porous concrete, grass pave, or gravel systems.	Many options for pervious pavement exist. Each comes with a different price point and its own set of considerations. Grass pave systems work well for overflow parking areas and pavers for smaller spaces. Pervious pave installations over clay soils should consider a gravel subbase to increase the retention area.

GREEN ROOFS

Green roof on a medical building in Webster, Texas. This roof has a shallow, engineered soil mix over a plastic water barrier to protect the roof. A variety of native plants were used to reduce maintenance needs.

An extension of a roof that adds water proofing, a drainage system, a lightweight growing medium, and plants.

Green roofs are one of the most picturesque GSI practices and create stunning views. These rooftop gardens have many engineering considerations due to the added weight load of the soil, water, and plants. These can all make retrofitting a building difficult, but any large building with a flat roof offers an opportunity for a green roof.

	Median Percent Removal Efficiency					
	Total Suspended Solids	Total Phosphorus	Total Nitrogen	Copper	Zinc	Bacteria
Wetlands	72	48	24	47	42	78
Rain gardens *(Bioretention)*	59	5	46	81	79	n/a
Infiltration *(includes pervious pavement)*	89	65	42	86	66	n/a
Open channels *(includes swales)*	81	24	56	65	71	-25

FIG. 8.7 Removal efficiency of various green infrastructure. (*Modified from*: Center for Watershed Protection, National Pollutant Removal Performance Database, version 3. https://owl.cwp.org/mdocs-posts/fraley-mcneall-_national_pollutant_removal_perf_v3/. 2016.)

CONTEXT AND SCALE

The degree of ecosystem services provided by green infrastructure is very much scale dependent. However, the ecological thread that connects the large- and small-scale green infrastructure landscapes, as previously discussed, are the natural processes that provide ecosystem services for humans that must live on the land. There is a fundamental scale dependency, and then, for the amount of ecosystem services that can be provided by green infrastructure. This scale dependency also affects how well given GSI BMPs work in rural, suburban, or urban environments. We cannot put a large self-sustaining prairie in a dense downtown, for example, although we may well be able to counter some of the impacts of dense urbanization by planting native grasses in medians or pocket parks or by using tree boxes with native trees.

BMPs in Context

A useful concept for thinking about where particular GSI BMPs do best is the Urban Transect [12]. The Transect was developed to partition the urban continuum into coherent segments amenable to particular management strategies. The idea is that urban form is tailored to each zone. So, codes for things such as building height, building setback, and sidewalk width will be customized for each zone. In fact, the New Urbanists have developed an entire coding system called form-based codes that is built on the Urban Transect.

The Urban Transect is an excellent guide for the placement of GSI BMPs. Location is not only about scale but also about context. Some GSI BMPs fit better in certain zones than in others. Some BMPs could be installed in any Transect zone, whereas others may fit well in one zone but would be completely out of place in another. BMPs provide the most function when used where they fit best. Fitness is a fundamental concept of ecological health and therefore resilience. A particular species or feature out of place usually means an ecosystem is stressed. An abundance of weeds, for example, suggests serious stress and that the system may be near a tipping point toward more degradation. Fitness also indicates that positive rather than negative feedback loops are present.

Fig. 8.8 depicts the continuum of the urban transect with GSI practices [13]. For instance, a detention basin with a stormwater wetland would not be appropriate in T-4 to T-6. Tree boxes or tree trenches would be suitable for the denser urban areas but likely would be overkill in a T-3 setting. There are GSI BMPs for every zone, but not every BMP is suitable for every T-zone. The darker colors of this graphic are where we hypothesize the greatest fit for each GSI BMP. The BMP will be most stable in its zone of greatest fit, and this is where it will contribute the most to system resilience. So good housekeeping (e.g., proper storage and use of fertilizers and pesticides) is a priority everywhere. Detention basins can take runoff from the higher T zones, but because they occupy a lot of space, they really belong in T-2 and T-3.

The T-1 Urban Transect zone corresponds to undeveloped land. Undeveloped land, particularly where

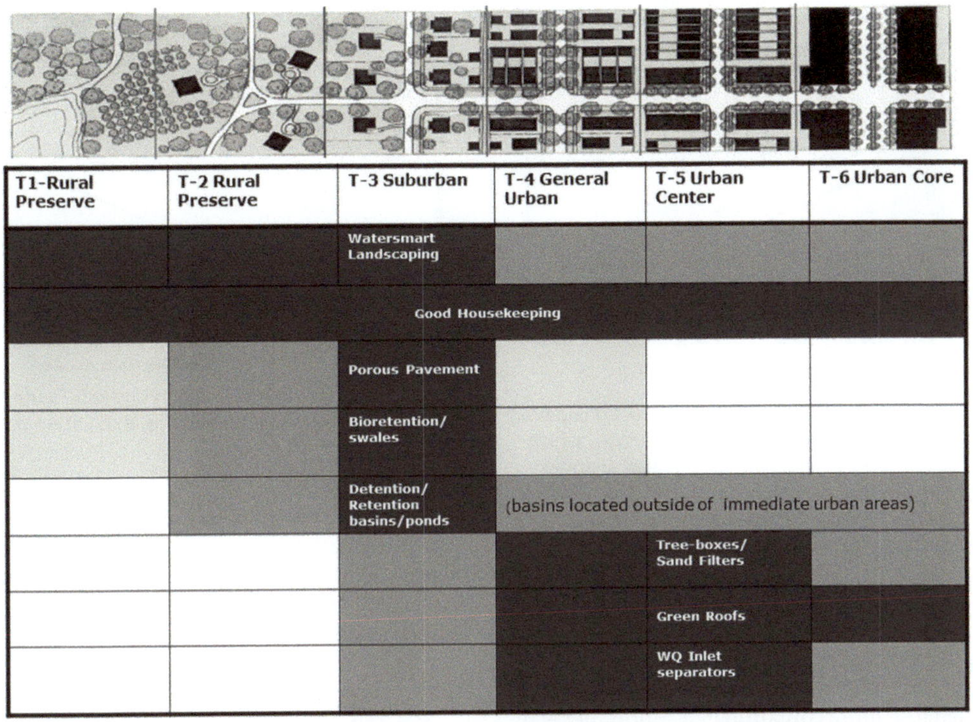

FIG. 8.8 This diagram of the New Urbanist Urban Transect overlaid with GSI BMPs depicts the zones where BMPs are most suitable and should be considered for installation.

intact ecosystems are functioning, is the area least in need of stormwater BMPs. This zone is where the macroscale GI practice of land preservation and restoration is best used, giving the best bang for the buck in terms of ecosystem services.

Watersheds and Walkability

The Impervious Cover Model (Fig. 8.9), developed by the Center for Watershed Protection, is a conceptual model that graphically summarizes a very large body of data validating the idea that stream health correlates very closely with percent impervious cover in the contributing watershed. Very significantly, the model posits critical thresholds at 10% and 20% impervious cover. Above 20% to 25% impervious cover, most streams will support only the hardiest aquatic species. Below 10% imperviousness, a stream and its watershed are considered "healthy," supporting a diversity of life. Recent research detects significant biotic degradation where watershed impervious cover is as low as 2%.

An unavoidable takeaway from this model is that urban development has an unmitigated negative impact on the environment and that the only remedy is to reduce density and install GSI wherever possible. A deeper dive into the impervious surface model, however, suggests that GSI is only part of the solution at higher levels of impervious cover, say above 40% to 60%. Standard suburban development—about 4 houses per acre—is roughly 30% to 40% impervious cover. Where development has more than 50% impervious cover, it would be very difficult to install enough BMPs to reduce impervious cover to anything approaching 25%, the empirically derived threshold for minimal watershed health, much less reaching 10%, the standard for a relatively healthy system. From a narrow environmental point of view, these higher-density, high-impervious-cover watersheds are just "throwaway" watersheds, where very little can be done to restore full watershed health.

But other factors come in to play at the higher end of the density scale, suggesting that these watersheds might be contributing to system resilience and not reducing it. Higher-density urban areas are much more likely to be walkable, an emerging indicator for urban resilience. Higher-density areas thus confer resilience by offering the most resilient form of transportation—the ability to walk from one place to another, in a short period of time.

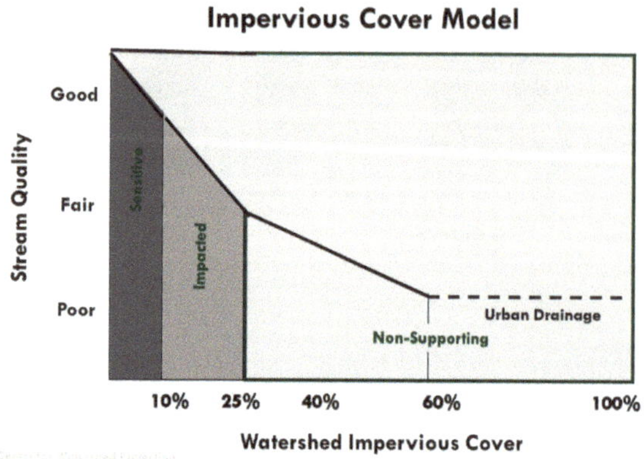

FIG. 8.9 Stream quality decreases as a watershed is paved over. This graphic summarizes hundreds of studies, which would more accurately be displayed as a scatter diagram. The intent of this graphic is to highlight the imperviousness thresholds. Natural diversity decreases from left to right. (*Modified from*: Center for Watershed Protection.)

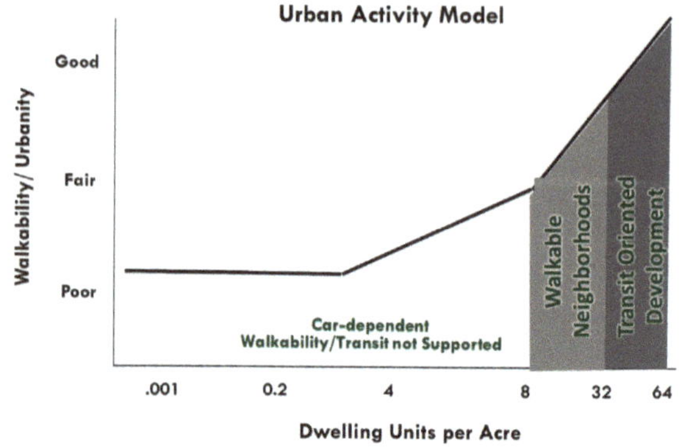

FIG. 8.10 The pattern of urban diversity is the mirror image of natural diversity as seen in Fig. 8.9. Urban diversity increases from left to right.

There is also a growing demand for walkable urban development. Walkability implies proximity, which implies high density, which implies more impervious surface, and therefore more contaminated runoff, per acre. However, the per capita impervious cover is markedly less in high-density developments than in low-density suburban developments. The ratio could be as much as 50 to 100 times per capita imperviousness and land consumed for suburban densities than for high-density, walkable areas [14].

The concentration of populations into higher-density urban areas, then, might result in markedly less pressure to develop low-impervious-cover watersheds, our very best and most effective kind of green infrastructure. Higher-density urbanization could thus be considered a beneficial form of development. Just a simple doubling of standard residential developments from 4 houses to the acre to 8 to 10 houses per acre could reduce nitrogen loadings by 20% to 40%, for a given number of houses (see Ref. [12]).

But how does this idea of higher-density square with the idea that "nature knows best"? First—it is an issue of scale, and the watershed scale is fundamental. We must consider the total loading of pollutants, not only the pollutant concentration of runoff from a lot or a neighborhood, important though that is. Nature is a system, not an individual lot. And second, we are interested in stability and resilience. Size matters, as does diversity.

Diversity of flora and fauna is part of what makes a natural landscape resilient, and the larger the tract, the more diverse and stable an area is likely to be. In turn, human diversity is also crucial for resilient communities. And urban diversity is very much dependent on density. Just as with the stream health curve in the Watershed Impervious Cover Model, we can construct an urban diversity, or perhaps better, an urbanity curve that is to some degree a mirror image of the stream health curve (Fig. 8.10).

We can construct an urbanity curve or model by changing the X axis units to density, in terms of houses or units per acre. The vertical scale becomes walkability. Just as with the thresholds for watershed health on the left of Fig. 8.9, we can posit some thresholds for urbanism on the right. At about 8 units per acre, there is just enough density to start to support some limited neighborhood business, such as a corner store. At about 30 units to the acre, there is enough density to support transit-oriented development, with its full complement of a diversity of shops and services within a neighborhood.

Diversity is thus a common currency for assessing the resilience of either natural areas (to the left) or human communities (on the right side).

Density of course is not sufficient to ensure walkable development takes place, but there can be no walkability without some density. It is not necessary to have Manhattan-level densities (in excess of 200 units/acre). Simply developing with narrow-lot, single-family-detached homes (8 units/acre) can greatly reduce runoff pollution compared with modern suburban densities. The sweet spot for maximum returns on density, in terms of availability of urban services, is at about 30 to 40 units/acre, corresponding approximately to three- to four-story development. This level of density supports practically all the urban amenities found in large, dense cities.

Combining the Watershed Imperviousness Model (Fig. 8.9) with the Urban Activity Curve (Fig. 8.10), we can construct a Watershed Health Curve (Fig. 8.11). One way to look at this curve is to think of the stream quality and urbanity curves as representing diversity—natural diversity on the left and urban diversity on the right. Diversity thus becomes a "common currency" for thinking about human and natural diversity at the watershed scale.

If diversity is a measure of stability, and current research suggests it is, then focusing on both urban and rural diversity, and maximizing both in their respective contexts, may be the safest path to infrastructure reliability and resilience [15,16].

The Watershed Health Curve allows us to see the value of compact development at the watershed scale.

The takeaway from this new Watershed Diversity Model is that there are four broad areas where natural solutions can be employed, but where contexts differ greatly.

1. Large intact ecosystems. These areas are the natural solution. Higher-density, walkable development relieves pressure on large natural systems.
2. Low-density suburban development. This is the area most in need to GSI-BMPs, and this is where there is room to easily employ practices such as stormwater treatment wetlands and rain gardens.
3. High-density walkability and transit-oriented development. Opportunities for GSI BMPs are more limited than in area 2, but well-placed BMPs such as green roofs can still contribute significantly to runoff reduction.
4. High-density development (above about 10 units to the acre) can itself be a BMP that reduces total pollutant loadings relative to low-density suburban areas.

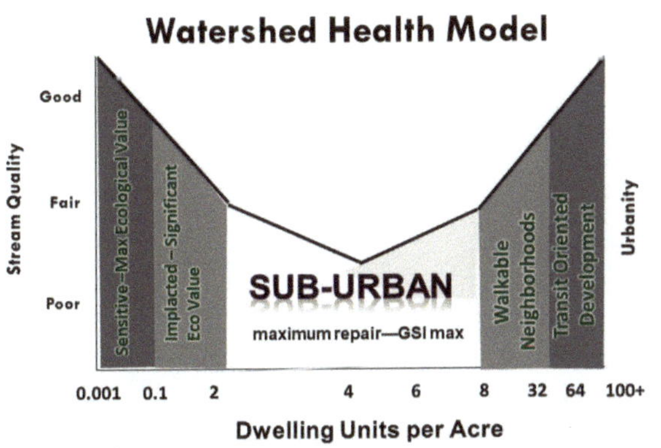

FIG. 8.11 Diversity provides a unifying theory of watershed health. Natural diversity or stream health increases to the left, with low human density. Urban diversity or urbanity increases with human density. The vast middle zone of 2–8 units to the acre, encompassing most development in the United States, is the zone of maximum per capita pollutant loading and thus the zone of maximum total loads. This is the zone where GSI can do the most good.

CONCLUSIONS

Nature provides us a largely untapped cornucopia of services. We are just barely beginning to scratch the surface of harnessing what nature freely provides. We know that nature knows best, we just do not quite yet understand how we can fully incorporate natural solutions into our overall infrastructure.

Large (over 5000 acres) tracts of healthy ecosystems have the most to offer in terms of services that replace more expensive and less reliable gray infrastructure. We understand so very little of the workings of complex ecosystems—all the more reason we urgently need to preserve these existing large tracts. They need to be preserved so that they can provide the services we need, but even more importantly for what we have yet to learn from these ecosystems. In a very real sense, these large tracts of functioning ecosystems are arks of wisdom and information that we need to take into the future, when we will no doubt know more about how to take advantage of ecosystem services. But even now, we do have some excellent examples of communities taking advantage of these services, even if it is only a very small piece of the GI that could be harnessed, for example, the case of New York City and the incorporation of GI into their water supply system.

Our best replication of ecosystem services occurs at the midscale of green infrastructure. We can incorporate green infrastructure such as stormwater treatment wetlands into drainage networks, to such a degree that emergent ecosystem properties begin to appear.

There are many opportunities for the incorporation of small-scale GSI in urban areas. These smaller-scale GSI BMPs cannot provide the full panoply of services that larger areas can, but they can make a very significant contribution to improving urban stormwater runoff, for example. It is very important to recognize, however, that no amount of small or even midscale GSI can make up for the loss of healthy prairies and forests.

It is not likely that natural solutions can replace all or even most of our gray infrastructure. But the more we can incorporate natural solutions into the overall system of infrastructure, the more resilient infrastructure will be. And very importantly, the more beautiful it will be. This is not something we can say very often about gray infrastructure.

REFERENCES

[1] E. McMahon, M.A. Benedict, Green Infrastructure: Linking Landscapes and Communities, Island Press, 2006.

[2] Texas A&M Agrilife Extension, Community Health and Resource Management. http://www.communitycharm.org/.

[3] J.W. Hanlon, Watershed Protection to Secure Ecosystem Services: the New York City Watershed Governance Arrangement. Case Studies in the Environment, April 2017. https://doi.org/10.1525/cse.2017.sc.400879.

[4] J. Jacob, F. Jaber, M. Sipocz, C. York, Stormwater Wetlands for the Texas Gulf Coast, second ed., 2012.

[5] City of Austin, Green Streets: An Introduction, September 2015. https://austintexas.gov/sites/default/files/files/Transportation/Complete_Streets/GreenStreetsWeb092115.pdf.

[6] U.S. Environmental Protection Agency. Learn About Green Streets. https://www.epa.gov/G3/learn-about-green-streets.

[7] R.G. Arendt, Conservation Design for Subdivisions: A Practical Guide for Creating Open Space Networks, Island Press, Washington, DC, 1996.

[8] Delaware Department of Natural Resources, Conservation Design for Stormwater Management, 1997. https://cfpub.epa.gov/watertrain/pdf/delawaremanual.pdf.

[9] H.E. Golden, N. Hohooghi, Green infrastructure and its catchment-scale effects: an emerging science, Wiley Interdisciplinary Reviews: Water 5 (2018) e1254, https://doi.org/10.1002/wat2.1254. 2018.

[10] City of Chicago, Chicago Green Roofs. https://www.chicago.gov/city/en/depts/dcd/supp_info/chicago_green_roofs.html.

[11] Center for Watershed Protection, National Pollutant Removal Performance Database, Version 3. https://owl.cwp.org/mdocs-posts/fraley-mcneall-_national_pollutant_removal_perf_v3/. 2016.

[12] A. Duany, E. Talen, Transect planning, Journal of the American Planning Association 68 (3) (2002) 245–266.

[13] J.S. Jacob, D. Crossley, Choices for Growth: Quality of Life and the Natural Environment. TAMU-SG-05-701, 2005.

[14] J.S. Jacob, R. Lopez, Is denser greener? A model for evaluating higher density development as an urban stormwater-quality best management practice, Journal of the American Water Resources Association 45 (3) (2009) 687–701, https://doi.org/10.1111/j.1752-1688.2009.00316.x. 2009.

[15] K.S. McCann, The diversity-stability debate, Nature 405 (2000) 228–233.

[16] S. Knapp, The link between diversity, ecosystem functions, and ecosystem services, in: M. Schröter, A. Bonn, S. Klotz, R. Seppelt, C. Baessler (Eds.), Atlas of Ecosystem Services, Springer, Cham, 2019.

CHAPTER 9

How Smart Land-Use Policies Help Avoid Future Headaches

SAMANTHA A. MEDLOCK, CFM • JAMES SCHWAB, FAICP, BA, MA

INTRODUCTION

Local communities across the nation contend with risks associated with aging and degrading infrastructure that will not withstand the chronic and sudden stresses associated with changes in demographics, extreme weather, and climate change. The scale of the problem is vast and often hidden from view, with more than one million miles of pipes delivering water, more than 640,000 miles of high-voltage transmission lines at full capacity, and more than 30,000 miles of levees that will require an estimated $80 billion in the next 10 years for maintenance and improvement [1]. The connection between land use and infrastructure is often overlooked, but community development plans and regulations drive decision making for nearly every infrastructure investment from regional and citywide scales down to the neighborhood, block, and lot.

Land-use planning as a discipline and a process provides a community vision of where and how to grow to enrich and protect quality of life. Plans and regulations typically take the form of community-wide comprehensive plans, capital improvement plans, land-use plans, and zoning documents that identify appropriate use for the range of types of development and spaces. These include designated areas for commercial, residential, industrial, or other uses and for parks, community services, and open space. Land-use planning should be an inclusive process with input from the full range of stakeholders, from residents and employers to public safety and other professional city staff to inform plans and designs to serve as the blueprint for community growth, as well as rebuilding following disasters. Land use is such an essential community function that a lack of planning can often be identified as a culprit for exacerbated disaster losses following major events. Moreover, communities often revisit land-use plans and building codes following such disasters to apply lessons learned in the form of enhanced codes and standards to guide redevelopment to reduce future losses.

PUTTING RISK AND RESILIENCE AT THE CENTER OF LOCAL LAND-USE POLICIES

Little more than a generation ago, risk and resilience were among the rarest topics to appear in a local comprehensive plan. Contrast that with the American Planning Association (APA) announcement in summer 2018 that its newest national planning award would be given for the best local resilience plan [2]. Growing awareness of the high costs of natural and other disasters, exacerbated by climate change and demographic shifts to high-hazard areas, has driven the planning profession in particular, and local governments more generally, to realize that designing communities for resilience is essential for long-term success and prosperity. Cities cannot continue to suffer devastation like that wrought upon New Orleans by Hurricane Katrina, the Jersey Shore by Hurricane Sandy, or Houston by Hurricane Harvey, nor can western states ignore the rapidly increasing costs of wildfires.[1]

In all these cases, land-use policies played a major role in defining the affected region's vulnerability. In all cases, infrastructure and development go hand in hand, for neither can exist without the other. In many communities, we are living with the legacy of past failures to understand those connections.

[1]Just for the decade beginning in 2010 through July 2018, NOAA [3] showed a grand total of CPI-adjusted losses in the United States of $657.7 billion. Both Katrina (in the previous decade) and Harvey individually topped the $100 billion mark. This was already more than 30% higher than the entire previous decade. This sharp upward trend in losses from major and catastrophic events is expected to continue.

Optimizing Community Infrastructure. https://doi.org/10.1016/B978-0-12-816240-8.00009-4

It is important to understand that risk has more than just physical dimensions. The physical dimensions obviously include threats to life, health, and property. Losses are inevitably tallied after disasters in lives lost, people injured, and structures damaged or destroyed. All of these have direct economic consequences, ranging from the costs to families of lost wage earners to medical costs for those injured or who become sick because of postdisaster conditions. However, the largest economic consequences of the physical destruction resulting from these events involve repairs, rebuilding, demolition of lost structures, and lost personal belongings and building contents. The latter costs necessarily include debris management and disposal, which in the worst disasters can involve massive tonnage of everything from ruined appliances to furniture to personal effects and vegetative debris like downed trees. The exact nature and proportions vary with the size, scope, and nature of the event.

In the modern world, however, development almost never exists without infrastructure, which includes roads, bridges, sewer and water lines, power lines, and communications systems, at the very least. The resilience of these systems, which includes elements of location and design, is a huge factor in determining the extent of disaster losses. Much of this infrastructure is publicly owned, but telecommunications and energy systems (electricity and natural gas) generally are not. Thus, losses will occur in both the public and private sectors, depending on the vulnerability and location of these systems. That vulnerability, in turn, will be determined to some extent by the nature of the hazards affecting the region. California, for instance, will focus much more on mitigating seismic and wildfire hazards, while the Gulf Coast and Southeast will be concerned with coastal storms. Each hazard has its own implications for land-use planning and infrastructure design.

However, losses are never limited to direct damage alone. Economic losses result from closed businesses, workers unable either to access jobs or to perform them because of illness or injuries, traffic congestion or delays because of destroyed bridges, power outages, and other interruptions stemming from infrastructure damage. Moreover, families dislocated from their homes must cope with the costs of short-term rentals, relocation, and adjustments to their normal commuting patterns. Lost public services, such as school closures, can also have detrimental impacts with economic consequences.

In addition to the physical and economic risks associated with natural hazards, communities and infrastructure owners and operators are facing increasingly complex liability risks where harm may be traceable to specific planning, siting, design, or maintenance decisions. As extreme weather events become more frequent, survivor claims against municipal defendants and infrastructure owners are also gaining greater traction with examples associated with Hurricane Harvey making their way through the courts at the time of publication [4–7]. Communities are well advised to integrate current risk data into their land-use planning and development standards to mitigate against future losses and to better manage potential legal risks that may arise as the severity and frequency of natural disasters increase.

In short, the rising overall costs of disasters ripples through all these component factors with serious implications for both present and future land use, compelling reassessment of land-use regulation at the local level.

Integrating Hazard Mitigation into Local Planning

One key response to this urgent need to reassess local land-use regulation came from the American Planning Association [8], with considerable amplification from the Federal Emergency Management Agency (FEMA), which sponsored the APA project [9,10]. The bottom line was almost deceptively simple. The Disaster Mitigation Act of 2000 (DMA) linked state and local eligibility for federal hazard mitigation grants to adoption of a hazard mitigation plan. FEMA reviewed and then approved plans meeting regulatory requirements established under DMA. Most of these plans have been prepared by emergency managers, but APA argued that a great deal of the effectiveness of such plans depends on the integration of mitigation priorities into land-use regulations and building codes, with the goal of guiding development out of harm's way.

Ideally, the premise for most local land-use regulations—which include zoning, subdivision ordinances, and similar legislation—emanates from the local comprehensive plan (alternatively known as a master or general plan, depending on the state). The APA report thus operated from the fundamental premise that the first target of integration for hazard mitigation priorities should be the comprehensive plan, but it did not stop there. It should also apply to other kinds of local plan making, such as area plans (e.g., downtown or corridor) and functional plans (e.g., transit); to the implementation tools for such plans, primarily land-use ordinances and building codes; to development work, such as site plan reviews; and capital improvements programs, which dictate priorities for public expenditures on infrastructure and public facilities. In short, the approach is holistic. Other APA projects, such as a subsequent report on post-disaster recovery [11], discuss other aspects of such integration including the need to address climate change. The National Oceanic and Atmospheric Administration (NOAA) has funded significant work on that last topic.

An important minority of states has incorporated some emphasis on hazard mitigation in the local comprehensive plan, either through incentives or through mandates. Notably, these include California, Florida, Colorado, Oregon, Iowa, and Washington. However, one can also find numerous local jurisdictions that are adopting, or have already adopted, an integrated approach even without state programs or mandates supporting it.

The point, ultimately, is that without such a focus on hazard mitigation throughout the planning process, many routine land-use decisions will escape scrutiny regarding their implications for resilience for both infrastructure and the built environment. The spotlight on hazard mitigation is essential to achieving meaningful community goals for increased resilience against future disasters.

Creating a Culture of Resilience

Those community resilience goals highlight an even larger purpose of this entire exercise: the creation of an underlying culture of resilience that will help institutionalize resilience as an operational concern of local government. However, as the National Academies has noted, "building the culture and practice of disaster resilience is not simple or inexpensive." [12].

It is not a one-shot effort. It requires deliberate and consistent leadership and a willingness to embed resilience as not only a priority, but also an assumption, of local governance. It requires public education and ongoing commitment.

However, it can be and has been done. Roseville, California, responding to numerous and varied hazard threats, has succeeded in presenting itself as a safe place to invest because of its initiatives [8]. Among other achievements, Roseville acquired Class 1 status in FEMA's Community Rating System, reducing its flood risks while gaining for its residents a 45% reduction in flood insurance premiums. Displaying some ingenuity, it has even used a herd of goats to chew down underbrush to reduce wildfire risks ahead of the annual fire season.

Cedar Rapids, Iowa, attracted the notice of the National Academies [12] study because it used the 2008 floods to educate citizens about their opportunities to reduce future flood risks and to establish a culture of preparedness that has responded well to subsequent flood threats [13]. In Cedar Rapids, this has included substantial and ongoing investments in flood control, buyouts, and other infrastructure and land-use approaches. These are two of the leading and most often-cited examples, but one can find others in the growing literature on local government resilience.

Finally, the National Institute of Standards and Technology expanded upon the theme of integration with a detailed description of how a community resilience plan can work with other community plans to respond to the need for resilience in the face of disasters and climate change [14]. Further discussion of some aspects of this study appears in the next section of this chapter.

RELATING INFRASTRUCTURE TO LAND USE

It is almost axiomatic, but almost as surely too simplistic, to discuss infrastructure by paraphrasing the well-worn movie line from *Field of Dreams*, "If you build it, they will come."[2] In reality, the cause-and-effect relationship between building infrastructure and new residential, commercial, or industrial development is more complex, interdependent, and organic than this catchphrase implies. Money for infrastructure is generally lean enough that new transportation facilities more often follow demand than precede it. On the other hand, state and interstate highways connecting different cities and metropolitan areas extend well beyond the current boundaries of developed areas and tend to encourage the extension of development along those corridors. The interstate highway system has long been cited as a factor in urban sprawl [15].

The physical presence of infrastructure, however, is not the only way in which development may be encouraged in a specific location. Equally important are public policies regarding infrastructure extension into new areas, including not only roads but also water and wastewater lines, power lines, and communications systems. If it is clear enough that the necessary

[2] In truth, the voice in the cornfields said, "If you build it, he will come," referring to Shoeless Joe Jackson. Nonetheless, it has been very tempting, however, for many people to quote the plural version in the context of development, usually referring to new or extended roads and highways.

infrastructure will accompany new development in a given area, developers generally have the signal they need—and may be able and willing either to find ways to support it, financially or in kind, or to advocate for public expenditures through various channels. There is generally nothing wrong with this process. It is the normal give and take of the development approval process and vital to encouraging community growth.

Where and How We Build

What is critical to the issue of resilience in this relationship between infrastructure policy and land use is the question of where we choose to build and how we build. Simply put, some locations are always more problematic than others with respect to hazard resilience. Floodplains are probably the most obvious issue, but other potentially hazardous areas include earthquake fault zones, liquefaction zones, steep slopes, and the wildland–urban interface. If we consider industrial and technological hazards as well, then chemical plants, refineries, railroads, and the like frequently need surrounding buffer zones and appropriate mitigation measures. These can usually be achieved through zoning codes and site plan reviews. This latter group generally requires significant infrastructure to serve its needs.

Natural hazards, however, generally afford the best opportunity to forestall questionable development through public policies that make development less attractive. In the most extreme cases, public policy can simply prohibit public investments that would aid development. At the federal level, the primary example would be the Coastal Barrier Resources Act [16], passed in 1982, which made certain barrier islands ineligible for federal expenditures and assistance, with the aim of preventing development in highly vulnerable (and physically unstable) locations. State and local policy can adopt a similar stance for hazardous or highly sensitive lands. Sometimes this is also achieved through purchases of development rights or direct purchase in fee simple, to ensure removal of such lands from consideration for future development.

Public policy can also limit future development by either not providing even the promise of adequate infrastructure connections. It can also aim to make development safer, for example, by ensuring adequate evacuation routes and appropriate building sites with strong code requirements. Regarding roads, for example, designing perpendicular routes that afford escape from either coastal or riverine flooding is a standard mitigation tool. So, too, is the requirement of multiple routes of egress in areas subject to hazards like wildfires [17]. In addition, appropriate land development practices, including landscaping and grading regulations (including road improvements), can minimize destabilization of slopes to help prevent landslides [18,19]. More importantly, however, zoning limitations on density based on slope ratios serve the purpose of simultaneously limiting the extent of infrastructure intrusion into hilly or mountainous areas vulnerable to erosion.

Land-use regulations can also serve to harden infrastructure against disaster damage. For example, development regulations can require undergrounding utility lines, either generally or in specific districts, to prevent power and telephone outages from high winds, ice storms, and similar events that can topple or overburden lines and poles. In many cases, this decision may be a function of cost–benefit analysis comparing the likelihood of such events against the costs of undergrounding utilities, as opposed to conventional aboveground construction, among other considerations [20].

Location and Design of Infrastructure

One particularly important facet of local decision-making in this regard concerns regulations for both subdivision design and planned unit developments (PUDs). Subdivision ordinances lay out the terms of engagement for developers wishing to subdivide plots of land into multiple lots, largely for residential housing. They establish a procedure for approval based on meeting standards for proposed development that include road layouts, infrastructure improvements, and the size and placement of structures and the lots on which they sit. Depending on the size of the subdivision, which can range from fewer than a dozen to thousands of homes, requirements for infrastructure hookups can be extensive and complicated.

APA and the Association of State Floodplain Managers (ASFPM) collaborated to examine how this process can properly address flood hazards [21]. The report noted that many communities will often need to go well beyond minimum standards of the National Flood Insurance Program (NFIP) because new subdivisions often occupy land involving creeks and streams not mapped by the NFIP because the land was previously not a priority for the NFIP, which focuses on developed areas in allocating floodplain mapping resources. It notes that communities may require developers to undertake such mapping as part of the process.

It also discusses why planning techniques such as cluster development can allow land within the floodplain to remain undeveloped while concentrating homes on higher land above base flood elevation. Clearly, infrastructure—both roads and utilities—will follow the placement of homes and thus also avoid areas of potential inundation. However, subdivision standards can also deal with issues such as floodwater retention and freeboard requirements.

Although the APA/ASFPM study focused on floods, communities can also find or adapt standards for subdivision design related to other designated hazardous areas such as landslide zones or the wildland–urban interface [22]. Colorado's Department of Local Affairs has established a website, "Planning for Hazards," that contains "Subdivision and Site Design Standards Model and Commentary," dealing with land suitability, subdivision improvement agreements, standards for hazard mitigation, and cross-references to zoning and other codes [23]. Most importantly, it takes a multihazard approach that includes geologic and wildfire hazards in addition to flooding. In neighboring Utah, which has dealt with both landslides and seismic threats, municipalities often require geotechnical studies of some soil conditions before development can proceed (see Ref. [24]). That obviously will also affect the siting and design of infrastructure intended to serve development where these requirements apply.

Maintaining Vital Infrastructure to Achieve Resilience

Planning deals with both prospective development and periodic or incremental redevelopment of existing built-out areas of the community. The foregoing section dealt in part with the need to plan for adequately resilient and disaster-resistant infrastructure in new development, but existing infrastructure requires ongoing, planned maintenance. In addition, like it or not, communities sometimes face the need for modest or substantial restoration of infrastructure after a disaster. Planning for such restoration or replacement before disaster strikes is one especially wise way to enhance community resilience. It is also critical in the 21st-century environment to incorporate resilience for future climate conditions into the planning of capital improvements.

RISK MANAGEMENT AND CRITICAL INFRASTRUCTURE

Efforts to better identify, manage, and mitigate risks associated with infrastructure begin with a robust and precise understanding of those risks. Following catastrophic events, notably major losses over the past 15 years, risk managers have benefited from significant advancements in the availability of risk data and analytics methodologies to produce more granular and actionable analyses to inform infrastructure project siting, configuration, and maintenance strategies. Communities, insurers, and investors can now achieve an unprecedented understanding of risks across the range of natural hazards through catastrophe models and climate impact modeling to include more reliably foreseeable impacts that may occur over the life cycle of the asset and manage risks across portfolios of assets. With advanced computing capability and scenario modeling available early in the planning stages, communities and infrastructure owners can adjust planning to account for future conditions against physical threats and foreseeable demographic and market changes.

Credit rating agencies are also taking advantage of greater availability of community asset and climate risk data to adjust their municipal credit rating factors to assess community creditworthiness and ability to meet financial obligations in full and on time [25]. Credit rating agencies are increasingly factoring in community resilience planning, insurance, and availability of liquidity to reduce dependency on federal aid into ratings, with implications for the cost of capital to communities.

Municipal bond issuers face risks associated with extreme weather and other effects of climate change in two ways. One is through the increasing costs of extreme weather events and uncertainties associated with federal disaster aid; the other is through the more gradual environmental changes, such as high-tide flooding and increased frequency of hot summers, that can affect land use and disrupt economic activity that increase costs, reduce revenues, or both [26]. Communities can take steps to increase ratings or prevent downgrades by engaging in resilience planning that takes future conditions into account, strengthening infrastructure, increasing community insurability, and ensuring adequate liquidity to weather disasters with greater resilience and less reliance on uncertain federal aid.

As risk information is increasingly available at more localized scales, community plans, standards, and development decisions can come under increasing scrutiny where they do not incorporate best-available data or resilience-based codes. For example, the City of Houston's lack of zoning laws [27], raised questions about whether land-use regulations could have reduced the widespread flood losses due to Hurricane Harvey [28]. By allowing thousands of homes to be built in floodplains and not providing for adequate drainage or flood risk reduction infrastructure, the City of Houston exacerbated flood losses associated with the storm [29]. Class action lawsuits were filed against federal, county, and city entities claiming that the City and Harris County bear joint responsibility for the release of floodwaters during the storm.

STRATEGIES AND TOOLS FOR COMMUNITY AND INFRASTRUCTURE RESILIENCE

As the first and last line of defense against threats and hazards, communities have the primary authority and responsibility for managing land use and for adopting and enforcing the range of codes and standards that drive demand for infrastructure. As infrastructure ages and failures occur, communities risk loss of essential services—transportation, power, potable water, wastewater treatment, and communications—with significant implications for public health and safety, as well as community economic vitality. Whether loss of services stems from a major storm event or the more chronic effects of climate change, communities and infrastructure owners need to anticipate and prepare for changing risks.

Land-use regulation is one of the most effective tools to ensure community and infrastructure resilience as it provides the opportunity to shape communities and assure adequate capacities for transportation, open space, drainage, and the distribution and consumption of energy. Local authorities over infrastructure can manifest at multiple nodes of the planning cycle, beginning with how land is subdivided and designated for particular uses. Site plan reviews, planned development processes, zoning amendments, requests for variances, and site inspection all offer important opportunities to ensure that development is consistent with requirements and the vision of the community as embodied in plans and standards.

Traditional planning strategies too often promote or even compel the use of "gray" infrastructure, particularly for stormwater and flood risk management, which has resulted in the proliferation of undersized and brittle infrastructure throughout the United States. These can include concrete-lined channels, culverts, pumps, and other drainage and urban flood infrastructure that are engineered to specific standards based on current conditions, but not as adaptable to changing conditions. These structures can also be overwhelmed in extreme weather events, resulting in greater damage when they fail. Flood risk infrastructure such as seawalls, levees, and dikes can attract additional development as they lower current flood risk, increasing the potential consequences from overtopping or failure, impacting populations that do not perceive the residual risk associated with these projects.

To keep pace with changing conditions, cities need to take new approaches to infrastructure that incorporate sustainability and resilience measures and provides for more gradual improvements to respond to extreme weather and other effects of climate change that can no longer be avoided. They also must reexamine the development codes and standards to incorporate approaches to reduce risk and cost. Approaches through use of floodplain freeboard, setbacks, and preservation of valley storage and conveyance are among widely adopted practices as they provide measurable reductions in risk and the cost of flood insurance. Guidance for building homes and businesses above minimum requirements through organizations like the Insurance Institute for Business and Home Safety have achieved significant traction in state and federal policy, both for their technical soundness and for the ways they are proven, storm after storm, to reduce loss and save lives.

Cities and infrastructure owners are also exploring the changes to governance and finance approaches that enable regional collaboration for planning, investment, and engagement with insurance and capital markets. When multiple jurisdictions engage in sustainability and resilience planning as an economic region, they can better design infrastructure projects with a longer-range planning horizon and devise finance approaches for both the project and its maintenance over its useful life.

CONCLUSION

Several lessons emerge from 21st-century community experience in addressing this interface between land use and infrastructure around disaster resilience. First is the need to recognize that losses encompass not only physical concerns, but also those stemming from economic losses and legal liability in connection with public policies related to development.

The second lesson involves an obvious, but complex, solution through the integration of hazard mitigation and resilience priorities throughout the local planning process, particularly the relationships between land-use policy and infrastructure. Much practical work has focused in recent years on a thorough integration that accounts for the impact not only of comprehensive planning but all types of plan making, as well as conscientious use of policy implementation tools such as zoning, subdivision ordinances, site plan review, and capital improvements programming. Holistic approaches to development regulation and guidance, baked into local governance, enable the creation of a culture of resilience both in local government and in the civic arena. The number of examples of this approach is growing at both the state level (e.g., Colorado) and in communities (e.g., Roseville, California, and Cedar Rapids, Iowa) and is being rewarded with recognition from major professional organizations like APA.

Finally, it is important to develop public-sector understanding of the importance of risk management for critical infrastructure. Major industry sectors such as insurers, investors, and credit rating agencies are paying increasing attention to these issues and influencing governmental behavior because of their increased scrutiny of community resilience. All these trends can be expected not only to continue, but also to grow.

REFERENCES

[1] American Society of Civil Engineers, 2017 Infrastructure Report Card, 2018. https://www.infrastructurereportcard.org.

[2] American Planning Association, National Planning Award Categories & Eligibility: Resilience Award. https://www.planning.org/awards/categories/#Resilience.

[3] NOAA National Centers for Environmental Information (NCEI) U.S, Billion-Dollar Weather and Climate Disasters, 2018. https://www.ncdc.noaa.gov/billions/.

[4] In Re Upstream Addicks and Barker (Texas) Flood-Control Reservoirs, Sub-master Docket No. 17-9001L.

[5] In Re Downstream Addicks and Barker (Texas) Flood-Control Reservoirs, Sub-master Docket No. 17-9002L.

[6] Val Anthony Aldred, et al., v. Harris County Flood Control District, et al., Docket No. 2017-57831.

[7] L. Abraham, et al. v. Costello, Inc. et al., Docket No. 2018-22747.

[8] J. Schwab, Hazard Mitigation: Integrating Best Practices into Planning, APA Planning Advisory Service, May 2010.

[9] Federal Emergency Management Agency, Integrating Hazard Mitigation into Local Planning: Case Studies and Tools for Local Officials, March 1, 2013.

[10] Federal Emergency Management Agency Region X, Integrating the Local Natural Hazard Mitigation Plan into a Community's Comprehensive Plan: A Guidebook for Local Governments.

[11] J. Schwab, Planning for Post-Disaster Recovery: Next Generation, APA Planning Advisory Service, 2014.

[12] National Research Council, Disaster Resilience: A National Imperative, The National Academies Press, Washington, DC, 2012. https://doi.org/10.17226/13457.

[13] KCRG-TV9, Cedar Rapids Sees Losses and Gains from the Flood of 2008, 2018. http://www.kcrg.com/content/news/Cedar-Rapids-sees-losses-and-gains-from-the-flood-of-2008-485483511.html.

[14] National Institute of Standards and Technology, Community Resilience Planning Guide for Buildings and Infrastructure Systems. NIST SP 1190, 2016. https://doi.org/10.6028/NIST.SP.1190v1. https://doi.org/10.6028/NIST.SP.1190v2.

[15] Transit Cooperative Research Program, The Cost of Sprawl—Revisited, TCRP Report 39, 1998.

[16] U.S. Congress, Coastal Barrier Resources Act (P.L. 97-348), 1982.

[17] J. Schwab, S. Meck, J. Simone, Planning for Wildfires, APA Planning Advisory Service, 2005.

[18] R.B. Olshansky, Planning for Hillside Development, November 1996.

[19] R.B. Olshansky, Regulation of hillside development in the United States, Environmental Management 22 (3) (1998) 383–392.

[20] Edwards, Kelcey, Inc/Exeter Associates, Inc, Maryland State Highway Administration Research Report: Cost Benefits For Overhead/Underground Utilities, 2003. https://www.roads.maryland.gov/opr_research/md-03-sp208b4c-cost-benefits-for-overhead-vs-underground-utility-study_report.pdf.

[21] J. Schwab, C. Berginnis, T. Turner, A. Read, N. Walny, Subdivision Design and Flood Hazard Areas, American Planning Association, 2016.

[22] National Fire Protection Association, NFPA 1141: Standard for Fire Protection Infrastructure for Land Development in Wildland, Rural, and Suburban Areas, 2017. https://www.nfpa.org/codes-and-standards/all-codes-and-standards/list-of-codes-and-standards/detail?code=1141.

[23] Colorado Department of Local Affairs, Planning for Hazards Implementation Project. https://www.colorado.gov/pacific/dola/planning-hazards-implementation-project.

[24] Highland City Public Works Department, Engineering Division. Minimum Requirements For Geotechnical Studies And Reports. https://www.highlandcity.org/DocumentCenter/Home/View/116.

[25] C. Flavelle, Moody's Warns cities to address climate risks or face downgrades. Bloomberg, November 29, 2017. https://www.bloomberg.com/news/articles/2017-11-29/moody-s-warns-cities-to-address-climate-risks-orface-downgrades.

[26] S&P Global, How Our U.S. Local Government Criteria Weather Climate Risk, March 20, 2018. https://www.capitaliq.com/CIQDotNet/CreditResearch/RenderArticle.aspx?articleId=2009147&SctArtId=450533&from=CM&nsl_code=LIME&sourceObjectId=10469131&sourceRevId=7&fee_ind=N&exp_date=20280319-20:06:24.

[27] City of Houston, Development Regulations: No Zoning Letter, 2019. http://www.houstontx.gov/planning/Forms/devregs/2019-coh-no-zoning-letter.pdf.

[28] S. Boburg, B. Reinhard, Houston's "wild west" growth, Washington Post (August 29, 2017). https://www.washingtonpost.com/graphics/2017/investigations/harvey-urban-planning/?utm_term=.6a974adab3ab.

[29] B. Fleming, The Real Villains in Harvey Flood: Urban Sprawl and the Politicians Who Allowed it. The Guardian. August 31, 2017, 10:13 AM. https://www.theguardian.com/commentisfree/2017/aug/31/real-villains-harvey-flood-urban-sprawl (citing S.D. Brody, S.E. Davis, W.E. Highfield, et al., Wetlands 28 (2008) 107. https://doi.org/10.1672/07-90.1).

Introduction

Imagine a world without buildings. It is very hard to do. Buildings are engrained in daily life and vital to the economy. Without buildings, the economy, the educational system, governments, and society break down. Hospitals, schools, police and fire stations, housing, office buildings, and stores all underly what it means to be a community. News reports following natural disasters often focus on two metrics as proxies for the level of impact—injuries and fatalities and the number or value of structures destroyed. This further illustrates the importance of buildings to community resilience.

The resilience strategies presented in the following chapters represent a microcosm of what it means to implement resilience holistically. Buildings are so ubiquitous that a failure to address their resilience almost certainly leaves communities vulnerable. The other systems covered in this book all rely on the functionality of buildings before, during, and after a disaster. Transportation networks are designed to move people and goods between buildings. Utilities require buildings to house their management functions, and their services are almost exclusively delivered to buildings. A significant portion of the nation's wealth is captured in buildings and real estate, and billions of dollars in assets flow through financial markets annually. Buildings, like most components of infrastructure, are intended to last for 30, 50, or even 75 years.

The challenges to achieving resilient buildings lie at the intersections. Architects and engineers have the skills to design buildings to withstand almost any anticipated risk, but they must weigh the likelihood of the risk with the cost of designing and implementing the higher level of design. Frequently, the risk tolerance, the risk awareness, or the availability of funding on the part of the owner or investor will dictate whether a project can be designed above the minimum level established by the building code. Within the building code development and adoption process, similar considerations are at play. At what level do building codes cost-effectively provide for a minimum level of protection? Defining what a minimum level of protection is (or in other terms, an acceptable level of risk) in both model code development and adoption at a state or local level present interesting social science and economics questions—questions that have not been well studied.

This section of the book explores the main contributors to resilient buildings: the investors and developers that influence how and where buildings are constructed; the designers that recommend and implement resilience strategies; and the building codes that provide the community's expectations for safety.

INVESTORS AND DEVELOPMENT

Resilience does not just happen. It requires investment of resources—particularly financial resources—to assess risks and implement measures to address those risks. For the private sector, those investments are made based on careful consideration of multiple factors. For commercial real estate, the resilience of a project (and the associated monetary implications) is just one factor in how and where projects are constructed and the value of existing properties. Chapter 10 by Nirmul and Scott examines how developers and investors consider resilience today and how the evolving landscape of risk will drive investment decisions in the future. As they reveal, if government policy and investments cannot keep pace with the investment community's level of concern, investments will flow elsewhere. Although not addressed in this chapter, the potential decreased investment in established cities that face significant risks may present another set of resilience challenges—the social and economic resilience of populations ill equipped to seek new opportunities in more resilient communities.

Other chapters in the book help complete the story on the role of finance in achieving resilience. A common theme across all these chapters is the need for signals that capture the value of resilience measures. These signals can come from a variety of sources including consumer or investor demand, insurance premiums, bond ratings, and government incentives. The National Institute of Building Sciences (NIBS) approach to Incentivization (covered in the introduction to Part VI) attempts to align all potential signals to determine the most cost-effective resilience strategy and prompt investment in its achievement.

Metrics that feed into existing decision-making processes have the best likelihood for early adoption and will open the door for more focused, in-depth analysis in the future. Initiatives like GRESB have started to raise awareness among the investment community on strategies that reduce environmental, social, and governance risk.

GRESB released its first assessment for real-estate companies and funds in 2009, establishing a benchmark for institutional investors looking to optimize the risk/return profile of their investments. In 2018, GRESB introduced an optional resilience module to begin capturing information on how real-estate and infrastructure companies and funds are preparing for potentially disruptive events and changing conditions, assessing long-term trends, and becoming more resilient over time. The module's main goals are:

1. Meet investor demand for information about the resilience of property and infrastructure companies and funds.
2. Provide more information about strategies used by property and infrastructure companies to assess and manage risks from social and environmental shocks and stressors, including the impact of climate change [1].

In the residential market, some communities have already seen a drop in home values due to risk exposure. First Street Foundation found that increased tidal flooding across 15 states between 2005 and 2017 resulted in $15.8 billion in lost home values. Florida experienced the largest loss, totaling $5.4 billion [2].

The Urban Land Institute (ULI) conducted a study to examine how the real-estate industry is currently assessing and mitigating climate risk—particularly physical risks and transitional risks such as regulatory changes, availability of resources, and attractiveness of location [3]. Fig. 1 captures some of these risks. The Task Force on Climate-related Financial Disclosures (TFCD) looked at similar risks across multiple industries (TFCD's activities were covered in the introduction to Part III) [4].

The ULI study found that physical risks may lead to increased insurance premiums, higher capital expenditure and operating costs, and a decrease in liquidity and value of buildings. Transitional risks may cause some locations to become less appealing, leading to the potential for assets to become obsolete. To address these risks, some investors have already undertaken efforts to develop tools and standards to support better pricing of climate risk. These efforts include the following:

- Mapping physical risk for current portfolios and potential acquisitions,
- Incorporating climate risk into due diligence and other investment decision-making processes,
- Exploring a variety of strategies to mitigate risk including portfolio diversification and investing directly in the mitigation measures for specific assets, and
- Engaging with policymakers on city-level resilience strategies, and supporting the investment by cities in mitigating the risk of all assets under their jurisdiction [3].

As investors make more informed decisions based on climate risk, they may foster what Keenan, Hill, and Gumber call "climate gentrification." They examined how higher elevation homes and neighborhoods in Miami-Dade County, Florida, may exhibit a consumer preference (resulting in a higher degree of value appreciation) due to flood risk [5]. As they recognize, "If investors/owners see a relative disadvantage or opportunity cost to their lower elevation properties, then this may be one of many other factors that lead to spatial relocation or the disposition of a particular asset ... The cost burden may be increased by virtue of a cycle of declining tax rolls and fewer and fewer tax payers." These impacts may further exacerbate the ability of vulnerable populations to achieve resilience without specific policies to counteract these shifts.

These changes in home values are corroborated by research by Bernstein, Gustafson, and Lewis who used data from Zillow to show that properties subject to sea-level rise sell at a 75% discount when compared with similar, less vulnerable properties. Although prospective owners or investors are considering long-term risk, short-term renters do not appear to exhibit such concerns [6].

These types of studies have almost exclusively focused on the impacts of flooding on consumer choices. In fact, a study considering broader risks seems to indicate that exposure to natural disasters (even in the near past) has no effect on property values [7]. According to the study, "Median home prices in cities in the top 20 percent (Very High) for natural hazard risk have appreciated 65 percent on average over the past five years and 9 percent on average over the past 10 years while median home prices cities in the bottom 20 percent (Very Low) for natural hazard risk have

	Category	Potential impact
Physical risks	**Catastrophic events** Extreme weather such as hurricanes and wildfires.	• Costs to repair or replace damaged or destroyed assets; value impairment • Property downtime and business disruption • Potential for increased insurance costs or reduced/no insurance availability
Physical risks	**Changes in weather patterns** Gradual changes in temperature and precipitation—such as higher temperatures, rising sea levels, increasing frequency of heavy rain and wind, and decreased rainfall—which are likely to exaggerate the impact of catastrophic events.	• Increased wear and tear on or damage to buildings, leading to increasing maintenance costs • Increased operating costs due to need for more, or alternative resources (energy and/or water) to operate a building • Cost of investment in adaptation measures, such as elevating buildings or incorporating additional cooling methods • Potential for increased damages from catastrophic events • Potential for increased insurance costs or reduced/no insurance availability
Transition risks	**Market** The possibility that markets vulnerable to climate change will become less desirable over time. Rising capital costs to pay for building and maintaining infrastructure to manage climate risks.	• Reduced economic activity in vulnerable markets • Reduced occupier demand for properties • Reduced asset value • Potential for increased real estate taxes
Transition risks	**Policy and regulation** Regulations to address climate change—e.g., climate risk disclosure, tougher building standards, carbon pricing, emissions caps, changes to subsidies—as well as changing policies for providing funding for infrastructure or rebuilding after major events.	• Increased cost of doing business due to new disclosure requirements and compliance measures • Increased taxes—both those resulting from public policies such as carbon taxes and those for funding adaptation infrastructure • Loss of subsidies or other funding opportunities • Additional capital investment to comply with stricter regulation
Transition risks	**Resource availability** Changes in the availability of key resources such as energy and water, including water scarcity.	• Increased costs and reduced net operating income due to higher prices for water and energy • Additional capital expenditures to adapt buildings to operate with reduced/alternative resources
Transition risks	**Reputation and market position** Growing stakeholder preference to work with companies incorporating climate risk into investment decisions, and consumer preference for real estate products incorporating climate mitigation.	• Risk to company brand and reputation if no action taken • Lower liquidity and/or reduced attractiveness of assets that have not incorporated climate mitigation

FIG. 1 Types of climate risk and their potential impact on real estate. (Source: Reprinted from, K. Burgess, E. Rappaport, Climate Risk and Real Estate Investment Decision-Making, Urban Land Institute and Heitman, February 2019. https://europe.uli.org/we-content/uploads/sites/127/2019/02/ULI_Heitlman_Climate_Risk_Report_February_2019.pdf.)

appreciated 32 percent on average over the past five years and 3 percent on average over the past 10 years." Presumably, the desire to live in such locations is stronger than the potential impacts of hazards. Although this may be the case in the near term, as the impacts of such events increase and if governments do not take measures to address these risks (including through building codes), the landscape may shift as Nirmul and Scott suggest.

DESIGNERS

In Chapter 11, Anderson examines how design professionals are a lynchpin for the implementation of resilience. Architects in particular possess the ability to look from the micro- to the macroscale, understanding the impacts of projects from the individual occupant to the surrounding community. Designers serve as trusted advisors to owners and developers, providing expert advice on how to address the variety of risks a project may face. Although Anderson provides insight from the perspective of an architect, her words should not be lost on the numerous design disciplines involved at all levels—from interior designers, lighting designers, and civil, mechanical, electrical, and plumbing engineers to landscape architects to urban planners.

As the need to address the risks a community faces grows, designers become an increasing part of the solution—a role that designers may not be able to shirk. The Conservation Law Foundation (CLF) examined the potential liability that designers may have based on the projects they design [8]. As discussed below, the existence of minimum building codes may not provide liability protection. Designers are subject to a standard of care established through contract, knowledge, regulation, custom, and foreseeability. Even unforeseen events that could be reasonably foreseeable could create an exposure. The design community should take this potential liability seriously and look for opportunities to proactively address the risks that their projects face today and may face into the future.

Professional organizations representing all aspects of the building industry came together in 2014 to begin examining how the industry can and should work collectively to enhance community resilience [9]. The Industry Statement on Resilience started with 21 signatories in May 2014 and has since grown to nearly 50 as of March 2018. The Statement recognizes that achieving resilience requires the engagement of designers, planners, manufacturers, contractors, operators, owners, and regulators. An industry-wide effort to advance education, advocacy, planning, research, and response is necessary. Since the initial signing ceremony, the signatories have made significant progress in addressing resilience—although there is still lots to do as illustrated in this book. The Statement signatories released a progress report in 2016 covering their efforts to date [10].

Groups such as the American Institute of Architects (AIA) and the American Society of Civil Engineers (ASCE) have undertaken initiatives to better prepare their members for a changing built environment. AIA has launched a series of continuing education courses focused on what designers should know about resilience and how to incorporate it into their projects and how to support their community [11]. Architects have also recognized how their expertise can serve communities before or after a disaster. The Disaster Assistance Program provides architects with the tools to provide postdisaster safety assessments, participate in the code development and adoption process, and engage their community in resilience-based decision-making. The Disaster Assistance Handbook provides architects with the latest knowledge on how architects can help communities prepare for and recover from adverse events [6]. The Center for Communities by Design supports probono work by teams of architects and other designers to support action to address key community challenges—including resilience [12].

ASCE has taken the lead on examining how standards and the design process must evolve to meet the challenges presented by climate change. Reilly and Ayyub touch on some of these efforts in Chapter 13 from the perspective of the individual engineer, but the implications for the design professions at large are significant. ASCE's Committee on Adaptation to a Changing Climate was formed in 2011 to evaluate the technical requirements and civil engineering challenges for adaptation to climate change. The Committee has produced numerous documents outlining a potential path forward including a manual of practice [13]. The 2015 ASCE publication *Adapting Infrastructure and Civil Engineering Practice to a Changing Climate* [14] outlines the complexities of changing risks and makes recommendations on an approach forward:

- Engineers should engage in cooperative research involving scientists from across many disciplines to gain an adequate, probabilistic understanding of the magnitudes of future extremes and their consequences. Doing so will improve the relevance of modeling and observations for use in the planning, design, operation, maintenance, and renewal of the built and natural environment. It is only when engineers work closely with scientists that the needs of the engineering community become fully understood, the limitations of the scientific knowledge become more transparent to engineers, and the uncertainties of the projections of future climate effects become fully recognized for engineering design purposes.
- Practicing engineers, project stakeholders, policymakers, and decision-makers should be informed about the uncertainty in projecting future climate and the reasons for the uncertainty, as elucidated by the climate science community. Because the uncertainty associated with future climate is not completely quantifiable, if projections of future climate are to be used in engineering practice, it will require considerable engineering judgment to balance the costs of mitigating risk through adaptation against the potential consequences of failure.
- Engineers should develop a new paradigm for engineering practice in a world in which climate is changing but cannot be projected with a high degree

of certainty. When it is not possible to fully define and estimate the risks and potential costs of a project and reduce the uncertainty in the time frame in which action should be taken, engineers should use low-regret, adaptive strategies such as the observational method to make a project more resilient to future climate and weather extremes. Engineers should seek alternatives that do well across a range of possible future conditions.

- Critical infrastructure that is most threatened by changing climate in a given region should be identified, and decision-makers and the public should be made aware of this assessment. An engineering-economic evaluation of the costs and benefits of strategies for resilience of critical infrastructure at national, state, and local levels should be undertaken.

BUILDING CODES

Building codes are an essential strategy for assuring that community goals surrounding the safety of its building stock and its ability to withstand a hazard event are achieved. As Davis and Ryan point out in Chapter 12, there are multiple requirements to effectively capture the benefits that codes provide. Like many of the other strategies outlined in this book, coordination is key. Development, adoption, and enforcement coupled with education and training and investment in personnel must be deployed with a broad vision. Multiple stakeholders contribute to the success (or failure) of building codes. Their development and implementation must engage product manufacturers and material suppliers, designers, advocates, contractors and builders, and regulators. Enforcement requires a robust structure at the state and local level. Product manufacturers must assure that their offerings comply with requirements in the code. Designers, contractors, and installers must receive education and training to assure they remain up-to-speed on the latest provisions. Researchers should be prepared to translate their findings into code change proposals that incorporate the latest knowledge.

Numerous studies have demonstrated the value that building codes provide. NIBS found that the 2018 International Building Code (IBC) and the International Residential Code (IRC) provide a national-level benefit of $11 for every $1 invested when compared with codes in place in 1990 [15]. A FEMA analysis from 2014 estimated approximately $500 million in annualized losses avoided in eight southeastern states because of the adoption of modern building codes [16]. Effective and well-enforced building codes in Missouri have reduced hail damage to homes by 10% to 20% on average [17]. And, in the 10 years following Florida's adoption of a statewide building code, the code's adoption and application reduced windstorm actual losses by as much as 72%, producing $6 in reduced loss to $1 of added cost [18].

Despite these significant benefits, many jurisdictions across the country have either not adopted building codes or not kept up with adopting newer editions. Many states set code baselines statewide, whereas in 21 states, local governments determine what, if any, building codes apply in their jurisdictions. Currently, five states, representing 12% of the population, have state building codes that are 9 or more years old and, where local governments determine code adoption, upward of 25% and 10% of residents in some Midwest and Gulf Coast states, respectively, use codes that are just as dated. Building code enforcement is essential to realizing the savings captured in the code text.

Again, the availability of common metrics helps tell the resilience story. The Building Code Effectiveness Grading Schedule (BCEGS) scores communities based on the codes in place and their commitment to code enforcement using a scale of 1 (exemplary) to 10. BCEGS incorporates information on the size of the building code enforcement budget relative to construction activity, the professional qualifications of building inspectors, and past code enforcement levels. BCEGS is used by the insurance industry to support underwriting of property insurance. Communities with better BCEGS scores have been shown to perform better during a natural disaster, exhibiting decreased insured losses and better protected properties, residents, and businesses. FEMA uses BCEGS data to track the rate of code adoption against the Building Sciences Program goal of increasing the percentage of communities with disaster-resistant building codes [19].

The Federal Alliance for Safe Homes (FLASH) focuses on educating and empowering consumers to support higher levels of disaster resilience. As part of this work, FLASH surveyed homeowners on the link between building codes and resilience. Most homeowners were "very" or "extremely" concerned about the impacts of natural disasters. However, most assumed that adequate building codes were in place and enforced in their community to protect them. Additionally, 68% indicated that they would be "extremely concerned" or "very concerned" if they learned that their community had no codes in place. These findings appear to indicate that consumers are not thinking about building codes because they assume they already have the necessary protections and that community leaders have taken the necessary steps [20].

Although most of the focus to date on building codes and resilience has been on the IBC and IRC, there is increasing recognition that codes covering the energy and water efficiency aspects of buildings support resilience. Organizations such as the American Council for an Energy Efficient Economy (ACEEE) [21,22], National Association of State Energy Officials (NASEO) [23], Institute for Market Transformation (IMT) [24], and Southwest Energy Efficiency Project (SWEEP) [25] have all developed resources focused on the energy/resilience nexus. The International Energy Conservation Code (IECC) and ANSI/ASHRAE/IES Standard 90.1: Energy Standard for Buildings Except Low-Rise Residential have become an important component of building codes and support achievement of national and local priorities—including social, economic, and infrastructural resilience. They enhance social resilience by reducing air pollution and associated health impacts, energy burdens, and exposure to pricing volatility—particularly for vulnerable populations. Energy codes serve to reduce urban heat island effects and the stresses on the grid associated with extreme temperatures. Energy efficiency contributes to passive survivability and the amount of time that facilities can remain operational without grid-supplied energy.

Regardless of the type of infrastructure—whether buildings, transportation, water, or energy—the latest codes and standards establish important minimum criteria that provide a foundation for community resilience.

REFERENCES

[1] GRESB, Resilience Reference Guide, 2018. https://GRESB.com/resilience.

[2] First Street Foundation, Rising Seas Erode $15.8 Billion in Home Value from Maine to Mississippi, February 27, 2019. https://assets.floodiq.com/2019/02/784113f9d16323db82f696b9c3b0874e-First-Street-Foundation-Mid-Atlantic-Press-Release-Immediate-Release.pdf.

[3] K. Burgess, E. Rappaport, Climate Risk and Real Estate Investment Decision-Making, Urban Land Institute and Heitman, February 2019. https://europe.uli.org/we-content/uploads/sites/127/2019/02/ULI_Heitlman_Climate_Risk_Report_February_2019.pdf.

[4] Task Force on Climate-Related Financial Disclosures, Recommendations of the Task Force on Climate-Related Financial Disclosures, June 2017. https://www.fsb-tcfd.org/wp-content/uploads/2017/06/FINAL-2017-TCFD-Report-11052018.pdf.

[5] J. Keenan, T. Hill, A. Gumber, Climate gentrification: from theory to empiricism in Miami-Dade county, Florida, Environmental Research Letters 13 (2018) 054001. https://doi.org/10.1088/1748-9326/aabb32.

[6] A. Bernstein, M. Gustafson, and R. Lewis. Disaster on the horizon: the price effect of sea level rise. Journal of Financial Economics. January 1, 2019. https://doi.org/10.1016/j.jfineco.2019.03.013.

[7] Attom Data Solutions, Home Prices Rising Twice as Fast in U.S. Cities with Highest Natural Hazard Risk than in Lowest-Risk Cities, September 2017. https://www.attomdata.com/news/risk/2017-u-s-natural-hazard-housing-risk-index/.

[8] Conservation Law Foundation, Climate Adaptation Liability: A Legal Primer and Workshop Summary Report, January 2018. https://www.clf.org/wp-content/uploads/2018/01/GRC_CLF_Report_R8.pdf.

[9] Building Industry Statement on Resilience, 2014. https://www.aia.org/resources/9336-building-industry-statement-on-resilience:56.

[10] American Institute of Architects and National Institute of Building Sciences, Preparing to Thrive: The Building Industry Statement on Resilience: Helping Communities Construct a More Certain Future, May 2016. https://www.nibs.org/resource/resmgr/Docs/WHRS_SignatoryReport_final.pdf.

[11] American Institute of Architects, AIA Resilience and Adaptation Online Series. https://www.aia.org/resources/205786-aia-resilience-and-adaptation-online-series:56.

[12] American Institute of Architects, Disaster Assistance Handbook, 2017. https://www.aia.org/resources/71636-disaster-assistance-handbook.

[13] American Institute of Architects, Center for Communities by Design. https://www.aia.org/pages/2891-center-for-communities-by-design.

[14] MOP 140ASCE Committee on Adaptation to a Changing Climate, B. Ayyub (Eds.), Climate-Resilient Infrastructure: Adaptive Design and Risk Management, 2018, https://sp360.asce.org/PersonifyEbusiness/Merchandise/Product-Details/productId/244232276.

[15] American Society of Civil Engineers, J.R. Olsen (Eds.), Adapting Infrastructure and Civil Engineering Practice to a Changing Climate, 2015.

[16] Multihazard Mitigation Council, Natural Hazard Mitigation Saves: 2018 Interim Report, K. Principal Investigator Porter, C. co-Principal Investigators Scawthorn, C. Huyck, Investigators: R. Eguchi, Z. Hu, A. Reeder, P. Schneider, Director, MMC. National Institute of Building Sciences, Washington, D.C. https://www.nibs.org/resource/resmgr/mmc/NIBS_MSv2-2018_Interim-Repor.pdf.

[17] Federal Emergency Management Agency, Phase 3 National Methodology and Phase 2 Regional Study Losses Avoided as a Result of Adopting and Enforcing Hazard-Resistant Building Codes, 2014.

[18] J. Czajkowski, K. Simmons, Convective storm vulnerability: quantifying the role of effective and well- enforced

building codes in minimizing Missouri hail property damage, Land Economics (2014).
[19] K.M. Simmons, et al., Economic effectiveness of implementing a statewide building code: the case of Florida, Land Economics (2018).
[20] Insurance Services Office, National Building Code Assessment Report: Building Code Effectiveness Grading Scale. 2019 ed.
[21] D. Ribeiro, E. Mackres, B. Baatz, R. Cluett, M. Jarrett, M. Kelly, S. Vaidyanathan, Enhancing Community Resilience through Energy Efficiency, American Council for an Energy Efficient Economy, October 2015.
[22] D. Ribeiro, T. Bailey, Indicators for Local Energy Resilience, American Council for an Energy Efficiency Economy, June 2017.
[23] NASEO, Resiliency through Energy Efficiency: Disaster Mitigation and Residential Rebuilding Strategies for and by State Energy Offices, March 2015.
[24] R. Meres, E. Makela, Building Energy Codes: Creating Safe, Resilient and Energy-Efficient Homes, Institute for Market Transformation, July 2013.
[25] C.H. Brinker, Energy codes are life safety codes, Builder Magazine (September 1, 2017). https://www.builderonline.com/building/building-science/energy-codes-are-life-safety-codes_o.

CHAPTER 10

The New Resilient Built Environment: Perspectives From Investors and Owners of Private Buildings

DEVESH NIRMUL, CEM, CSDP, LEED AP O+M • JOHN SCOTT, BOMA FELLOW, RPA

INTRODUCTION

The built environment's vulnerability to environmental changes is a function of both potential impacts and its ability to withstand those impacts. The successful mitigation of vulnerabilities depends on how well stakeholders and shareholders alike can anticipate, prepare for and mitigate impacts, and communicate to the economic marketplace that risks are within a manageable range. In other words, for investments in the built environment to remain "sustainable and resilient" over their life cycle, they have to continue to signal to the investment community that they (1) sustain their high-performance attributes such as energy efficiency and (2) are able to absorb environmental impacts with minimal additional cost all while still providing for attractive returns.

The reality today is that human communities, their economies, and the natural environment are experiencing stresses and shocks that alter the socioeconomic-political system's current paradigm for effective anticipation, preparation, and mitigation of impacts. Whether it is a question of how fast we are able to adapt, how much investment (both public and private) we are able to make or not, or the extent to which intermarket and legislative forces are aligned effectively enough, it is prudent that we identify the gaps, missing links, and/or opportunities to enhance our ability to adapt and manage risk. The environmental change scenarios within this dynamic include, but are not limited to, climate change, pollution, other human-induced deterioration of the environment, and our current economic system's over dependence on fossil fuels—which bring their own set of vulnerabilities of price, availability, and access. This chapter focuses on the role of private investment and ownership in the built environment, particularly on climate change risk as experienced through (1) extreme storm events and sea-level rise and/or (2) vulnerability of the energy infrastructure that is both contributing to increased climate change risk and the source for the solutions that help mitigate such risks and help socioeconomic systems be more resilient in the face of current and future climate impacts.

Although the United States macroeconomy continues to incur more and more defensive spending associated with the impacts of climate change, the microeconomy through the decisions of individual building owners and investors is compelled to make strategic investments or in some cases divestments to maintain profitability and manageable levels of risk or perceived risk. It is within this strategic investment arena that innovation is occurring or will need to occur in the marketplace. In the true sense of the meaning of "investment," owners and investors have to derive value from resiliency investments either through (1) risk/return evaluation of existing assets and the creation of risk mitigation preparedness strategies starting with no/low-cost investments, (2) financial incentives such as reduced insurance premiums, lower interest rates or resiliency financing capability, and terms that maintain healthy cash flows and reduce operating costs all while enhancing property value, and/or (3) enhanced value that results from high-performance, differentiated products/services, or technologically leading-edge and synergistic

Optimizing Community Infrastructure. https://doi.org/10.1016/B978-0-12-816240-8.00010-0

investments (e.g., battery storage that provides both reliable and zero-emission back-up power during disasters/extreme events and subsequent reduced business loss insurance premiums for tenants, but that also can help building owners save during summer and winter peak energy consumption periods and reduce operating costs significantly). The caveat is that even with the promise of innovation to enable effective adaptation, the investor/owner community is challenged by a built-in inertia for status quo and inadequate levels of resiliency investments in the marketplace, whether that inertia is derived from market demand or the more short-term orientation around immediate profits and sales.

This chapter will walk through current-day examples of mainstream, environmentally-induced economic vulnerability scenarios and explore both existing and potential adaptation strategies within the investment demand, development, operations, and reinvestment continuum.

THE INVESTMENT–REINVESTMENT CONTINUUM

The authors interviewed institutional investors, property managers, owners, and tenants within the office, retail, and multifamily building sectors to determine what has changed over the last 20 or so years within the commercial real estate investment–reinvestment continuum and what we can expect in the future. They sought to identify how environmental stressors are manifesting themselves and being managed within the economic framework that underpins life cycles within the built environment. They specifically inquired about (1) drivers of concern whether they be shareholder/stakeholder or cost-driven, (2) concerted strategies to reduce risk and cost, and invest in mitigation of current/future impacts, and (3) the challenges that still remain in reconciling the asymmetries in the market's ability to effectively price and mitigate risks across this continuum/life cycle.

The continuum is broken-up into particular stages of the built environment (re)investment life cycle to focus on each relevant actor and their level of response to market and legislative resiliency drivers within this continuum (see Fig. 10.1). We have attempted to identify all shareholders/stakeholders, who through their voices and actions (or lack thereof), determine the market's resiliency drivers. For clarity, we assigned each shareholder/stakeholder within each continuum a symbol that represents their reactions or actions taken, if any, based on the specific dynamic resiliency scenarios we are exploring. The following key summarizes what each symbol signifies:

● A "Solid Circle" signifies the perception and/or reality that risk is being effectively managed, enabling a healthy investment–reinvestment lifecycle.[1]

○ A "Hollow Circle" reflects concern/awareness around the built environment's ability to address environmental stressors and shocks.

⬇ A "Down-Arrow in a Circle" represents the perception or reality among stakeholders/shareholders that current risk is not being effectively mitigated in the marketplace. As such, these institutions/individuals are sending a signal to the market that (re)investment may not happen or will be too risky and/or offer too little reward for the investment required to mitigate risks.

⬆ An "Up-Arrow in a Circle" represents successful mechanisms deployed by stakeholders, and/or marketplace recognition of such mechanisms, to mitigate risk to invite reinvestment within the next 20 years or some other specified time period.

⇧ A "Hollow Up-Arrow" represents early and critical need–based public and/or private investments to improve the built environment to more effectively absorb current and future impacts. In their current iteration, such investments may not be adequately addressing perceived and/or real risk. A good example would be an energy-intensive sea-level rise/flooding mitigation solution that relies heavily on power from the electric grid, a system that is often not available during an extreme event.

X An "X" signifies that the authors did not have enough information about the activities or perceptions of particular market-complementary stakeholders that the owner/investor contributors may interact with in the marketplace.

Figs. 10.1–10.3 provide some illustrative "templated" examples of resiliency scenario dynamics based on what the authors have observed in the marketplace. The actual relationship dynamic between the actors within each of these continuums can vary considerably across different geographies and real estate markets. The authors view this dynamic framework more as a tool for assisting a group of stakeholders within a particular region or community to figure out how each other's action/lack of action regarding vulnerability and

[1]It is very possible in a world of imperfect information and multiple drivers of investment and risk that while some stakeholders/shareholders within a particular vulnerability/resiliency scenario are raising concerns and even considering divestiture of real estate assets, others are not putting the same weight or any weight on perceived/real risks.

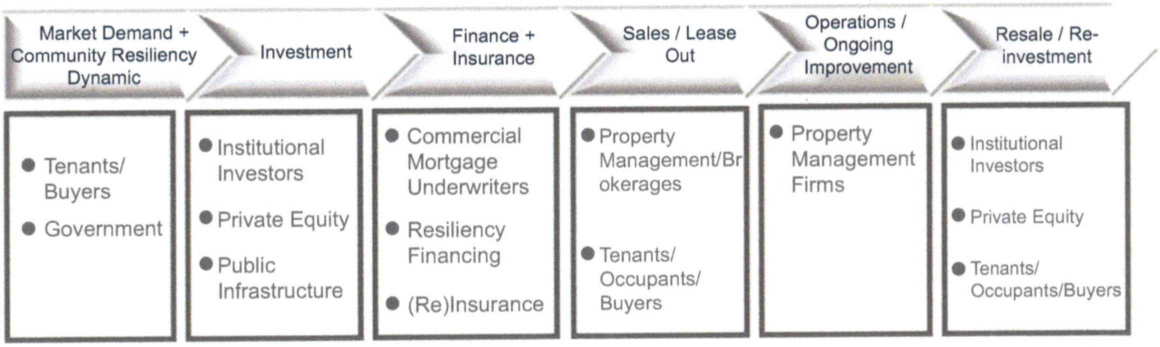

FIG. 10.1 Environmentally resilient built environment investment–reinvestment life cycle/continuum + stakeholder/shareholder mapping.

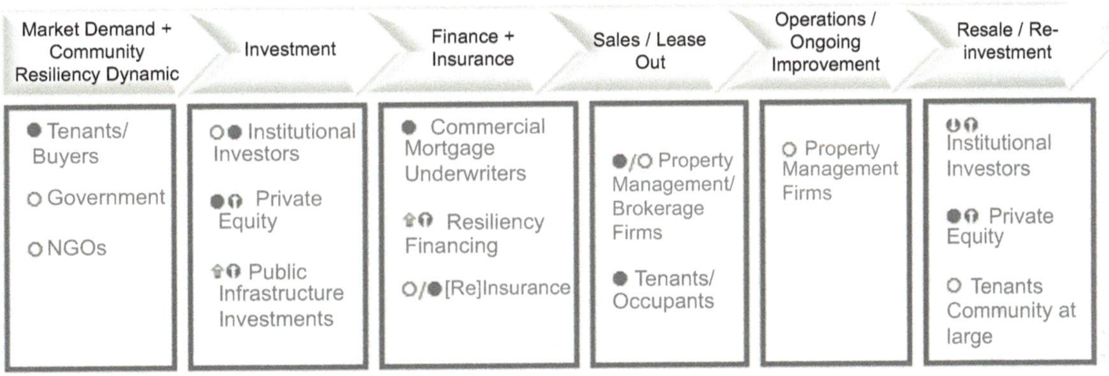

FIG. 10.2 Current-day environmentally vulnerable coastal urban built environment investment–reinvestment life cycle/continuum (e.g., cities like Miami, Boston, and New York).

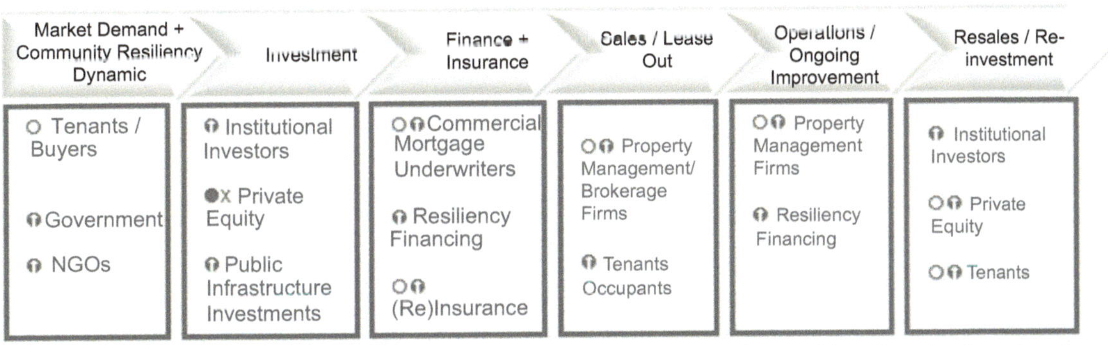

FIG. 10.3 Current/future environmentally resilient coastal urban built environment investment–reinvestment life cycle/continuum (e.g., city like Amsterdam or future of Miami, New York, or Boston).

resiliency investments impact the other actors' perception or reality around whether risk has been mitigated or not. For each specific shareholder/stakeholder the authors have interviewed, separate investment–reinvestment continuums are developed where necessary to illustrate both (1) specific investments/actions taken and (2) the interplay between actors within a particular resiliency scenario.

The following matrix provides context for some of the resiliency scenarios the authors explored with owners, property managers, finance, and insurance companies within the commercial real estate space. The

TABLE 10.1
Key U.S. 2018 National Climate Assessment Built Environment Infrastructure Impacts: Ramifications for Private Real Estate Owners, Tenants, and Stakeholders in the Larger Community.

2018 National Assessment "Key Message for Infrastructure"	Potential Impact on Vulnerable Private Commercial Owners and Buildings	Other Potentially Implicated Sectors of the Economy/Society
City government agencies and organizations have started adaptation plans that focus on infrastructure systems and public health. To be successful, these adaptation efforts require cooperative private sector and governmental activities, but institutions face many barriers to implementing coordinated efforts.	• Without effective coordination around minimal adaptation standards, policies and/or financing/incentives, privately owned dense urban building stock in low-lying coastal and windstorm-prone regions, could, without warning, lose value and become defunct and a strain on the local economy • Businesses and real estate owners will seek locations/markets that have a manageable/predictable level of climate risk	• Loss of local, regional, and national private companies that create economic value • Loss of substantive portion of the urban tax base • Decreased future reinvestment in the community especially in light of investors having taken losses either through (1) unexpected or severe impacts or (2) loss of capacity in the insurance and reinsurance markets
Extreme weather events are affecting energy production and delivery facilities, causing supply disruptions of varying lengths and magnitudes and affecting other infrastructure that depends on energy supply. The frequency and intensity of certain types of extreme weather events are expected to change.	• Increased business interruption losses that derive from both direct impacts to buildings (grid energy-dependent systems) • Indirect impacts to public infrastructure that private business, their employees and customers depend on for being able to successfully engage in commerce (ex. roads, bridges, water/sewer services, etc.)[2]	Same as above
Higher summer temperatures will increase electricity use, causing higher summer peak loads, while warmer winters will decrease energy demands for heating. Net electricity use is projected to increase.	• Businesses and real estate owners will seek locations/markets, energy purchasing and conservation strategies that offer a hedge against rising energy costs • Geographic regions that have already successfully integrated buildings into the capacity markets, through incentive structures, currently have a head start over other regions	• Increasingly, cities and regions will see energy cost and reliability become more and more of either a hindrance or opportunity for economic development • Increasingly and in concert with decreasing costs for clean/renewable energy generation and storage, current paradigms for reliable energy provision through a central plant–based utility model could become less attractive • State and local governments will be challenged to look at alternatives to dependency on investor owned utilities and policies of public utility/service commissions that are incompatible with climate change.

Changes in water availability, both episodic and long-lasting, will constrain different forms of energy production.	• Businesses and real estate owners will seek locations/markets, energy purchasing and conservation strategies that offer a hedge against rising energy costs	Same as above
In the longer term, sea-level rise, extreme storm surge events, and high tides will affect coastal facilities and infrastructure on which many energy systems, markets, and consumers depend.	• These impacts are already being felt in coastal vulnerable cities and are driving up costs for building owners and thus deterring investors from continued investment in such regions • Some ownerships (e.g., high-end residential) are implementing low/no-cost measures to reduce liabilities, insurance costs and maintain adequate insurance coverage	Cities that are effectively planning for adaptation to coastal risks (Miami for example), have introduced new standards for new construction to enable effective adaptation over the next 20 to 40 yrs. The remaining challenge for cities is existing buildings. Energy benchmarking/disclosure policies are being mobilized in major metro areas around the country. Do the energy resiliency and operational savings that may result from such programs provide sufficient incentive for broader-scope resiliency investments? What other tools and funding mechanisms can be mobilized to encourage more comprehensive retrofits (e.g., PACE, resiliency-focused investment trusts, mutual insurance companies, etc.)?
As new investments in energy technologies occur, future energy systems will differ from today's in uncertain ways. Depending on the character of changes in the energy mix, climate change will introduce new risks as well as opportunities.	The trend within the real estate industry is toward more sustainable buildings and ultimately buildings that work in harmony with and enhance the natural environment. For energy this translates into moving beyond energy conservation to energy generation and net-positive buildings. For owners and investors, new economic opportunity for energy investments are emerging.	Current-day energy utility generation and distribution business and governance models are being/will be challenged to remain effective and profitable. Those utilities that have evolved their business models to interact with buildings and communities for the provision of grid capacity and resiliency will be better positioned to be relevant and part of the resiliency solutions for today and tomorrow.

Adapted from the Fourth National Climate Assessment [1].

matrix (Table 10.1), references some of the built environment infrastructure impacts from the latest National Climate Assessment [1] and articulates the observed/potential ramifications for building owners, investors, operators and the larger community.

OWNER/INVESTOR COMMUNITY INTERVIEWS

The authors have organized the primary source content for this chapter to provide multiple complementary vantage points that build upon each other, spanning the spectrum of the *assessment of perceived and real resiliency risk* to the *tools and drivers for and investments in tangible energy and climate resiliency*. We start with a macro focus from a portfolio institutional owner and investor followed by a multiple owner/building-level perspective through the eyes of a commercial real estate/sustainability consultant. We then shift to owner representatives/owner advisors through an international commercial property management firm's overview of the approach to integrating sustainability into a diverse managed portfolio followed by insights into resiliency benchmarking and investments that are driven by a premier North American multifamily property management firm on behalf of condominium owner boards. We then expand our perspective by getting insight into the characterization of resiliency risks and incentivization of investments to mitigate such risks by an overview of the climate-resilient business model and market focus of a mutual commercial property/business insurer and a well-established "Green Real Estate Investment Trust (REIT)" backed sustainability and resiliency investment firm.

INSTITUTIONAL PORTFOLIO OWNER PERSPECTIVE: PRINCIPAL REAL ESTATE INVESTORS

Fostering a Culture of Resiliency Within a Vertically Integrated Commercial Real Estate Investment Firm

Principal Real Estate Investors (Principal) is the dedicated real estate asset-management group of Principal Global Investors. Principal manages investments through a vertically integrated platform, incorporating all disciplines of commercial real estate across the spectrum of public and private equity and debt investments. As a vertically integrated firm, Principal is incentivized to align the high-level investment objectives of their clients with the operating protocols of each building in their portfolio. As such, assessing climate risk and enhancing the resiliency of buildings has emerged as a priority for Principal.

To address the challenge that climate change risk poses to their investment management objectives and performance targets, Principal employs multiple approaches across the investment/reinvestment continuum that are referenced in this chapter (see Fig. 10.4).

Management Strategy/Corporate Culture

Principal Real Estate Investors utilizes a unique environmental, social, and governance (ESG) framework, the Pillars of Responsible Property Investing (PRPI) initiative, to help drive asset management and fiduciary governance and deliver positive financial and environmental results. This framework leverages Principal's capabilities and supports client objectives with the goal of making buildings sustainable, smart, productive, resilient, and profitable.

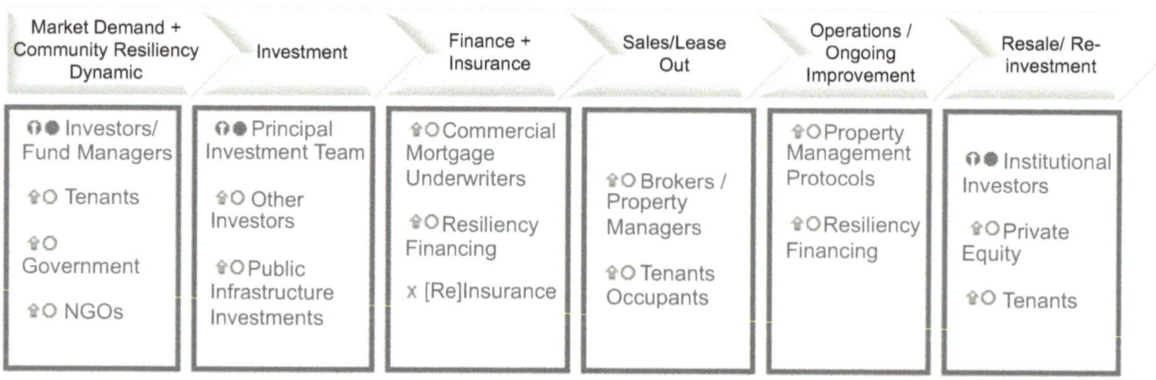

FIG. 10.4 Principal Real Estate Investors impacts the resiliency dynamic of all phases of the investment/reinvestment continuum. (Based on input from Jennifer McConkey, Devesh Nirmul, and John Scott.)

The PRPI philosophy integrates ESG and resiliency aspects within every stage of their investment process:

- Potential investments are evaluated using a method that incorporates ESG considerations, including a formal review of utility performance, certifications, sustainability programs and policies, and climate risk and resilience features.
- Operational assets are continuously monitored to identify and capitalize on ESG value-creation opportunities and risk reduction efforts.
  - Properties participating in the PRPI initiative are required to track utility performance monthly, implement operational best practices, and engage tenants in building-level sustainability efforts.
  - A new pilot effort will assess property-level climate risk with the goal of strengthening Principal's ability to identify and manage these risks portfolio-wide.
  - A broadened appraisal scope requires the review and valuation of high-performance, energy-efficient, and resilience attributes of properties.
- Principal discloses and promotes building-level sustainability and resiliency attributes to potential buyers.
- Mortgage lending policies require the inclusion of ESG and resilience factors, such as the tracking of building-level sustainability attributes and the monitoring of exposure to natural hazards.

Principal seeks to engage all stakeholders in their ESG strategy through their Responsible Property Investing working groups for debt and equity, which consist of representatives from all major disciplines within the organization. These stakeholders determine and execute the portfolio-wide ESG strategy in partnership with third party property management firms. In practice, ESG and resilience requirements and preferences are shared throughout management via a top-down and bottom-up approach. Fund managers convey the ESG strategy, including resiliency requirements, through the management structure down to asset managers. From the bottom-up, portfolio managers that are in touch with the more "resiliency-sensitive" millennial demographic and community-wide sentiments (e.g., in major metropolitan areas) send the signal of tenant and occupant preferences for sustainable and resilient buildings up to management, fund managers, and investors.

Tenant Engagement

High-performance buildings and their operators must effectively engage with tenants to achieve a unified culture of high performance that propels buildings to achieve even greater levels of sustainability (increased energy efficiency, resiliency, comfort, health, productivity, etc.). Energy consumption has traditionally been a logical and tangible performance metric to engage tenants. As part of an effort to extend energy conservation to energy resiliency and overall climate resiliency, Principal has developed a comprehensive tenant engagement program as part of the PRPI initiative. Through this initiative all property teams are expected to educate tenants on building-level sustainability efforts and encourage them to participate in a wide variety of engagement programs. Principal has also initiated the inclusion of questions in their annual tenant satisfaction survey to obtain first-hand information on tenant preferences and awareness of property-level sustainability efforts. Moreover, they are coordinating with brokers on marketing strategies that include a resiliency focus. Through these initiatives, Principal expects to see engagement and embracement across the spectrum of tenants, enabling the firm to move ahead with a more holistic approach on resiliency.

MULTIPLE OWNER/BUILDING-LEVEL PERSPECTIVE: A COMMERCIAL REAL ESTATE SUSTAINABILITY CONSULTANT

A Perspective on Strategy and Energy-Climate Impact Mitigation Cobenefits of Owners and Tenants in the Commercial Property Sector

The authors interviewed Brenna Walraven on her current engagement of commercial ownership clientele. Following is a summary of the relevant Q and A:

(1) *Does energy consumption weigh into the resiliency management equation?* It can, but not necessarily. There often can be positive resiliency benefits when doing energy efficiency retrofits—for example, energy efficiency investments that improve reliability and reduce risk of failure (a new energy management system for example) would reduce the risk of a building "being off line" and thus unoccupiable. Another, could be where a large retrofit (with significant savings potential) could allow for infrastructure improvements that help "pay for" the resiliency investments—so a five or six component retrofit with significant savings could, for example, help pay for electrical panel protection from water via pumps or water proofing.

(2) *What about the role of renewables?* From the context of property assessed clean energy (PACE)

financing, which makes investing in renewables more feasible and can be used for resiliency infrastructure, both with large upfront cost challenges, these investments can be enabled. PACE is increasingly being enabled across the country such that this becomes a viable option for implementing synergistic deeper retrofits, renewables, and infrastructure improvements.

(3) *How about more extreme storms and blackouts? Generators are the default response now, what about investing in more sophisticated resiliency options (solar with battery back-up and other alternatives)?* What is different today from the past is that typically solar and storage are not set up to produce power for the entire building—and storage is not set up to replace or even supplement a building generator. What we are starting to see is that these projects could produce base building power and sustain a building through some duration in a dramatic weather event. A more comprehensive approach should be delivered as we start to see more distributed energy models being implemented—community solar is an example where the community simply agrees to buy the power at a discount but also has better resiliency in storm scenarios.

(4) *What steps are being taken as part of your (1) build-out specifications and (2) ask of the building owner to ensure a resilient and efficient business operation during your lease term?* I have clients now that are indicating tenants are asking about not just sustainability or certifications but really vetting the landlord on their approach to sustainability.

(5) *What will the new resilient built environment look like?* More damage and disruption will need to occur before deeper focus and investment happens within the real estate sector. The larger, most sophisticated investor owners have gradually been making improvements because they will increasingly see the need and because they will face the costs and penalties of inaction in the form of property damage, lost rents (insurance only goes so far in this regard), tenant failures, insurance premium increases, etc. For example, Prologis and USAA Real Estate are really focused on this, but again these are the most sophisticated owners—which make up less than 15% to 20% of the overall market. The best opportunity is if there are greater partnerships between cities, counties, states, owners, and operators to include real estate owners and government agencies such as the Department of Defense and General Services Administration (GSA) and associations such as Building Owners and Managers Association International and the Urban Land Institute.

PORTFOLIO BUILDING MANAGER PERSPECTIVE: COLLIERS INTERNATIONAL

Driving Climate Resiliency from the Momentum Generated through Energy Efficiency

Colliers International, a leading international commercial property management and brokerage firm, with a diverse portfolio of ownership clientele and building types (office, industrial, retail, mixed-use, etc.) and classes (A through C) has had the opportunity to provide leadership on the cutting edge of sustainable buildings for both new construction and major renovation. By virtue of its diverse portfolio, it has also had the opportunity to see where the market or legislative environment has not been able to deliver solutions for properties that are not positioned for resiliency both from an energy risk and climate impact risk perspective.

Furthermore, the value proposition of energy efficiency has proven itself in the Class A properties where (1) tenant demand for high performance, (2) institutional long-term investor appetites for sustaining high performance within their assets, and/or (3) larger societal and cultural influences as exhibited through city sustainability policies and grassroots efforts for encouraging more sustainable built environments (e.g., the green building movement) have motivated ownerships to continuously strive for higher and higher levels of performance. However, due to structural dysfunctionality such as the landlord-tenant split incentive challenge [3] or imperfect pricing/valuation of high performance in the microeconomy, the inability to factor in risks and future liabilities effectively, and/or to take on the upfront costs of high-performance investments, many of the Class B and C properties and ownerships are not able to transition to more efficient and resilient property portfolios. Despite this divergence on high performance, Colliers International and other globally diverse real estate management firms are well positioned to be change agents and innovators for the entire spectrum of ownerships. Ultimately, firms such

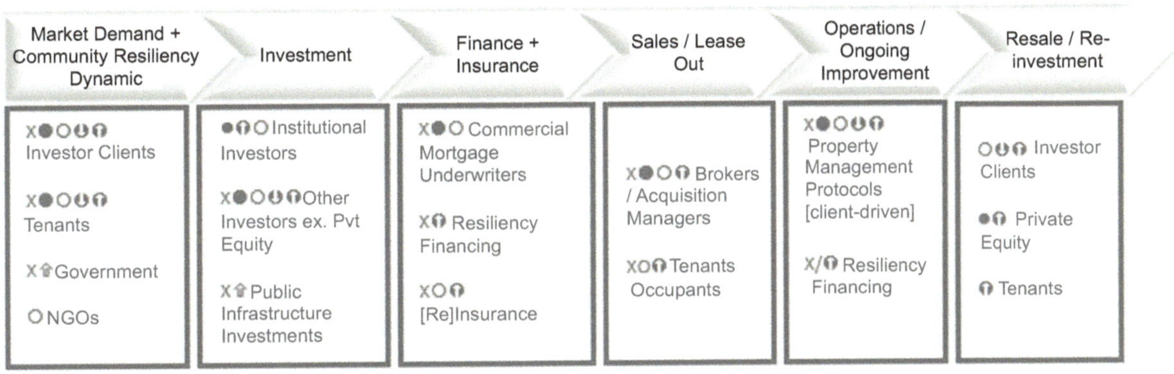

FIG. 10.5 Commercial property management and brokerage firm leveraging the concern and behavior of high-performance actors in the marketplace to drive less-motivated ownerships and stakeholders to value climate risk and actions to mitigate risk strategically.

as Colliers are motivated by expanding their clientele base and retaining their current clientele. Differentiating their offerings and looking out for the best interest for their owners invites a strategic focus on high performance and resilience (Fig. 10.5).

CASE STUDY: PARK TOWER: COLLIERS IMPLEMENTED AND TENANT-DRIVEN HIGH-PERFORMANCE TRANSFORMATION

Challenge: Purchased in 2006, Park Tower is a 33-year young property. Most systems had exceeded their design life and began declining to the point of diminishing returns on investment. Park Tower no longer met the rigid perimeters of its GSA tenants who require the building to be EnergyStar rated.

Strategy: While maintaining optimal tenant comfort, the management team investigated and chose the best performing and most energy-efficient equipment to replace the existing, outdated equipment. Updates included the following:

- Modernized elevators
- 2–750-ton high-efficiency chillers
- Variable speed drives for each floor's air handlers
- State of the art energy management system
- 5600 new high-efficiency light fixtures
- Motion sensor lighting
- New transformers for each floor
- New fire panel and building rewiring
- 750 kVA generator
- Energy-efficient roof coating

Results: Through implementation of these strategies, the building achieved the following results:

- Average EnergyStar score rose to a 99% rating in 2013
- Energy usage reduced by 47%. The January 2007 electric bill totaled $155,324 or 44,970 kwh/daily; whereas, in January 2014, the bill was $73,704.98 or 23,902 kwh/daily.
- The savings also equate to a $13 million increase in Park Tower's value.
- Receipt of $32,611 in rebates from Tampa Electric

Energy Resiliency—Park Tower Multitenant High-Rise Office in Tampa, FL

The Park Tower high-performance transformation from a deferred maintenance approach to what is now a Leadership in Energy and Environmental Design (LEED) Gold Certified property and the oldest high-rise office tower in downtown Tampa is a great example of how a fundamental economic driver, in this case, the energy performance standards of a GSA tenant that was occupying 43% of the building's leasable space, has resulted in not only one high-performance property, but a push by the owner of the property to replicate the same transformation across their portfolio.

Although the impetus for the transformation in Park Tower came from a focus on energy, it was a focus that had significant bottom-line impacts to operating costs and ultimately building valuation, not to mention the priority performance metrics of tenant retention and occupancy rates. Similar drivers within the context of climate change impacts can also be a source for transformation, especially if they are supported by incentives from both the insurance and financial markets.

Protecting the Long-Term Investment Horizon of Institutional Investors

Addressing the challenge of climate impacts in the most vulnerable urban dense communities has become a business reality for Colliers International in places such as coastal South Florida. The flooding, windstorm, and related impacts on business operations and productivity of tenants in these properties has alarmed these long-term investors and now divestiture of these assets is a very real possibility. Acting in their client's best interest, Colliers has made it a priority to actively engage the government and nonprofit stakeholders on strategies for successful mitigation of current-day climate impacts and near-term impacts. Colliers is specifically working with South Florida municipalities such as Miami-Dade County, the City of Miami, and the City of Miami Beach. As an example of resiliency leadership, the City of Miami has legislated sea-level rise projection-based codes for new construction. These are a great approach for new construction; however, there is a need for incentives and cost-effective investments to mitigate the property and performance risks facing existing buildings.

The synergy across the spectrum of *energy efficiency, clean energy, windstorm, and sea-level rise impact mitigation* presents an opportunity for building owners and investors to develop a plan for implementing a logical sequence of building improvements, whereby operational cost savings from each step help to finance the subsequent step. A building like Park Tower having significantly improved its energy cost competitiveness is in a better position to pursue clean energy, windstorm and, where necessary, sea-level rise mitigation measures. With access to financial products, incentives, and public–private partnerships, these organic paths to greater resiliency may be formalized and replicated in the marketplace.

MULTIFAMILY PORTFOLIO MANAGER PERSPECTIVE: FIRSTSERVICE RESIDENTIAL

A Market-Transforming Approach to Engaging Condominium Owners on Sea-Level Rise and Windstorm Vulnerability/Impacts and Driving Resiliency Investments in Southern Florida

The high-rise coastal condominium properties in Southern Florida and the owners of the individual units in these buildings are literally on the front lines of climate change adaptation and climate/energy resiliency. Properties in Miami Beach and Naples already have to prepare for and/or mitigate impacts from both sea-level rise and windstorms. Through the lens of the largest residential property management firm in North America, FirstService Residential (FSR), we explore the impacts, particular ownership dynamics, risk mitigation measures, investment/financing strategies, and evolution of the adaptation to climate risk from the recent past into the future within this sector (Fig. 10.6).

Coastal Condominium Marketplace: Market Drivers and the Relevance of Resiliency

The primary driver of the development and high occupancy of coastal condominium properties is the desire of owners to live or have a property on the coast/beach. As long as a building developer is able to secure a

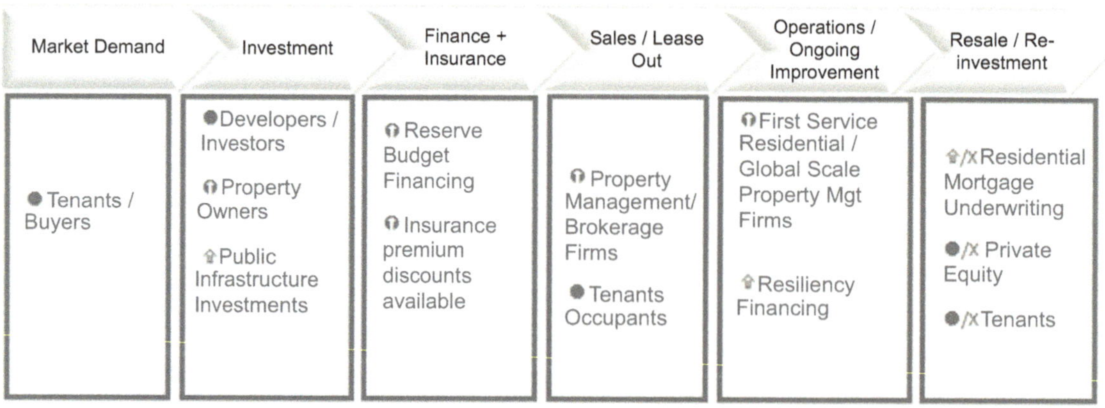

FIG. 10.6 Florida condominium marketplace climate risk mitigation investment–reinvestment continuum. (Based on input from David Diestel, Devesh Nirmul, Chris Normandeau, and John Scott.)

master-insurance policy for these properties and other fundamental building-based lending criteria are met (e.g., low rate of property delinquencies, etc.), individual owners are able to qualify for mortgages. Although this demand factor continues to drive development, it is clear that the frequency and severity of particular impacts such as flooding from sea-level rise, building envelope damage from windstorms and electricity outages are interrupting the livability of these properties and the finances needed to both pay for insurance protection and the actual improvements/repair to keep the buildings operating normally.

The Evolving Condominium Risk-Reserves Management Paradigm

The long-term outlook that condominium owners have for the buildings they reside in has resulted in normal operating protocols including establishment of a reserve fund. The reserve budget is typically determined via an initial reserve study that looks at the useful life and replacement costs of major equipment and building features. A key influential variable in this system of useful life preservation of a building and its components is in the choice that condominium boards make around whether to fund a reserve account or implement one-time assessments whenever repairs/replacements are required. According to FSR, the real estate market tends to favor and place a higher value on buildings that prepare for both known and unknown repair/replacement costs through a regular funding of their reserve accounts. From the insurance angle, there are both flood and wind-insurance premium reductions available for qualifying measures that are implemented. FSR has also seen, over the last 10 years, an 81% increase (6.1% compound annual growth rate) (measured across Broward, Miami-Dade, and Palm Beach Counties) in the amount of annual budget dedicated to reserve budgets. Looking forward, this trend is expected to increase [4].

The Role of and Value Proposition of Property Management Firms in Managing Risk

- Benchmarking (energy and water consumption and costs as well as other costs in a condominium's budget) across buildings in a portfolio to identify what attributes are driving low and high performance [4].
- Negotiating power on behalf of owners to secure vendors and cost-effective services when they are needed (e.g., frequent impact events like king tides and major disaster scenarios like hurricanes)
- Preparing and response planning and action for high-impact events like hurricanes
- Managing operations for resilience (back-up generator testing, etc.)
- Balancing market trends with the specific nuances of a particular building and board

Condominium owners who are aware of climate risks (whether labeled as such or not) currently budget for important upgrades such as high-impact windows, bottom-floor equipment relocation, dune restoration, and/or seawall infrastructure. The following are some specific examples extracted from real-life scenarios at South Florida FSR-managed condominiums. Average payback numbers for investments are provided where applicable.

Example 1: Dune restoration and sea wall installation.

Minor investments in dune restoration and sea walls for coastal condominiums has resulted in savings in the realm of \$75,000 per condo association—or about 75% of annual flood insurance premiums—at a cost of less than \$50,000 for an average simple payback of less than 1 year in flood insurance savings alone.

Example 2: Bottom-floor equipment relocation—Miami Beach beachfront high rise.

A condominium owners association reduced their flood insurance premiums from \$138K to \$107K per year by changing the elevation of the lowest covered piece of equipment, the costs of which were negligible.

Example 3: Pumps for removal of floodwaters.

Investments in pumps (at a cost of about \$10,000 to \$20,000) for removal of floodwaters resulted in a \$5000–\$10,000 reduction in flood insurance premiums—an average simple payback in the realm of 3 to 5 years based on flood insurance premium savings alone.

Example 4: Energy performance benchmarking and investments—example of climate change mitigation and adaptation synergy.

FSR's efforts—and benchmarking the energy consumption and cost of condominiums against each other within the South Florida region—have increased awareness of owners and helped to push down energy and water consumption and cost. This has resulted in reduced greenhouse gas emissions and greater ability to potentially reallocate energy and water cost savings to climate impact resiliency investments and/or energy-security type resiliency investments (for example, enabling minimal/critical operations during windstorm-induced electric blackouts). Annual savings in energy and water bills has reached up to 30% (and higher in some instances) freeing up tens, or even hundreds of thousands of dollars for associations to invest elsewhere.

PRIVATE NEW CONSTRUCTION AND RENOVATION RESILIENCY FINANCING: COUNTERPOINTE SUSTAINABLE REAL ESTATE—HANNON ARMSTRONG

PACE in Florida: An Innovative Catalyst for Resiliency Investments

The State of Florida, by virtue of its geography, faces an entire spectrum of resiliency challenges from windstorm risk, sea-level rise, and an energy infrastructure that is susceptible to storm-induced interruption in service. The ability of a privately owned property/building to weather one or more of these climate change impacts will depend on (1) when it was built (e.g., pre- or post-Miami-Dade wind code) or (2) whether or not it was renovated to code in the recent past or (3) the ability of the building owner to save for and invest in resiliency measures. To address the risk and uncertainty that underlies the level of windstorm and energy resiliency of privately owned properties across the state, the Florida legislature enacted PACE legislation in 2010. Twenty-five other states and the District of Columbia have their own version of PACE programs as well [5].

PACE attracts private investors to direct capital to specific building-owner investments that result in either energy savings, clean energy generation, or wind mitigation through building envelope improvements (roofs, windows, doors, etc.), the cost of which is repaid through financing at a fixed rate and fixed term where repayment is made on the building owner's tax bill as a non-ad valorem tax assessment. PACE financing was born in California, which is also facing its own set of climate-related resiliency challenges (droughts, fires, etc.). In California, PACE applies to water conservation, energy conservation, clean energy, earthquake risk, and most recently fireproofing investments.

Counterpointe Sustainable Real Estate and its partner Hannon Armstrong (CSRE/HASI), a leading "Green" Real Estate Investment Trust (REIT) has been actively working with property owners around the country on financing new construction and renovation projects through PACE. The objective of this type of financing is to (1) ensure owners are making the necessary renovations to effectively withstand windstorm, fire, or other relevant types of climate-induced natural disaster impact risks based on a state's PACE program, (2) encourage owners to not leave valuable sustainability and resiliency retrofits on the table due to not having sufficient capital or operating budgets to make investments or having an ROI threshold that voids the possibility of high-performance wind/fire/seismic impact, energy/water efficiency, or clean/energy-generating projects, and (3) encourage owners/investors to push the envelope and leapfrog into an enhanced-resilient scenario by giving them the ability to invest in net-energy positive buildings that can withstand natural hazards.

Fig. 10.7 provides some examples of projects CSRE/HASI has invested in around the country that illustrate the ability of a financing tool like PACE to (1) accelerate the uptake of resiliency measures, (2) enable owners and investors to preserve and even enhance their financial position, (3) increase the well-being, comfort, productivity and/or health of building occupants/tenants, and (4) reduce the built environment's impact on the natural environment or in some cases complement and enhance or regenerate the natural environment.

PACE as a Catalyst for the Mainstream Banking and Finance Market to Motivate and Incentivize Resiliency Investments

Commercial PACE financing, through (1) its ability to finance both soft and hard costs, (2) insistence on minimum performance requirements and recognition of operational cost savings that result from high-performance investments, (3) underwriting that focuses primarily on the leverage available in the property for both new construction and retrofits, (4) non-ad valorem special assessment tax-based financing methodology, and (5) ability to have the repayment of financing be passed through from building owners to tenants is demonstrating a model for the real estate financial marketplace of how private capital can be allocated to enable private building owners to shift to greater resiliency while eliminating the high upfront cost or unplanned for costs of high-performance building investments.

PRIVATE INSURER PERSPECTIVE: FM GLOBAL

A Mutual Commercial Insurance Firm with a Built-in Incentive for Advancing Resiliency Investments

FM Global, as a membership or mutual company, has an obligation to its stakeholders/policy holders to make sure they are acting responsibly and working with companies that are like-minded with respect to mitigating risk: everyone who writes/buys a policy with the firm is invested in the overall organization and wants it to perform well financially. This

Southern Oaks Rehabilitation & Nursing Center — **Pensacola, FL**

Project: Impact Windows, Impact Roof & Elevators

Financing: PACE

Transaction: $527,860 – 25 years

Property: Assisted Living Facility

Built in 1978, the Southern Oaks Rehabilitation & Nursing Center replaced the original roof and 267 windows to conform with the updated Florida Building Code and help reduce insurance premiums. In addition, the owner upgraded elevators to modernize the building and decrease energy usage.

Hilton Ocala — **Ocala, FL**

Project: HVAC Chiller

Financing: PACE

Transaction: $634,995 – 30 years

Property: Hospitality

Located near Ocala International Airport and Paddock Mall, this 196-room full-service Hilton hotel completed a chiller replacement using PACE financing. The owner expects to save nearly $700,000 over the useful life of the improvement.

Hyatt Centric — **Sacramento, CA**

Project: Ground Up Construction

Financing: C-PACE

Transaction: $10.5MM – 25 years

Property: Hospitality

The Hyatt Centric Downtown Sacramento is a ground up construction of a 172-room, 11-story upscale full-service hotel adjacent to the Golden 1 Center in the heart of Sacramento. The project is the redevelopment of the Hotel Marshall, a landmark on the city's register of historic sites. Seismic strengthening, lighting, elevators, water conservation and HVAC were financed.

Papavero & Piazza Apartments — **Fontana, CA**

Project: Solar

Financing: PACE

Transaction: $1MM – 25 years

Property: Affordable Senior Housing

Two apartment buildings with 210 units for very-low income senior citizens in Fontana. The owner financed over $1MM in solar systems with lender consent from the U.S. Department of Housing and Urban Development. The solar systems are expected to save over $5.3MM in utility savings over the life of the system.

FIG. 10.7 PACE financing case studies. (Based on input from Eric Alini and Devesh Nirmul, Hannon Armstrong Sustainable Real Estate/Counterpointe SRE.)

orientation is supported not by the mainstream actuarial approach to assessing and pricing risk but instead through an "engineering" approach. The built environment assets and operations of each of FM Global's mutual partners are assessed for risk exposure from the location/site level all the way down to the materials employed in the construction of buildings. This type of risk assessment, specific risk mitigation recommendations, and ultimately policy writing and premium pricing results in a much more resilient group of commercial businesses and real estate owners.

Deconstructing the FM Global Approach: Miami Metro Area Tropical Storm Risk and Building-Level Insurance

A Miami high-rise office building seeking a policy with FM Global would have its structure and operations assessed according to standards that FM Global has determined through analysis carried out at its research campus in New England. The engineering approach is accompanied by algorithms that are set up based on geography and physical protections in-place (for example, flood exposure; condition of walls, doors, and roofs; location of critical equipment; ability to mitigate business operational loss during electric outage; etc.). In the global, regional, and building-level context, FM Global assesses risk and mitigation solutions and rewards both community-wide actions and building-level actions within the framework of global, regional, and building-level risk assessment. Despite available reinsurance capacity, FM Global would adhere to building-level standards that adequately offset risk according to FM's own standards. Fig. 10.8 illustrates a typical global- to building-level risk assessment-pricing-mitigation recommendations spectrum that an FM client/member company may be subject to.

For a real-world example, the authors have encountered a South Florida industrial property that was seeking full property insurance coverage that had met actuarial compliance and location risk acceptance but was refused based on *engineering identification* that fire suppression systems were not adequate. The authors conclude that resiliency-based insurance coverage will follow this pattern. Fig. 10.8 demonstrates this alignment that is currently in practice at FM Global.

> **Insured Asset in Post-Irma and Maria Puerto Rico: Assessment of Interdependencies**
>
> In its assessment of member-insured facilities in Puerto Rico, FM Global determined that even if a facility is hardened and is able to withstand wind impacts, if it did not have access to power, clean water, or ingress/egress, they may not be not able to operate in the poststorm environment.

The Empirical Experience of FM Global's Strategy

Based on the last couple of years, the reinsurance market is not drying up as a slew of alternative sources of capital (e.g., pension funds and hedge funds) have flooded into the insurance market. This capacity plays a part in what FM Global can charge clientele for property and business loss coverage. Nonetheless, there is an overall environment of rising rates for business loss coverage and given FM Global's insistence on members' implementation of baseline risk mitigation measures, it has been able to maintain competitive rates. FM Global tends to generally perform better than the rest of the market because of its aggressive approach to risk mitigation and the selection of partners (insured members). In its post-Maria analysis for example, FM Global reported that actual losses tended to be less then half of what was predicted by simulation models.

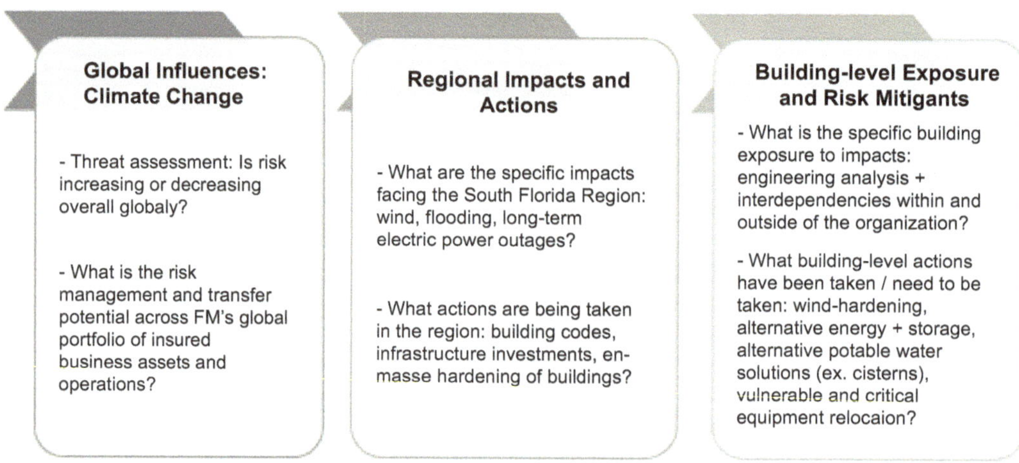

FIG. 10.8 FM Global's macro- to microlevel risk assessment, pricing, and mitigation recommendations spectrum. (Based on input from Devesh Nirmul, John Scott and Christopher Wegman.)

A Mutual Insurance Company's Legacy for an Age of Unprecedented Change in Risk and Impacts from Climate

FM Global's business model, relative to the mainstream insurance marketplace, seems to provide a more resilient or climate change-adapted approach to ensuring manageable risk so that despite the timing and magnitude of impacts we are currently managing or will have to manage in the future, real estate assets and the business operations tied to those assets can perform within a tolerable level of risk and predictable resilience. The authors suggest that the insurance industry, government, and non-governmental organizations such as the National Institute of Building Sciences invest in more research and stakeholder engagement into translating the business model of this type of company into the larger real estate insurance and finance marketplace.

CONCLUSION

Within the framework of market, legislative and community-wide stakeholder drivers of the recognition of climate change risk and impacts and actions to mitigate this risk, there is no doubt a new and growing focus among some stakeholders/shareholders in the marketplace on resiliency. There is still, however, the complex reality private building owners and investors have to reconcile on a daily basis—making investment decisions and interacting with relevant actors in a world of imperfect information and differing levels of value placed on both current-day impacts and uncertain, yet probable, future impacts of climate change.

On the leading edge of this spectrum of inaction-recognition-action, the authors have documented a "balance-sheet" level recognition of perceived and real climate risk across institutional investors, portfolio-level real estate owners and managers, leading mutual commercial business and property insurance companies, resiliency and sustainability finance and investment firms, and climate-impact vulnerable states and municipalities such as Amsterdam, New York, and Miami, and the promise of increased climate-resiliency investments among private building owners. This is being motivated by those actors in the market through application of their own "investment decision-making filter" or data/engineering-based assessment of challenges and opportunities. They are recognizing at the global and regional level, identifying at the property level and, in some cases, strategically getting ahead of current-day impacts of climate change through specific property investments. In the middle of this spectrum, government-level and real estate representatives/advocates are formalizing multistakeholder and client-focused awareness raising (e.g., regional planning councils with resiliency committees or emerging multistakeholder efforts like the government-based Southeast Florida Climate Change Compact or sustainability and resiliency advisory departments within global property management firms), processes for assessing risk, interactions, and tangible actions/investments.

Finally, there are a number of actors that, while they may be aware of risks and current and potential future impacts, they may have not yet witnessed the market, legislative or stakeholder forces that would drive a change in investment behavior. The opportunity for accelerating the uptake of meaningful recognition, calculated assessment, and strategic investments around climate change vulnerability, risk, impacts and solutions will be realized by the actions of the types of stakeholders and shareholders highlighted in this chapter as part of an overarching effort to improve the entire marketplace's awareness, understanding, and commitment to take action for enhanced resiliency from the global to regional and right-down to the asset/building-level.

REFERENCES

[1] U.S. Global Change Research Program, Fourth National Climate Assessment, 2018. https://www.globalchange.gov/nca4.

[2] Risk and Insurance, Climate Change as a Business Interruption Multiplier, 2018. http://riskandinsurance.com/climate-change-as-a-business-interruption-multiplier/.

[3] Rocky Mountain Institute, The Piece of Paper that Could Make or Break Commercial Net-Zero Energy, 2018. https://rmi.org/piece-paper-make-break-commercial-net-zero-energy/.

[4] FirstService Residential, The Definitive Guide to Florida Condominium & HOA Operating Spend, The Benchmarking Guide, 2017.

[5] PACE Nation, PACE Programs Near You. https://pacenation.us/pace-programs/.

CHAPTER 11

The Role of Designers and Other Building Practitioners in Advancing Resilience

ALLISON HOADLEY ANDERSON, FAIA, LEED AP

INTRODUCTION

Communities need a variety of advocates and policymakers, influencers, and implementers to become resilient, but they also need designers to contribute their best ideas to address hazards and the current and future impacts of climate change. Architects, engineers, and urban designers play key roles in creating a built environment that can adapt to changing conditions, withstand hazards, and recover rapidly from disruptions. Design professionals are charged with protecting the health, safety, and welfare of the public, but advancing resilience calls for more ambitious goals: to understand and create innovative visions for the future, test policy with built examples, and transform desired community values such as reduced risk, improved economic security, and environmental sustainability into a physical framework.

Addressing the challenges of resilience requires designers to assume multiple roles. As a resource for the community, they help residents envision the future and encourage broad participation in its creation. As an adviser to clients, they share information about predicted future conditions that may affect the project, provide designs to meet the anticipated service life, and recommend adaptation or mitigation strategies. They use professional skills to design adaptation measures: conceptualizing, testing, and implementing projects that can mitigate risks.

RISK AND RESILIENCE

Calculating risk is defined by two factors: the hazard event itself and the likelihood that people and property will be harmed. Every new project is designed within an established context, bounded not only by geographical proximity to hazards but also by political districts, existing infrastructure, the physical environment, and sociocultural characteristics. Context affects public health and environmental quality—is the community prosperous, vibrant and connected or underresourced, aging, and struggling—distinctions that determine the community's risk profile. High-risk neighborhoods have characteristics that create more vulnerability to the consequences of hazard events and result in effects that are devastating, rather than simply inconvenient, when disaster strikes.

Community vulnerabilities may be social, physical, or economic in nature. Social vulnerability includes a population with low educational attainment, limited communication abilities, and an unusual percentage of elderly or young residents or other conditions that rely on assistance from outside the community. Physical vulnerability denotes structural deficiencies in the community, for example: housing is built in the floodplain or on steep slopes; access is limited, leading to potential isolation; or there are inadequate services such as unreliable power and water or a lack of sources of fresh food. Economic vulnerability includes income inequality leading to imbalances in recovery or pervasive resource limitations that prevent residents from preparing for and recovering from disasters.

DESIGNERS AS COMMUNITY RESOURCES

Architects clearly understand their responsibility for addressing physical vulnerability against the impacts of disasters (high winds, flooding, fire), but they also have a critical role to play in addressing social and economic vulnerability. The impact of a disaster has the greatest consequences for residents with the fewest

Optimizing Community Infrastructure. https://doi.org/10.1016/B978-0-12-816240-8.00011-2

resources to recover, but good architecture and urban design can reduce human vulnerability by reinforcing an ecology of support within a neighborhood. Active commerce and busy street life, density, family and friends living nearby, well-maintained infrastructure, and green space are signs of a healthy community, one where even fragile residents can participate in community life.

Neighborhoods that support social interaction fare better in a crisis. Architects may have limited responsibility for many urban concerns, but as members of a community with access to many of the people in power they have a voice in directing attention and investment to help neighborhoods in need. Consistently recognized among the most prestigious professions, architects can clearly outline the parameters of a problem and propose solutions. As a resource in their community, designers can provide pro-bono services to initiate projects, facilitate groups envisioning their future, encourage community participation and social equity in decision-making, advocate for sound development policies that consider public health, safety, and welfare, and share resilience and adaptation lessons with community leaders.

Architects are skilled at envisioning the future. Speculative, conceptual, preliminary, visionary—these words describe the first steps in every design project, whether at the community scale or an individual site. The ability to imagine a different future, one that improves the present condition and has plausible steps to implementation under real world conditions, is a hallmark of the design professions. A project resides briefly between the spaces that already exist and those to come, between the inconceivable and the inevitable.

Most projects begin with research, starting with programmatic information pertaining to the need, quality, budget, and time: conversations about performance, priorities, and tradeoffs, differentiating what can be jettisoned and what must be fulfilled. Research also includes the analysis of a site to determine its opportunities. A standard design process usually begins with research on a myriad of topics ranging from existing conditions, future trends, complicating factors, precedents, materials, and assemblies; these same parameters are directly applicable to designing for resilience.

After the programming phase is defined, the architect begins to shape alternate scenarios. The ability to generate multiple solutions to a single problem has tremendous value for communities; designers know that there is not a single way to respond to a challenge, even a highly repetitive issue such as flooding. Exploring alternatives and evaluating the potential benefits and drawbacks of each idea reinforce the validity of the preferred option, the one that is finally selected as the best of many possible answers. These design talents are applicable beyond the systems of a building to the broader scale of the community: circulation systems (transportation), envelopes (armoring), and services (infrastructure).

The difference is that resilience planning must envision alternate scenarios, not just in form, but also in impact. Will sea level rise by 30 in. or 60 in. higher by 2100? Will the loss of barrier islands increase storm surge? These questions lead to inquiries about the impact upon human systems, such as the following: How often will flooding affect transportation networks? Will critical functions be stranded or inundated? Are there ways to protect against sea level rise? These questions spawn further ones to ask about the client's acceptable level of risk. Is a homeowner willing to build higher, and how much higher? Are waterfront property owners willing to experience isolation due to flooding, and for how many days each year? Is it economically feasible to build seawalls, breakwaters, pumps, elevated roadways or other structures to mitigate this problem? Is managed retreat an option? Is a natural barrier along the coastline feasible? Each community will have different responses to the desired performance against hazards, and the acceptable levels of exposure and risk. They will also have different resources to apply to the solutions.

Every community has a distinct set of constraints that will result in a unique design solution, but only if individuals from the community participate in the decision-making. Without the perspectives of a wide range of residents, there may be an attitude that a proven solution can be applied again at a different location, in a different context. In practice, the issues are never identical, stakeholders never advance the same concerns, and the context is never quite the same as the last community.

Resilience projects require widespread participation to look beyond individual property lines and consider the needs of the entire community. Working with large groups of stakeholders to find consensus is a staple for architects experienced in public or institutional projects, but the ability to solicit ideas and opinions from a community at risk has two further purposes: to enlarge the range of possible solutions to a hazard using vernacular knowledge; and to build adaptive capacity within the community by educating residents about hazards, impacts, and available resources.

Community participation helps the design team to understand multiple perspectives in solving knotty problems and addressing social, political, and economic equity issues. The ability to work across race, age, and background has the potential to bring energy, innovation, and interdisciplinary endeavors to adapt to the threats. Author and activist Jane Jacobs wrote, "Cities have the capability of providing something for everybody, only because, and only when, they are created by everybody [1]."

There is a critical difference between consulting the public through desultory "information sessions" (a hallmark of public planning processes) and giving residents real power to affect the design and policy solutions. The people who live within a community are often the best resource for insight into how it works, what is important, and where gaps exist. They can help the design team to understand what gives a place its own unique culture and identity. The community members' role is not to provide technical expertise in transportation, retail service, or cultural heritage, but they can have a profound effect on a project through anecdotal evidence on these subjects. In resilience planning, comments about existing housing, public space, and city services provide valuable information for improvements.

Without a comprehensive strategy to improve infrastructure's performance against today's hazards and the growing impacts of climate change, operations will continue to deteriorate. The need for infrastructure investment is critical now, as most systems are burdened by deferred maintenance, and the anticipated need will become more pressing. A World Bank study found the expected cost of adapting to a 2°C warmer world by 2050 is in the range of $70 to $100 billion per year [2].

A 2°C limit to climate change may be a vanishing possibility [3]. More worrisome than the cost of making infrastructure resilient is the shrinking timeframe to plan and implement mitigation and adaptation measures. The first myth about climate change—that it will occur over decades and its impacts would not be felt for years—has been categorically disproven. The second myth—that only the broad outlines of climate change impacts are known, and specific risks at a given location are too uncertain to plan for—is also losing validity. There is no rationale that supports further delays. Uncertainty about the intensity and frequency of future impacts may reduce the appetite to take any action until "more is known," an approach that can further increase society's exposure to hazards, degrade the natural ecosystems that temper climate stress, and delay projects that could strengthen adaptive capacity if they were built. The caveat is that uncertainties may

Case Study: Rebuild By Design, Bridgeport, CT

Rebuild by Design was a competition to improve resilience in New York, New Jersey, and Connecticut following Hurricane Sandy. Ten interdisciplinary design teams were selected to participate in a 3-month research phase and a 5-month design phase. Philanthropic support from the Rockefeller Foundation and other organizations provided financial incentives and organizational support, and the U.S. Department of Housing and Urban Development funded the implementation of selected projects.

During the research phase, teams were introduced to a variety of stakeholders as they toured the shattered landscapes, learning from a wide range of perspectives, from narratives of racial segregation and forced relocation to stories of heroism and cooperation during the extended power outage. Each viewpoint helped the designers to understand weaknesses in the existing built environment and the uneven social fabric of the region.

During the design phase, each team developed a design opportunity and implementation strategy for a climate-adapted future by working within a specific community with support from local residents and government officials. The challenge was to create solutions that not only addressed future disasters, but also benefited residents in their everyday life. The team led by unabridged Architecture with Waggonner and Ball and the Gulf Coast Community Design Studio was assigned to Bridgeport, Connecticut, a highly vulnerable city at the confluence of several rivers and the Long Island Sound. The team focused first on the South End neighborhood, where residents were isolated for several days following the storm by flooding along the Interstate-95 and Metro North corridor.

Bridgeport has a long history of innovation, a rich tradition of welcoming immigrants, and a struggling economy marred by environmental contamination, losses in manufacturing, and barriers to expansion. Rebuild by Design advisors encouraged the team to seek out nontraditional community groups to help shape the design proposal. Engagement activities included developing tools to communicate across language barriers with adults in an English as a Second Language class; a winter walk the length of the Pequonnock River through the city; a teen bike repair clinic and bike ride; a tour of the municipal waste-to-energy plant and discussion with manufacturing entities nearby; meeting with providers of community services to public

Continued

housing residents; and planting vegetables at a community garden. The team heard about daily challenges the residents navigated, as well as the consequences of Hurricane Sandy.

One of the most contentious—and useful—meetings was held in March 2014 in the semirestored community room of the Marina Village public housing complex, which had not reopened following the storm. Former residents' lives had been transformed by this event, and yet they had very little access to information regarding the future of the complex. They were understandably angry, uncertain about the opportunity to return to their homes, and anxious to have a timeline for rebuilding. The Marina Village meeting was not an appropriate time to present grand visions. Instead, the design team listened to these residents, a group who had not had a strong voice in the first year of recovery. There were several revelations: this neighborhood has the lowest rate of automobile ownership in the state; public transit requires an average of two transfers to reach essential services such as daycare or fresh food; there were a high proportion of elderly and medically disabled residents; there were no public places in the neighborhood to recharge a cell phone, hook up a child's nebulizer, or get warm. Many residents refused to go to a shelter after the storm because they were fearful of what might happen in an unfamiliar place, such as the possibility that families would be separated, belongings stolen, or children would be exposed to unwelcome cultural differences.

Their stories of daily tribulations and disaster events had a powerful impact on the final design proposal, inspiring the team to design a mixed use development to fill gaps in neighborhood services, house the most vulnerable residents in high performance buildings with backup power and water, create a workforce training center to support a burgeoning green economy, and establish a familiar neighborhood spot where people would feel welcome in a crisis, with access to electricity, food, and medical care.

The culmination of the Bridgeport team's design process was the "All-Scales" workshop, which brought community members from the disparate meetings together with representatives from government agencies for a public dialogue about transportation, housing, infrastructure, economic development, and other topics. The design team collected the salient points from this workshop and developed a final design proposal that highlighted the edges of the city, reconnected isolated neighborhoods, and pinpointed locations for resilience centers. The proposal succeeded in garnering $75M from HUD including funding from Rebuild by Design and the National Disaster Resilience Competition. The community engagement process touched the lives of hundreds of Bridgeport residents, building the networks and community strength that will be necessary for a more comprehensive recovery in the next disaster (Fig 11.1).

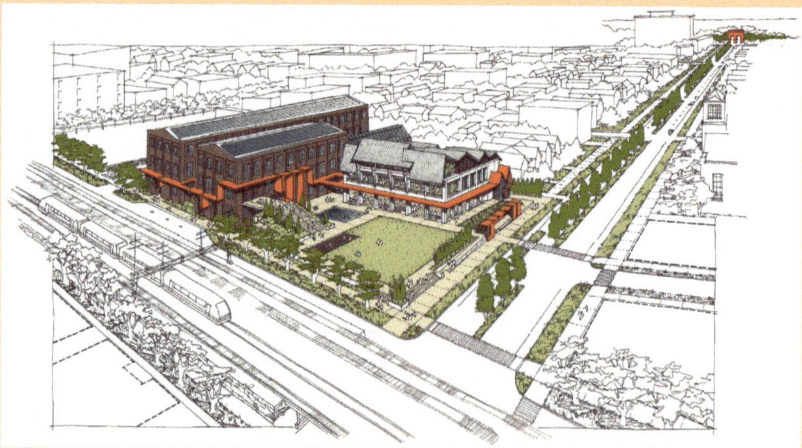

FIG. 11.1 Resilience Center, Bridgeport, CT, Rebuild by Design. The South End neighborhood was isolated during Hurricane Sandy by flooding along the Interstate-95 corridor. Resilience Centers were proposed to fill gaps in current neighborhood services and create familiar places for residents to shelter during hazard events. (Credit: WB/unabridged with Yale/Arcadis.)

not diminish over time. There is no endpoint predicted for climate change.

The typical project planning phase for a major infrastructure project is on the order of 25 years from conception to construction. In the transportation sector, planners must consider the proposed geographic alignment in relation to foreseeable hazards, the existence of alternate routes, the effect of the roadway on future development patterns, environmental impacts, and linkages to other parts of the transport network, any part of that may also be at risk. The standard service life for a new thoroughfare, bridge, or other transportation asset is 100 to 150 years. Before the onset of rapid climate change, planners had more time to establish goals for a study, inventory environmental conditions, and assess potential impacts before scheduling the project funding. The accelerating frequency and intensity of events; the increasing operational costs from heat, flooding, and deferred maintenance; and the escalation of construction costs inspire a sense of urgency.

Early adaptation saves money and lives, but limited funding and shrinking timeframes are working against effective responses. So is a lack of knowledge about appropriate adaptation measures. Many communities are operating with outdated or insufficient policies that prioritize bigger over better, dispersed over compact, and extensive over intensive. Architects can use their role as experts in the built environment to advocate for sound development policies to mitigate carbon emissions: compact developments with higher density and mixed uses to reduce vehicle miles traveled; alternative transit options such as pedestrian sidewalks, bicycle paths, and public transit; and energy standards for buildings that exceed national codes. Designers can promote green building certification, with its attendant energy use reductions and lower resource use. Urbanists might recommend a comprehensive plan with development limits and green belts to restrict horizontal community growth, enhance conservation and green space, protect critical habitats, and restore urban forests, or the transfer of development rights to lands appropriate for higher density. Engineers might propose placing limits on future development within the floodplain and ordinances that require the first inch of stormwater to be retained on site.

Design professionals can advocate for adaptation policies through their service on committees, preservation boards, zoning commissions, economic development groups, and emergency planning opportunities. If an architect contributes to a community's Hazard Mitigation Plan, they can propose modifications to existing buildings and structures to help them better withstand the forces of a hazard or encourage the removal of structures from hazardous locations; they can support opportunities for natural resource protective measures and evaluate structural protections such as seawalls or levees. If a designer is nominated to the local hospital board, they can support funding for relocating vulnerable equipment to higher floors, floodproofing critical functions, and improving thermal performance.

There are many ways to influence community perceptions about climate change and adaptation. Any member of a community may invite speakers to a local library, host movie nights and book discussions on impacts, donate resource materials to schools, write an editorial for the local newspaper, or speak at community forums, but if an architect presents the material—an expert not given to doomsday predictions or exaggerations—it may be more readily accepted by the public. Discussions on topics related to the built environment will raise the architect's profile in the community and may lead to commissions for projects with mitigation and adaptation measures.

DESIGNERS AS CLIENT ADVISERS

An architect's responsibility to their client is to serve them competently, exercise unbiased judgment, and safeguard their trust. The Canon of General Obligations demands that members of the profession "maintain and advance their knowledge of the art and science of architecture, respect the body of architectural accomplishment, contribute to its growth, thoughtfully consider the social and environmental impact of their professional activities, and exercise learned and uncompromised professional judgment [4]."However architects also have a standard of reasonable care, the skill that is ordinarily used by other architects in the same locality. This rule implies that practicing in hazard-prone areas requires the architect to apply more advanced technical knowledge than if they practice in a community with less risk.

There is no such thing as a "natural disaster." Hazards are natural events, but a hazard becomes a disaster when human systems expose people and property to those risks. Development in high-risk locations creates greater vulnerability for populations and investments, but planning ahead reduces that risk. Information about the future of a project site must be shared with the client in a way that helps them to act judiciously, and this responsibility belongs to the architect. What are the hazards that threaten to impact the site: sea level rise, storm surge, extreme heat events, intense storms, tornadoes, fire, earthquakes?

Understanding the observed and anticipated hazards that may impact a project is the first step in advancing the resiliency of a project. Architects are practiced in analysis at the outset of a project, but hazard analysis is a predictive science, a future benchmark against which designers measure performance. Sources for accurate scientific information on the types of hazards and their geographic range include the National Climate Assessment, which aggregates data within the United States; in the United Kingdom, the Climate Change Risk Assessment Evidence Report presents regional information on hazards.

Sites with the greatest susceptibility to hazards may be designated as "sending areas" characterized by a planned reduction on the population. This can be accomplished through policies such as better incentives to build elsewhere, a strict refusal to "grandfather" reconstruction of houses following a disaster, and systematic buy-outs; it can also occur through market-based mechanisms such as steeply rising insurance costs. "Receiving areas" are places where the population can relocate safely with the physical and economic advantages to support more density and better living conditions. It is essential that architects are involved in the delineation and design of both areas so that public safety efforts do not trade one type of hazard for another or further segregate one group from another.

The inland migration of the floodplain during the next century is the hazard with the most rapid rate of change, and the one that will most profoundly reshape cities. A map illustrating a project site in relation to anticipated sea level rise and storm surge is a critical data point, whether the site lies within the floodplain (subject to a greater risk of flooding), or outside the floodplain (with the possibility of higher density development). A map of the future can adequately describe these threats, but plotting sea level rise and storm surge on the façade of an existing community landmark is an even more graphic illustration, an emotional "eye-level" view of the current and future conditions (Fig. 11.2). An equally effective diagram might depict the road network with current flooding hotspots, next to a diagram of the same network affected by 30 or 60 in. of sea level rise; the map will illustrate the future barriers to movement, resembling a labyrinth with no exit.

Flooding combined with overdue maintenance is capable of isolating a site or neighborhood, interrupting power and water supplies in an emergency, and severing emergency assistance, direct consequences that may lead investors to choose an alternative site

FIG. 11.2 Seaside Village, Bridgeport, CT, Rebuild by Design. The photo of a neighborhood listed on the National Register of Historic Places with observed and predicted flood levels superimposed on the façade. (Credit: WB/unabridged with Yale/Arcadis.)

within the same community. However, there are hazards that cannot be avoided within a wide geographic region, such as the number of degree days over 90°F, the extent of droughts, or a propensity toward wildfire. These hazards affect an area from which it may be difficult or impossible to retreat, making it necessary to adapt the built environment to protect against these hazards rather than avoid them. Becoming aware of climate trends early in the site selection and programming phases ensures the professional has a clear understanding about which hazards pose a threat to give the client information about the available alternatives.

Critical to this discussion is an understanding of the project's anticipated service life. Commercial buildings have a service life of about 25 years, so a site located in an area that will have a 1% per year chance of flooding by the year 2100 may have little impact on the design. Residential projects typically have a 75-year lifespan, during which much of an owner's wealth is tied to the ability to convey the property to another owner at the end of their tenure, making robustness and continuity important. Institutions and governments specify longer periods of service life, from 100 to 150 years, and these clients may demand the project experience minimal interruption from hazard impacts. (Some installations require continuity of service for mission-critical operations, but these clients will come prepared with their own hazard data.)

Resilient design considers the anticipated service life of the structure and the effects of hazards and climate change over that time, aligning these so that the building has a better chance of survival. The client's expectation of durability has both environmental and economic repercussions, and every new construction project should evaluate the impact of material selection. The U.S. Green Building Council (USGBC) has calculated that the service life of a building must be at least 60 years to fully account for the maintenance and replacement of the materials and resources based on carbon emissions, depletion of atmospheric ozone, depletion of nonrenewable energy resources, and other categories [5]. A long service life is a marker of sustainable design, but one jeopardized by climate change.

Decisions about what and where to build reside with the client. Understanding the potential hazards and the client's expectations for function and tenure should lead to a discussion of specific client concerns: a commercial developer may require flexible workplaces and convertible retail space so that tenants can recover quickly after events. Governmental clients may demand redundant utilities and high-performance envelopes so that operations continue uninterrupted. Projects with multifamily housing may need to shelter children or elderly residents to ensure they remain safe through the event. Stewards of historic buildings may need to prepare for hazards without altering the visual appearance or compromising the original materials.

Communication is the key to uncovering the client's concerns about risk and vulnerability. Many clients believe that architects need to improve their understanding of the commercial drivers for development; that architects must understand more than the client's building, they need to understand the client's business [6]. Linking the client's desired return on investment, operational timelines, and the direct and indirect cost of interruption to the architect's brief may be slightly unusual, but it forms a trustworthy basis for communication between the architect and the client and demonstrates the value of a savvy design team.

Case Study: Marine Education Center

The University of Southern Mississippi Marine Education Center in Ocean Springs, MS, teaches children and adults about coastal ecosystems along the Gulf of Mexico. The program's field trips, summer camps, teacher training, college programs, and citizen science research were interrupted by Hurricane Katrina in 2005, when the center's concrete structure on the beachfront was inundated. The process of rebuilding honed the desire for a building that could withstand the potential for catastrophic storms along the Mississippi coast, including high winds, storm surge, and intense rainfall; it is also expected to meet the challenge of rising temperatures in outdoor education. A new facility would need to accommodate sea level rise and reduce storm recovery time so that research and education missions could continue with minimal interruptions.

The administrators' first decision was to retreat from the water's edge. A new site was selected with a ground elevation of 19 ft above sea level (ASL), a significant improvement over the previous site at about 1-foot ASL. The flood zone at the new site is Zone X (0.2% annual chance of flood hazard). The building was constructed with a finished floor elevation of 20.5 ft ASL, a decision reached early in the design discussions to address the anticipated 30 in. of sea level rise by 2100, a term that matches the anticipated service life of this project. The high ground encompassed two narrow peninsulas, restricting the floor plate to articulated segments to avoid steep slopes and wetland areas. A suspension bridge connects the two sides of the site across a natural drainage area.

This center, focused on stewardship of the marine ecosystem, had a clear expectation that its construction should not harm the environment. The design team used a precautionary list for materials based on sustainable guidelines, expanding it to rule out additional materials based upon their effect upon the marine environment. For example, zinc—often used in countertops, roofing, cladding, and hurricane fasteners in wood construction—is toxic to many species of juvenile fish and shellfish that spawn in the marsh areas adjacent to the site. An early preference for a zinc roof was changed to specify a steel roof with an enamel coating that resembles zinc but does not have the same negative effect on runoff.

Stormwater from the roof is collected in trenches beneath the drip line, which promotes groundwater infiltration instead of releasing it directly into the marsh. This process also reduces the temperature of the runoff, a common development stress upon marine life. The infiltration trenches, permeable driveways, and gravel parking areas filter contaminants and provide extra capacity for stormwater to be absorbed on site. The permeable surfaces have multiple benefits, including reduced site temperatures, an important design goal for this client.

When first encountered, the project site was thickly wooded with a mix of pines and hardwoods. The client and design team worked to retain much of the forest cover for several reasons: to reduce temperatures in outdoor gathering areas and shade the buildings; retain the native habitat for birds, amphibians, and small mammals; and reinforce the impression that the buildings have a deep and lasting connection to the site. However, keeping the trees required evaluating two competing adaptation measures: minimal clearance between the buildings and the treeline allows the trees to provide shade and gives occupants an intimate view of the outside, but in a high-wind environment, trees (especially pines with their brittle root systems) can blow over easily and cause damage to the building. In conversation with the client, the benefits of the narrow perimeter (shading, stormwater uptake, habitat, views) outweighed the concerns (collateral damage, extended construction

Continued

schedule). The team identified hazard trees and selectively removed them, but this decision has the potential to cause interruptions over the next 80 years as the forest grows, matures, and deteriorates.

The goal of reducing energy demand was achieved through a robust building envelope, highly reflective roof, efficient mechanical system, and a system of window controls that allow occupants to open windows when a green light indicates conditions are favorable. These conditions include times when the mechanical cooling is turned off, humidity is low, and outdoor temperatures are within a client-specified range. In this facility, users dress for the weather; educators, students, volunteers, and administrators spend a great deal of their day outdoors and returning to an overcooled indoor environment is not desirable, so the range of acceptable indoor temperatures is wider than the standard thermal curves. People get dirty on purpose here.

This structure relies on steel and heavy timber framing to meet the high winds likely in this region. The resulting structural design exceeds the building code requirements for lateral design pressure, providing a comfortable redundancy to handle most conditions. Although the site lies within Exposure Category C (according to the International Building Code, Surface Roughness C is characterized by open terrain with scattered obstructions and all water surfaces in hurricane-prone regions), the -surrounding forest absorbs and deflects some of the wind. Windows and doors are designed with impact-rated glazing to meet missile level E that prevents a 9-lb 2×4 propelled at 80 ft/s from breaching the glazing or frame [7].

Material selections can have climate change mitigation benefits as well as adaptation benefits. The building foundations are helical piers, augured into place with grout to withstand the lateral and uplift forces (tension) as well as the gravity loads (compression). This system was selected primarily because of the reduced damage to tree roots, but it uses less concrete and therefore emits less carbon.

The project's focus on resilience was not limited to materials; it also built flexibility into the spatial programming. Offices are "loose fit," interchangeable rooms of one size that work for several functions: one administrator, two staff members, or a small conference room. The laboratories have identical water, gas, and power resources, and are distinguished by furnishings rather than forms. There are open spaces such as the bullpen office in the administrative area, exhibit hall at the entrance, and shared porches at the labs, that serve one function now, but can adapt to different programming needs in the future (Fig 11.3).

FIG. 11.3 At the University of Southern Mississippi Marine Education Center in Ocean Springs, MS, laboratories for education and citizen-led science projects are built on high ground elevations and maintain a narrow perimeter to the existing coastal forest to reduce wind exposure. (Credit: Casey Dunn, photographer. Lake Flato Architects with unabridged Architecture.)

DESIGN FOR ADAPTATION

Architects and other building practitioners are responsible for preparing for the future through their design actions. Most of the time, professionals take action in service to their clients, but can at the same time incorporate benefits that extend to the community at large, fostering community interaction and reducing the potential for maladaptation that can increase vulnerability. In the quest for resilience, designers must plan for flexibility to deal with changing conditions and pursue measures that have benefits with immediate returns.

The need for adaptation is extensive but there are limited funds for community-scale infrastructure and modest budgets for project-scale protections. Therefore, the solutions that succeed in being implemented are those that confer multiple benefits. At the community scale, permanent coastal barriers can also connect previously disjointed neighborhoods with sidewalks and greenspace. Designs for urban flood storage might be used for recreation during dry periods, as illustrated in the popular Dutch "water plaza" concept. In a suburban neighborhood, greenways for riverine flooding may plant a high-density tree canopy to help with stormwater uptake and cool adjacent streets and structures. A planned evacuation route could become a corridor for public transit. A new floodwall might integrate all of these, forming a connective, multimodal path with tree-lined edges and open space; overlaying programmatic functions with infrastructure can make it easier to leverage funding from local, state, philanthropic, and federal sources.

Private owners also search for ways to maximize their return on investment, whether they build a single-family home or a downtown, multistory, mixed-use development. Adaptation measures are easiest to justify when one element does the work of many, or an investment pays back on a daily basis. Sun-shading devices might also shield intense rainfall. Passive House standards reduce heat gain or loss and also reduce the size of mechanical equipment. Installing solar panels and a battery backup can power a building during a hurricane and also reduce operating costs for lighting, receptacles, and appliances. Reducing water infiltration improves indoor air quality, discourages mold growth, and improves respiratory health. These are "low regrets" options that yield benefits even in the absence of a hazard event, and "win–win" measures that manage several climate risks at once.

Adaptive measures reduce the direct cost of repairs and sheltering during displacement. But what price can be placed on the cost of interruption versus an accelerated recovery? For a grocery store, there might be immediate—and long-remembered—value to being the last store open and the first to reopen. For homeowners, strengthening their primary investment against storms lowers the risk of mortgage default and allows them to return to work more quickly. If every new coastal residence included a slightly higher initial investment, such as additional nailing and an ice and water shield beneath the primary roofing plane and installing impact-rated windows, homeowners could avoid up to $5 in high-wind losses for every $1 invested [8]. Going beyond code in flood zones to raise the floor elevation 1–4 ft above the flood insurance rate map requirement can provide annual savings on insurance costs that may pay back in just a few years.

Balancing the client's wishlist against the myriad possible solutions is difficult, as is fulfilling stakeholders' needs within the budget and planning for future uncertainties. In public projects, a cost–benefit analysis (CBA) quantifies the proposed benefits to determine if they outweigh the estimated costs. In FEMA Hazard Mitigation projects, when the ratio of benefit to cost is greater than 1.0, the prospective project is of sufficient value to justify the cost [9]. The CBA methodology was implemented in the 1920s to evaluate whether federal waterway infrastructure proposed by the U.S. Army Corps of Engineers was economically viable, serving as an important tool to convince the legislature and taxpayers that investments in flood control have immediate benefits and avoided future costs.

CBA begins by determining the reference condition, or what would realistically happen if no action is taken by looking ahead 20 years, 50 years, or more. Then, the CBA evaluates the positive and negative effects of a project in comparison to the "do-nothing" framework: does it improve outcomes? Does it prolong the viability of at-risk structures and institutions or artificially encourage investment in hazard areas? The CBA includes categories such as life cycle costs, environmental value, social value, economic value, robustness, flexibility, and difficulty of implementation. Although an architect's standard training does not include this type of public pro forma, demand for this service is growing, and the rudiments are easy to grasp.

Designers of the public realm are responsible for the social infrastructure—churches, parks, libraries, and civic spaces of all kinds—the places where people gather, make friends, learn, and build connections. These shared spaces, or even a well-designed street, foster the type of community interaction that can save lives. How can architects design to overcome inequality arising from the physical environment?

Eric Klinenberg writes, "The dangerous ecology of abandoned buildings, open spaces, commercial depletion, violent crime, degraded infrastructure, low population density, and family dispersion undermines the viability of public life and the strength of local support systems, rendering older residents particularly vulnerable to isolation [10]." In the Chicago heat wave of 1995 many of the fatalities were people with medical troubles, who lacked family nearby, had limited access to transportation, or with limited social contact. Spatial attributes also contributed to higher mortality rates: dilapidated streets with vacant storefronts and empty lots made residents feel unsafe, a lack of commercial functions such as banks, food stores, and restaurants gave people no reason to linger on the streets, and poor street lighting and badly maintained roadways further reduced the watchful observers on the street.

Community health has spatial solutions to economic woes: appropriately scaled streets with sidewalks, lighting, and street trees, streets that connect to mixed-use corridors, and streets that encourage walking foster social contact. Neighborhoods with adequate parks and public facilities will economically outstrip those lacking in these essential public goods. Neighborhoods that fill the gaps of retail vacancy before they become an epidemic of closures and boarded-up storefronts can often halt the wholesale abandonment of the downtown. Appropriate density, not overcrowding; diversity of housing types and levels of quality; and the retention of long-term populations in place promote relationships among residents and opportunities to connect. After Katrina, or Sandy, or any number of catastrophic events, the spatial characteristics of the neighborhoods where residents looked out for one another and joined forces to assist in the recovery embodied these principles.

DESIGN RESPONSES TO CLIMATE RISK

Architects and other design professionals have proven their value as strategic partners in mitigation to slow the continued acceleration of climate change. Residential and commercial buildings in the United States were responsible for 39% of total energy consumption in 2017 [11]. Progress has been made toward sustainable energy goals, with an annual energy intensity falling more than 2.8% in 2015 due to a market-based transformation in building design, improvements in the industrial sector, and gains in renewable energy [12]. Adopting stronger building efficiency standards could reduce global energy consumption by 14%, which is critical because energy is the most dominant contributor to climate change [13]. Design professionals have been outspoken advocates and leaders in climate change mitigation, and can take on a similar role in adaptation.

There are spatial characteristics that heighten vulnerability and cause maladaptation, which increases vulnerability to climate variables and undermines the ability to adapt in the future. Examples of maladaptation include poorly selected adaptation measures: levees that protect one community at the expense of a neighboring locale; floodgates that deflect storm surge but impound rainwater to cause inland flooding.

Initiatives labeled "climate change adaptation" range from effective projects that match designs with long-term predictions to projects that are the equivalent of uninformed guesses about future outcomes. There are occasions when it is difficult to discern a good project from a bad one. This may create a false sense of security, place people in harm's way, or use scarce resources on a project with dubious returns. There is no reason for delays in addressing climate change, but addressing the observed and predicted effects must attempt to reduce the potential for maladaptation.

In the hypothetical example of building a levee, one conclusion may be commonly held about the height of sea level rise during the project's service life, but rapid ice melt scenarios may result in greater impacts after the resources (money and material) have already been spent. Options for modifying the levee, which is too low to shield residents from flooding, are limited: scab on a higher wall if the foundations permit or build a higher levee adjacent to the first. Both options have squandered at least some of the value of the initial investment and made people more—rather than less—vulnerable.

Knowing exactly what to do is often a difficult choice that depends on a number of "what if" scenarios. New Orleans struggled with this following Hurricane Katrina—did the levees surrounding the city contribute to a false sense of security? Would people have elevated their houses out of the floodplain had there been no levees?

There is physical evidence that New Orleans would have developed very differently without levees. The natural levee along the Mississippi River deposited sediment over thousands of years, making the French Quarter the highest ground in the city. In this area, the most prevalent typology was the narrow, shotgun style house, raised 2 to 4 ft above the street level for ventilation and to avoid periodic inundations. Buildings in the French Quarter did not flood when the levees failed after Hurricane Katrina, the result of being located on a high riverbank and vernacular design choices.

The French Quarter fared well in comparison to the Gentilly neighborhood, one of the areas of the city with the lowest elevation and one of the last to be developed in the post-World War II boom. Many of the houses in the Filmore, Mirabeau, and Gentilly neighborhoods are "ranch houses," a style imported from the dry American West with hallmarks including slab-on-grade foundations, brick veneer facades, and detailing with an uneasy geographic provenance. These houses were constructed on filled wetlands kept dry by the intervention of city pumps. However, those same city pumps removed much of the groundwater, causing subsidence, exacerbating flooding, foundation settlement, and underslab problems. Without the artifice of the levees or the hubris of transforming wetlands into building sites, it is unlikely that this neighborhood would have been developed at all. It certainly would not have developed in this pattern.

Maladaptation has the potential to cause negative effects both locally and on neighboring developments. The construction of a floodwall can increase scouring and erosion at its base or displace impacts to other locations and cause higher flooding elsewhere. Construction of maladapted infrastructure also has an *opportunity cost*, the loss of a potential gain from other alternatives when a single option is chosen. A floodwall may reduce the natural protective value of the existing ecosystem, such as sandy water bottoms with high friction and roughness, or cause the disappearance of marsh grasses that absorb wave action.

Maladaptation can unevenly impact different sociocultural groups. Wealthy residents have a degree of immunity due to their ability to reserve the best lands by zoning, buying, and occupying them with large lots and single-family residences. They may also have access to better knowledge about the community and prospective changes, as there have been few established channels to engage local communities in policy-making. This opacity is reinforced by not accurately communicating risks and hazards to potential new residents, especially those buying properties with repetitive losses, which may be less costly. As transactions in high-risk areas become dicey, with legal and moral consequences, efforts must be made to eliminate the categories of "winners" and "losers" and reduce exposure for everyone, with clear standards for reporting risks. Democratic nations foster an expectation of fairness and architects can promote wider distribution of information and involvement.

Another example of maladaptation is having a monoculture of economic activity in which the majority of industry in an area is threatened by the same climate-related hazards. Florida's late-20th century explosion of coastal tourism, where great swathes of the state's beaches and maritime forests were bulldozed for resorts, condominiums, and vacation houses, made developments more susceptible to impacts from hurricanes and coastal storms because communities no longer had the natural protections of the dunes and forest floor. Residents also suffered from the lack of diverse economic activity. In the aftermath of storms or other disasters, such as an oil spill, the loss of income from tourism had devastating effects, and the extended recovery period caused economic stress on residents who depend on tourists for jobs and income.

One final note on maladaptation. Much of climate change and adaptation policy is made at the federal level but carried out by local governments. If the objectives or background data are not clear, local officials may not support adaptation measures, fail to enforce guidelines, or modify projects so that the planned efficacy is reduced. The potential for maladaptation can be reduced when there is minimum ambiguity in the proposed outcomes, there are incentives and support for participation, and a clearly defined system of checks to ensure fidelity with the policy's intent.

Infrastructure is the same as every other design problem: the cost to make changes and improve the final product is relatively low early in the design process, increases as the project is detailed in the contract documents phase, and rises further when it enters construction. Once a project is substantially complete, the ability to make changes decreases, often to zero. Architects, engineers, and other building professionals understand the importance of programming the proper performance goals at the beginning of a project. Innovative projects relying on new technology or unproven methods of construction take the chance of not meeting the expected quality or performance requirements; this is one reason they may not pass regulatory reviews or receive funding. In adaptation projects, the most common question is as follows: will the project be too big, or not big enough? Achieving success in these circumstances requires analysis, prediction, consideration of different scenarios, and the ability to select the solution with the capacity to withstand rapidly changing conditions.

Instead of finding the "best" solution, adaptation favors the most robust solution: one that is the least sensitive to future climate uncertainty, avoids irreversible actions, and has a high margin of safety to meet greater-than-predicted impacts. Planning for expansion is one way of future-proofing a project. Traditional expansion occurred horizontally, as roadways and engineering works sprawled beyond their original boundaries to accommodate a growing number of users. More of the population is moving to urban centers and coastlines and empty land for expansion is unavailable or prohibitively expensive; expansion has turned vertical, a shift that requires forethought to provide the additional structural capacity before the demand exists.

Retrofitting large-scale projects to resist threats after they are constructed is not an easy task. Extending an existing levee or floodwall to resist overtopping has traditionally been done with manpower and sandbags, a solution that seems primitive in this era. Technical solutions can be limited and costly, so having a wide variety of tools to tackle the problems is needed. For solutions to flooding, we must thank the Dutch.

The history of flood control in the Netherlands is a centuries-long pursuit of safety in a low-lying land. Dikes and levees transformed the landscape of the fen, the bog, the *moeras,* into an economic powerhouse, and a fertile proving ground for techniques to manage sea level rise, storm surge, and flooding. "Living with water"

Case Study: Room for the River

A recently completed project along the Waal River near Nijmegen moved an existing dike 350 m inland to construct an overflow channel, effectively widening the river at what had been a bottleneck at the river's bend. Flooding along the Rhine River in 1995 caused 250,000 residents to evacuate when dikes threatened collapse, but the new dike wall is higher and features sloped concrete edges with cobblestone plazas and steps; materials that can be submerged during high-water events and return to service quickly.

The project reduced high-water levels while at the same time providing more recreational space, a linear path along the river, and space for a growing city. The new channel is 150 to 200 m wide by about 3 km long, forming an urban water park with bicycle and walking paths, cafes, and a marina. At times of high water, people sit along the quay edges because they enjoy seeing the river from a place of relative safety. Shoals are visible in the summer, but not in winter. The project created space for new development; dredge materials from the overflow channel were used to raise a 3 hectare on the new island by 2 m, about 75 cm above the dike. These new programmatic functions did not ignore the past; efforts to preserve the historic character incorporated cultural elements, such as a fortress, into the design [14].

"It is not easy to do a multi-disciplinary project—what does historic preservation or salamanders mean to the hydraulic objective? [15]" Citizens informed the design team about what was meaningful, and their participation allowed many people to lay claim to the landmark project. The design team of over 250 people was led by Royal HaskoningDHV from their headquarters in Nijmegen, taking on the roles of project management, engineering, and environmental consulting, with a team of consultants including landscape architecture and modeling. Their expertise was mirrored by the city's own consultants who collaborated in managing stakeholders and implementing the project to ensure the spatial quality added value for the residents.

Integrating so many variables requires starting with the big issue—in this case, hydrology—and working out the less critical but still important performance and functional goals as the process unfolds. In this project, the nation's responsiveness to local concerns combatted initial resistance and created an attitude of pride in helping to make the Netherlands a little safer (Fig. 11.4).

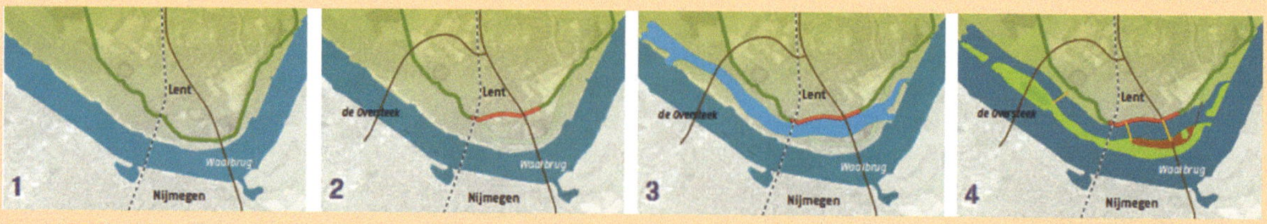

FIG. 11.4 Transformation of the river at Nijmegen. *Ruimte voor de Rivier*, Waal River, the Netherlands. 1. The original condition, a bottleneck in the Waal River that caused flooding. 2. The dike was moved 350 m inland. 3. A high-water channel creates an overflow area for more capacity. 4. The island was raised using spoils from the dredging, and bridges connect to the island for new development. (Credit: Royal Haskoning DHV.)

in the Netherlands includes creating additional storage capacity in secondary channels and lowered floodplains, high-water channels for overflows, and by relocating dikes to allow more space for water to flow without threatening houses, farms, and commercial spaces. The resulting projects are not simply engineered responses to water movement, but incorporate transportation, housing, recreation, agriculture, and urban form.

CONCLUSION

Expanding the design team to solve impacts from climate change produces a richer solution, layered with cultural meaning and commercial functions, stoutly engineered and robust against uncertainties. The design team benefits from experts who are not typically included in the design of cities. Designing for environmental sustainability requires a landscape architect but should include food producers, land conservancies, ecologists, biologists, and market owners. Designing for economic stability requires governmental leaders but also small business owners, logistics companies, retail consultants, chamber of commerce directors, workforce trainers, manufacturers, and artists. For quality of life, the parks director should no longer have the sole voice, but committees can include kayakers, young adults (to combat "brain drain"), social workers, housing advocates, and others. The floodplain manager is essential to promote adaptation, but so are scientists, transportation professionals, energy companies, and the emergency manager. Architects, often considered the last of the generalists in the production of form and space, must sometimes fight for a seat at the table.

The physical planning of cities is shifting from a technical process jealously guarded by government staff and planning experts to an integrative and participatory decision-making process. A design team with broad experience, just as a widespread constituency in engagement, can better resolve competing ideas and interests. It removes ownership of an idea from a single source and invites everyone to influence the formal and spatial qualities of a solution.

The results are more transparent (a common planning buzzword), but applied to urban design, transparency effectively means that the plans have been more rigorously evaluated and will be more widely accepted. The network of people involved in creating the plan will each have a role in its implementation and the elements will continue to stimulate discussion and debate around the alternative futures for the city.

Adapting to climate change is a wicked problem, not because climate change is evil but because it is difficult to solve due to incomplete information, uncertain future impacts, and entrenched policy frameworks that are resistant to change. It is also because there is no constant state predicted for the climate, no endpoint at which the problem will be "solved [16]." Resilience projects often straddle multiple jurisdictions, so implementation may be hampered by regulatory processes, barriers to stakeholder cooperation, and political conflicts. These concerns make it even more critical to widen the group of people charged with adapting to climate change and transform an ethical and theoretical dilemma into strategies that can be shared until every community has access to proven adaptation measures. Architects and building professionals should lead the way.

REFERENCES

[1] J. Jacobs, The Death and Life of Great American Cities, Vintage, New York, 1992, p. 238.

[2] World Bank, Economics of Adaptation to Climate Change - Synthesis Report (English), 2010. Washington, DC. Retrieved from: http://documents.worldbank.org/curated/en/646291468171244256/Economics-of-adaptation-to-climate-change-Synthesis-report.

[3] Carbon Brief staff, Six Years Worth of Carbon Emissions Would Blow the Carbon Budget for 1.5 Degrees, 2014. London, November 18, 2014. Retrieved from: https://www.carbonbrief.org/six-years-worth-of-current-emissions-would-blow-the-carbon-budget-for-1-5-degrees.

[4] The American Institute of Architects, AIA Code of Ethics and Professional Conduct, 2018. Washington, DC. Retrieved from: http://www.aiacc.org/2016/10/26/aia-code-ethics-professional-conduct/.

[5] The United States Green Building Council, Whole Building Life Cycle Assessment. LEED BD+C: New Construction v.3-LEED 2009, 2009. Retrieved from: https://www.usgbc.org/credits/new-construction-core-and-shell-schools-new-construction-retail-new-construction-healthcar-9.

[6] Royal Institute of British Architects, What Clients Think of Architects: Feedback from the 'Working with Architects' Client Survey of 2016, 2016. London. Retrieved from: https://www.architecture.com/-/media/gathercontent/working-with-architects-survey/additional-documents/ribaclientsurveyfinalscreenwithoutappendixpdf.pdf.

[7] ASTM International, ASTM E 1996 Standard Specification for Performance of Exterior Windows, Curtain Walls, Doors and Storm Shutters Impacted by Windborne Debris in Hurricanes, 2009. West Conshohocken, PA.

[8] National Institute of Building Sciences, Natural Hazard Mitigation Saves: 2017 Interim Report, 2017. Washington, DC. Retrieved from https://www.nibs.org/news/news.asp?id=381874.

[9] Federal Emergency Management Agency, Benefit-Cost Analysis, 2018. Washington, DC. Retrieved from: https://www.fema.gov/benefit-cost-analysis.

[10] E. Klinenberg, Heat Wave, University of Chicago Press, Chicago, 2002, p. 91.
[11] Environmental Information Administration, How Much Energy Is Consumed in U.S. Residential and Commercial Buildings?, 2018. Washington, DC. Retrieved from: https://www.eia.gov/tools/faqs/faq.php?id=86&t=1.
[12] The World Bank, Tracking SDG7: The Energy Progress Report 2018, 2018. Washington, DC. Retrieved from: https://openknowledge.worldbank.org/handle/10986/29812.
[13] United Nations, Sustainable Development Goals, Goal 7: Affordable and Clean Energy, 2018. Retrieved from: http://www.undp.org/content/undp/en/home/sustainable-development-goals/goal-7-affordable-and-clean-energy.html.
[14] ClimateWire, How the Dutch Make 'room for the River' by Redesigning Downtown, 2012. January 20, 2012. Retrieved from https://www.scientificamerican.com/article/how-the-dutch-make-room-for-the-river/.
[15] M. Hillen, Telephone interview with Royal HaskoningDHV team member on the Nijmegen Ruimtevoor de Rivier project, on 7 August 2018 by the author, 2018.
[16] H.W.J. Rittel, M.M. Webber, Dilemmas in a general theory of planning, Elsevier, Policy Sciences 4 (1973) 155–169, https://doi.org/10.1007/bf01405730.

CHAPTER 12

Building Codes: The Foundation for Resilient Communities

CINDY DAVIS, CBO • JAMES TIM RYAN, CBO

Building codes serve as an important component of a community's strategy for achieving resilience. Although code adoption alone does not assure resilience, a community without up-to-date codes and strong enforcement is unlikely to be considered resilient and may suffer dire consequences should a hazard event occur.

As identified in numerous studies, building codes provide a mechanism to avoid considerable loss of lives and property. The Insurance Institute for Business and Home Safety (IBHS) found that homes built to strong high-wind codes adopted following Hurricane Andrew received significantly less damage in Hurricane Charley than those built before the adoption of such codes [1]. The frequency of insurance claims was reduced by 60%, and the claim was 42% less severe when a loss did occur. Researchers at Louisiana State University found that if stronger building codes had been in place, wind damages from Hurricane Katrina would have been reduced by 80% [2].

The Congressionally established National Institute of Building Sciences (NIBS) found that adopting the latest edition of the building codes as compared with the 1990 editions provides a benefit of $11 for every $1 invested [3]. NIBS also found that adoption and implementation of the International Wildland/Urban Interface Code (IWUIC) would provide $4 of benefit for every $1 invested [4].

This chapter provides an introduction to codes through their history and how they are currently developed. It addresses current challenges to their effectiveness including assuring the adequacy of resources provided to departments charged with their enforcement. Finally, it touches on why strong building codes are necessary, but not sufficient to assure community resilience.

HISTORY OF CODES

Building codes have a long history. They can be traced back to around 1800 BC when the Code of Hammurabi provided the basis for the first known performance code. A performance code generally does not provide specifics on how to achieve the performance but rather simply calls out what the expected performance is. The performance expectations in the Code of Hammurabi as documented by the Avalon Project at Yale Law School [5] include the following:

- If a builder build[s] a house for someone and complete[s] it, he or she shall give him or her a fee of two shekels in money for each sar[1] of surface.
- If a builder build[s] a house for someone and does not construct it properly, and the house which he or she built fall[s] in and kill[s] its owner, then that builder shall be put to death.
- If it kill[s] the son of the owner, the son of that builder shall be put to death.
- If it kill[s] a slave of the owner, then he or she shall pay slave for slave to the owner of the house.
- If it ruin[s] goods, he or she shall make compensation for all that has been ruined, and inasmuch as he or she did not construct properly this house, which he or she built and it fell, he or she shall reerect the house from his or her own means.
- If a builder build[s] a house for someone, even though he or she has not yet completed it; if then the walls seem toppling, the builder must make the walls solid from his or her own means.

As might be expected, general performance criteria may not be sufficient to explain exactly how a building should be constructed to achieve the expected performance criteria. When a code provides a clear explanation of how to achieve that performance, it is called a

[1]A sar is an ancient unit of measurement equal to an area 12 cubits square or approximately 30 square feet.

Optimizing Community Infrastructure. https://doi.org/10.1016/B978-0-12-816240-8.00012-4

prescriptive code. Modern-day building codes fall more in line with this type of code, as they, in most but not all cases, describe how something must be constructed. This is especially true for the model code known as the International Residential Code (IRC) where prescriptive methods comprise almost the entire code. These too can be traced back to antiquity, with the first known prescriptive code residing in the Biblical book of Genesis [6] where God gave Noah very specific instructions on how to construct the Ark:

> *Make yourself an ark of gopher wood; make rooms in the ark, and cover it inside and outside with pitch (Genesis 6:14). And this is how you shall make it: The length of the ark shall be three hundred cubits, its width fifty cubits, and its height thirty cubits (Genesis 6:15). You shall make a window for the ark, and you shall finish it to a cubit from above; and set the door of the ark in its side. You shall make it with lower, second, and third decks (Genesis 6:16).*

Fast forwarding through history, building codes, as a general rule, followed disastrous fires, becoming more refined with each one. Documented by Ellen Vaughan and Jim Turner [7], beginning in ancient Rome (4 AD) and then in Boston (1631), London (1666), Chicago (1871), Baltimore (1904), and Cleveland (1929), building codes have evolved throughout history.

In the United States, the development and implementation of modern codes began with the publication of the National Electrical Code (NEC) by the National Fire Protection Association (NFPA).[2] This was followed by the publication of the National Building Code after the San Francisco Earthquake of 1906 [8].

As the evolution continued, the United States, by the middle of the 20th century, ended up with three primary construction codes, all of which referenced a number of standards. These three model code groups were generally geographically based, with the seismic focused provisions in the International Council of Building Officials (ICBO) codes on the west coast, high wind-related codes in the south and hurricane-prone areas governed by the Southern Building Code Congress International (SBCCI), and snow load-prone regions in the northeast governed by the Building Officials and Code Administrators (BOCA) codes.

Economic pressures from interest groups, business, and industry, who were tired of the redundancy and expense in designing to three different national standards, finally resulted in the consolidation of these three model code groups into one national model code organization, the International Code Council (ICC).

Although the International Building Code (IBC) is primarily used throughout the United States, it is important to understand that there are many codes and standards developed by a number of organizations that play an important role in the resiliency of communities. ICC and NFPA are just two of the organizations that develop codes and standards that are in use throughout the United States and internationally.

DEVELOPING TODAY'S MODEL CODES

Different codes are developed by different organizations, and each has their own set of rules for how their particular code is developed. The Whole Building Design Guide summarizes the different development processes, which are also provided in Fig. 12.1 [9].

The ICC [10] uses a governmental consensus process that meets the principles of the US Standards Strategy [11] and Office of Management and Budget (OMB) Circular A-119, the Federal Participation in the Development and Use of Voluntary Consensus Standards and in Conformity Assessment Activities [12]. Additionally, the governmental consensus process complies with the National Technology Transfer and Advancement Act of 1995 (Public Law 104–113) [13].

In the governmental consensus process, the final decisions on content are determined by governmental voting members with no vested financial interest in the outcome of the proposal. Industry, scientists, trade associations, and other professionals are not left out. Proposals may be submitted and advocated on by any interested person or organization. The process also includes review by a balanced committee that includes user interests, general interests, or producer interests. However, one-third of all committees must be public safety officials. Committee members must disclose any conflicts of interest and cannot vote on proposals that present a conflict. The committee then makes a recommendation to the full governmental membership. That recommendation may or may not be upheld by the governmental members.

This process of submitting proposals, reviewing by committees, and voting by governmental members occurs in 3-year cycles with half of the codes being evaluated the first year and half the second year. Currently, no code development occurs in the third year of a cycle, but that is being evaluated for possible

[2]NFPA and National Fire Protection Association are registered trademarks of the National Fire Protection Association, Quincy, MA. All rights reserved.

PROCESS NAME	DEFINING CHARACTERISTICS
ANSI Consensus	• Consensus must be reached by representatives from materially affected and interested parties • Standards are required to undergo public reviews when any member of the public may submit comments • Comments from the consensus body and public review commenters must be responded to in good faith • An appeals process is required • Standards developed by ANSI accredited organizations that follow this process are considered American National Standards
Government Consensus	• Participation in the development of the codes, including code hearings, is open to all at no cost. • Anyone can submit a code change proposal or make a public comment • Code committees must consider all views before voting • Final decisions are made in an open hearing by government officials
State or Local Amendments to Model Codes or Standards	• In some jurisdictions, amendments may be made to the model codes or standards • The actual process for amendment may vary considerably however, most processes offer some level of public/stakeholder engagement • Proposed amendments may be made by or through an advisory board for consideration by the jurisdiction's legislative body during the adoption process • The legislative body may have delegated the development, review and incorporation of amendments to a committee that ultimately adopts the code for the jurisdiction
State or Locally Developed Code	• If a jurisdiction chooses to develop its own code, a Council or Committee is usually charged with such development. • The development process usually relies on administrative procedures or another approved process to assure public/stakeholder due process

FIG. 12.1 Common code and standard development processes. (Source: Whole Building Design Guide, Codes and Standards Development, 2016. http://www.wbdg.org/resources/codes-and-standards-development.)

change. Fig. 12.2 provides a flowchart of the current ICC process [14].

NFPA produces codes and standards as well. Probably the most recognized code produced by NFPA is the NEC.

All NFPA standards [15] are revised and updated every 3 to 5 years, in revision cycles that begin twice each year. Normally a standard's cycle takes approximately 2 years to complete. Each revision cycle proceeds according to a published schedule, which includes final dates for each stage in the standards development process. The four fundamental steps in the NFPA Standards Development Process are as follows (Fig. 12.3):

1. Public Input
2. Public Comment
3. NFPA Technical Meeting (Tech Session)
4. Standards Council Action (Appeals and Issuance of Standard)

Irrespective of which code is being used, the takeaway should be that model codes and standards development in the United States embraces and encourages participation by all interests and persons without payment and without membership requirements. The processes are open and transparent, and the better the process is understood by the general public, the more participation will increase. The result will be better technical standards for guiding and creating a resilient built environment.

LOCAL AND STATE ADOPTION OF CODES

Although having a written set of criteria for buildings to be designed and constructed to is important, the widespread applicability of codes is dependent on adoption at the state or local level. Codes provide important functions for the adopting communities. They establish the community's acceptable level of risk and minimum quality requirements for housing and buildings. No matter where someone goes within a community they rely on the adoption and enforcement of the code to

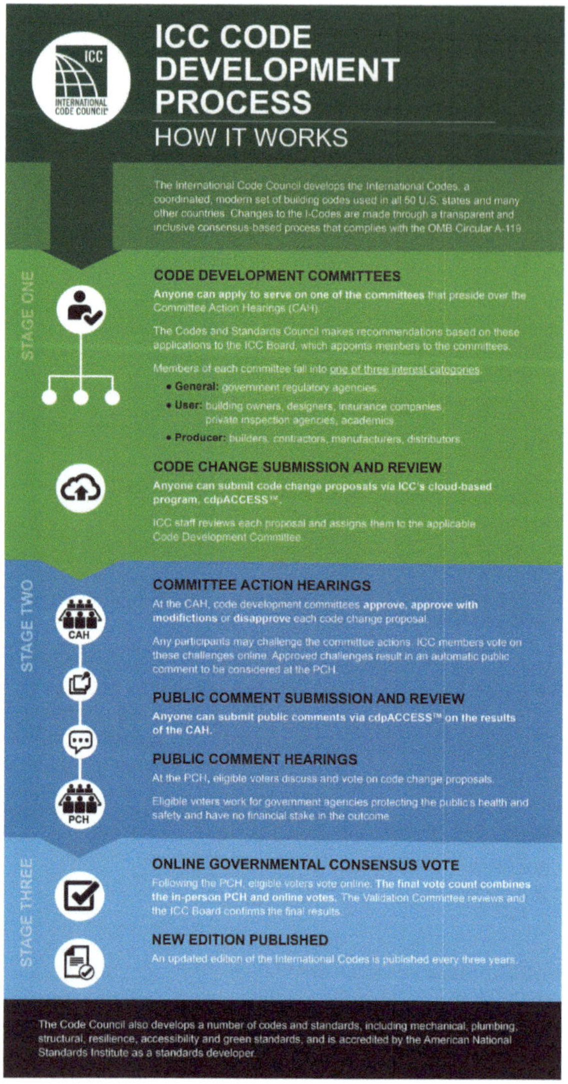

FIG. 12.2 The International Code Council's code development process. (Source: International Code Council.)

support their safety. They also understand that surrounding buildings are designed and constructed to not adversely impact their building.

Codes also serve an important consumer protection function. Individual homeowners or building owners are unlikely to have the knowledge to identify the requirements necessary to protect their investment from hazard impacts. They also do not have the ability to adequately inspect their property throughout the construction process to verify that such requirements are being met. Codes and the departments that support their enforcement centralize this function at a jurisdiction level where specialized knowledge can be developed and consumers can look to for such expertise.

Developing and adopting codes within individual jurisdictions does come with its challenges. The development of model codes and standards in the United States has evolved to the point of creating minimum safety requirements based on building science and the most technically advanced construction methods and materials. However, the culture, methodology, and process for adopting and enforcing codes vary substantially between jurisdictions nationwide. Although the advancement in the development of building codes has occurred, it must be recognized that most, if not all, state and local jurisdictions adopt amendments to the published model codes.

Amendments to the published model codes identify changes that reflect community priorities. In most adoptions, amendments tend to lessen the acceptable level of safety between published editions of the model codes. Such amendments are usually based on cost of construction concerns, geographical and technical evaluations, and political influences. Furthermore, such amendments create inconsistencies and challenges for designers, contractors, and enforcement officials.

Another major factor that impacts the level of safety being provided within communities is the lack of adoption of the most current editions of nationally recognized model codes and standards. Many jurisdictions throughout the nation choose to remain on older editions of model codes or choose not to adopt or enforce model codes. This substantially increases the level of risk to natural and man-made disasters for these communities.

Although it may not be critical to communities to immediately adopt the most recently published edition of a model code, jurisdictions should stay current within two cycles of the most current edition to support incorporation of new technologies and practices and the latest research findings into their building stock. Furthermore, it is critical that all jurisdictions have policies, procedures, and methodologies to continually update to the more current codes and standards.

Most states have statutes dealing with the adoption of codes and enforcement. Not all states have mandatory statewide code adoption and enforcement, particularly in home rule states. In these states, such statutes are boiler plate requirements that do not ensure code compliance at the local level. Furthermore, such state statutes often have exemptions for one- and

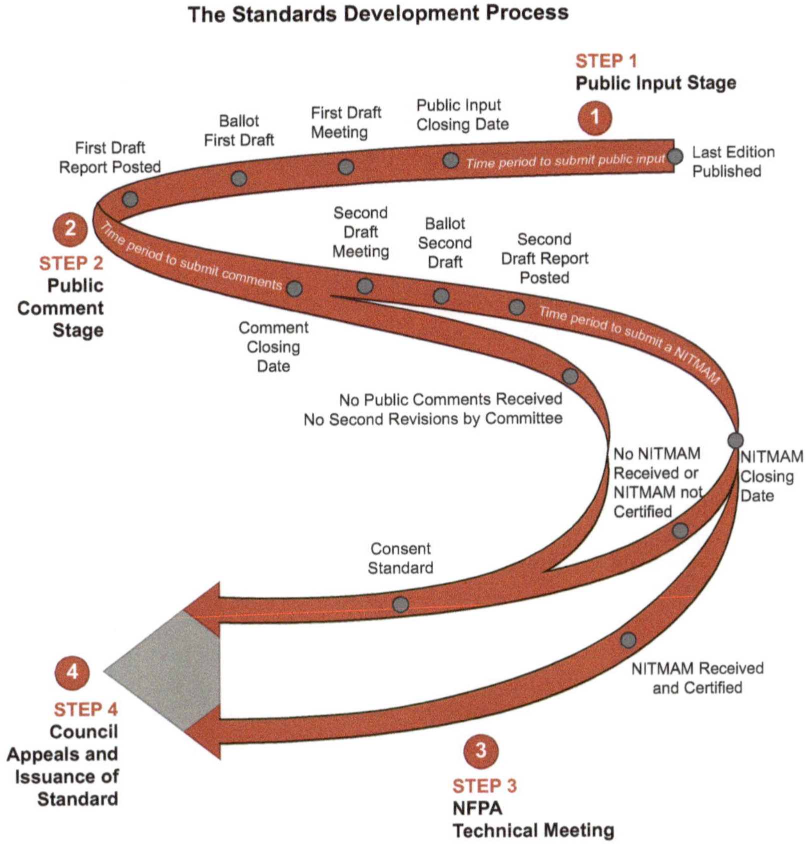

FIG. 12.3 The NFPA standards development process. (Source: Copyright © 2019, National Fire Protection Association. Additional information on the NFPA standards development process to can be obtained through the NFPA web site at www.nfpa.org.)

two-family dwellings. These types of exemptions are also applicable to licensure laws for design professionals.

THE ROLE OF CODES IN ADDRESSING EXISTING BUILDINGS

It may be intuitive to expect that new construction will meet minimum standards that enable a structure to withstand at least some forces expected when natural disasters occur. However, not all buildings are new. So how do codes influence existing buildings?

In many jurisdictions that choose to enforce a model building code, a reference is made either to a model code for changes to existing buildings and minimum levels of maintenance or to a state or local code that addresses requirements for existing buildings. The primary reason is to acknowledge the need and desire for existing buildings to be maintained and upgraded without the burden of having to bring the building entirely up to the requirements for new construction. Contemporary model codes for existing buildings provide multiple options, which incentivize building owners to maintain and update code compliance for their buildings.

The other area that model codes address is the continued maintenance of existing buildings. For example, the International Property Maintenance Code (IPMC) provides specific regulations pointed to property owners to maintain their buildings and properties. Such codes require owners to maintain their buildings related to structural stability; minimum levels of fire and life safety; basic levels of sanitation and environmental conditions; etc.

Unfortunately, some jurisdictions do not provide the same level of emphasis on property maintenance as they do for new construction or at all. Enforcement in some jurisdictions is minimal, that is, inspection of exterior property only; inspection based only on complaints from residents; fire prevention inspections only in common areas; etc.

Some states, such as Ohio and Kansas, have passed state laws, which inhibit enforcement officers from making routine inspections of rental properties. These laws were established to provide protection to residents from invasion of privacy and unlawful search and seizure. Code enforcement officers can only make inspections if the legal resident invites them to inspect their property or rental space.

In 2014, Virginia implemented the mandatory use of the Virginia Rehab Code (based on the International Existing Building Code) for all existing buildings. What are the benefits of an existing building code and why was it made mandatory?

The language from the Code of Virginia [16] sets forth the reasoning behind the law:

In accordance with Section 36-99.01 of the Code of Virginia, the General Assembly of Virginia has declared that (i) there is an urgent need to improve the housing conditions of low and moderate income individuals and families, many of whom live in substandard housing, particularly in the older cities of the Commonwealth; (ii) there are large numbers of older residential buildings in the Commonwealth, both occupied and vacant, which are in urgent need of rehabilitation and must be rehabilitated if the state's citizens are to be housed in decent, sound, and sanitary conditions; and (iii) the application of those building code requirements currently in force to housing rehabilitation has sometimes led to the imposition of costly and time-consuming requirements that result in a significant reduction in the amount of rehabilitation activity taking place.

The General Assembly further declares that (i) there is an urgent need to improve the existing condition of many of the Commonwealth's stock of commercial properties, particularly in older cities; (ii) there are large numbers of older commercial buildings in the Commonwealth, both occupied and vacant, that are in urgent need of rehabilitation and that must be rehabilitated if the citizens of the Commonwealth are to be provided with decent, sound and sanitary work spaces; and (iii) the application of the existing building code to such rehabilitation has sometimes led to the imposition of costly and time-consuming requirements that result in a significant reduction in the amount of rehabilitation activity taking place.

The evolution of the Rehabilitation Code in Virginia is depicted in Fig. 12.4 [17].

It is not hard to grasp the understanding that if buildings are not maintained, they will begin a process of decay and dilapidation that is the beginning of physical disorder in a community. Societal and community ills directly related to dilapidated buildings include the following:

- increased crime including drugs, arson, shootings, gang activity, etc.,
- declining property values,
- declining tax base,
- health and safety hazards, and
- homeless encampments.

In *Vacant Properties: The True Costs to Communities* [18], the data show that in Austin, Texas blocks with vacant buildings had 3.2 times as many drug calls to police; 1.8 times as many theft calls; and twice the number of violent calls. Furthermore, a study by the National Vacant Properties Campaign [19] shows that more than 12,000 fires break out in vacant structures each year in the United States resulting in $73 million in property damage annually. Most are arson.

A 2016 Study by Wichita State University's School of Public Affairs, Public Policy and Management Center, found that surrounding neighborhoods of foreclosures experience an average decline in price per home of $8667, and the aggregate national property value decline was estimated at $352 billion [20].

None of the above bode well for a healthy resilient community. It makes sense then for a community to "incentivize not penalize" those wishing to improve an existing structure.

According to the NIBS white paper [21] titled *The Role of Existing Building Codes in Safely, Cost-Effectively Transforming the Nation's Building Stock*, in 2005, 13 state governments and 13 localities in other states adopted the International Existing Building Code (IEBC) for the first time for enforcement and use in existing buildings. By 2013, 21 state governments and 18 localities in other states had adopted the IEBC. The authors of the paper opine that it is extremely encouraging to see that, by 2016, the IEBC is effective statewide in 23 states plus the District of Columbia and localities in 18 other states are using the code. Meanwhile, New Jersey and Rhode Island have their own statewide existing building code.

Using either the model IEBC or a custom version of a rehabilitation code created specifically for a community is an effective way to help ensure the resiliency of a community.

THE IMPORTANCE OF BUILDING DEPARTMENTS

As stated, the procedures for the development of nationally recognized, contemporary model codes and

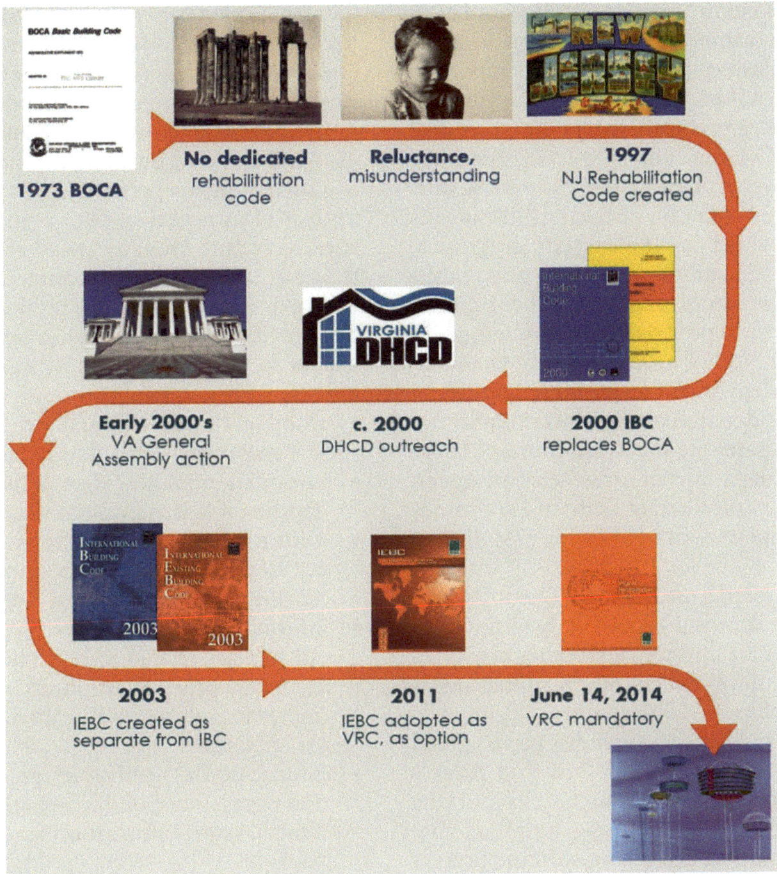

FIG. 12.4 Evolution of the Virginia Rehabilitation Code. (Source: Virginia Department of Housing and Community Development.)

standards in the United States are well established. However, the level of enforcement and compliance with codes is inconsistent and varies jurisdiction to jurisdiction. It is unfortunate that the enforcement of building safety regulations is still not included in the conversations related to "public safety." Such conversations are usually reserved for police and firefighting operations. The capacity of building departments to provide technical expertise to residents and business owners of a community is easily overshadowed by reactionary responses when things go wrong.

It should also be noted that as a public safety department, the building department generally covers most, if not all, of their costs from user fees, generally in the form of building permits. Thus, one of the most important departments of the community is generally self-sustaining and often requires little in the form of tax revenue to operate. In fact, in many cases, jurisdictions are using fees generated by building permits to subsidize the general fund, which in some states can be seen as an illegal tax. Clark County, Nevada, settled a lawsuit for $1.2 million because the county comingled permit fees with general fund dollars. By state law in Nevada, permit fees can only cover the true cost of processing a permit [22].

Additionally, the code official workforce is facing significant challenges. A 2014 study by NIBS for ICC found that 30% of the code official workforce expected to retire within 5 years and 80% within 15 years [23]. These code professionals will be leaving the workforce without an equivalent number of younger professionals already involved in the profession to take their place—only about 15% of the respondents are under 45 years old, with only about 3% under 35. These findings are particularly concerning as more than half of the code departments have nine employees or less.

Complacency with the adoption and enforcement of model building codes, in some jurisdictions, results in the lack of resources being provided. Further, and even more detrimental to achieving safe buildings in some jurisdictions, is the lack of political will to support an efficient and viable building code compliance culture.

Jurisdictions that are considered to be successful such as the States of Virginia and New Jersey and jurisdictions such as Overland Park, Kansas, and Seattle, Washington, all have common characteristics such as providing a highly skilled and trained staff; progressive code adoptions; strong commitment to fair and consistent enforcement procedures; and a culture of political and executive support to provide for safe buildings.

Jurisdictions that enforce model codes generally do so by designating a department or agency to administer the codes. Do not underestimate the powerful resource the agency or department is for a community. Those employed by these departments or agencies are generally highly skilled professionals with an enormous amount of knowledge that can aid in the planning of any project.

All too often, these departments are overlooked at the beginning of an approval process only to find that large amounts of time, money, and other resources were devoted to the development of a plan that needs to be changed due to code requirements.

In fact, NIBS published a white paper titled *Engaging Code Officials Early in the Process to Achieve High-Performance Buildings* [24], which calls out the importance and benefits of engaging building, fire, mechanical, and plumbing officials early in the process of any project. Today, despite evolving delivery models, the project team's interaction with code officials continues to occur late in the design process. This late involvement often contributes to project delays and increased costs because the project team's identified design solutions do not comply with adopted codes or standards. Too often, project teams only engage the regulatory agencies at time of submittal. Yet, code officials in jurisdictions across the nation have found that identifying concerns early provides the best opportunity to rectify them.

Although not specifically called out in the paper, it is implicit that early involvement would naturally allow the identification of potential resilient elements of the project which, if identified early, would be relatively cost-effective to introduce. Conversely, trying to apply the same resilient features retroactively after significant design calculations have been employed may cause an owner to be not as receptive to resilient building features.

The NIBS white paper focuses on high-performance buildings, but certainly, early involvement is beneficial for any project no matter the size. One overlooked resource for citizens within the community is using the building department as a resource when planning a weekend "DIY" project. Homeowners can reap enormous benefits and avoid costly mistakes by consulting with their building department when planning a project.

Often, homeowners are reluctant to obtain a permit for fear of increased taxes or perhaps out of pure defiance, as private property issues are always a hot topic. However, the building department is in place to help ensure a vibrant, safe, and resilient community. They can provide homeowners with important information related to natural disasters as well as common safety hazards such as the following:

- Flood elevation information for those properties that may be within a flood plain.
- Information on how to make a structure resistant to flooding if it is within a flood prone area.
- Information on materials to avoid and safeguards to put in place for structures and properties prone to hurricanes or in high-wind regions.
- Information on materials to avoid and safeguards to put in place related to vegetation around a home if in an area prone to wildfires.
- Important information related to proper venting of gas appliances.
- Information on consumer product safety information or recall of building materials.
- General safety information for decks and swimming pools.

Funding of building departments is a critical benchmark in determining resiliency as identified by the Alliance for National & Community Resilience (ANCR) [25]. Although building departments generate a substantial amount of monies from building permits, plan reviews, and inspections, these monies should be used for the purpose they were intended—to ensure a safe built environment for the community. However, in some jurisdictions, this revenue is not directly earmarked to support the operations of the building department and may be dispersed to other departments that generate little or no revenue, thereby reducing the resources available to the building department, resulting in a potential reduction of community resiliency.

An enterprise funding plan is much like a private business. The revenues generated by the building department are used directly to fund the operations of the department. The problem with this type of budgeting is that the financial health of the department is directly related to the level of construction activity within the community. When activity is low or stagnant, the department may have to downsize their staffing, which is the largest cost to the department. Experience and institutional knowledge are lost.

The risk with these types of funding methods is that building safety departments can be underfunded based on the level of impact they are providing for the community. They lack personnel, training and education, and other resources needed to properly execute the administration and enforcement of building codes. Furthermore, building safety departments usually do not have budget strength to withstand the ebbs and flows of the construction industry. These departments are usually understaffed when construction activity is robust and overstaffed when construction activity is low.

Building departments need to have other methods of securing funding for their activities. This starts with establishing the level of importance to public safety by the jurisdiction. In contrast, the fire service receives funding through several grants from the Department of Homeland Security and FEMA. Many of these grants are for basic education and training of fire service personnel. Training and education is one of the primary concerns for the men and women charged with enforcing building codes, particularly with the ever-changing construction technology and publication cycles of the codes and standards. This begs the question why building departments cannot receive grants like those received by the fire service. Like the US Fire Administration, should there be a US Building Administration?

All of this information is important to current homeowners to ensure their safety along with the long-term viability of their home. What is often overlooked is the impact that inspections and permit review will have in the future. It is often difficult to sell a property if the proper permits and inspections have not been obtained from the jurisdiction. Although that may be a frustration on the end of the seller, it is an important protection to the buyer. Someone's child or grandchildren will be moving into a home that was once owned by someone else. It is important for communities to protect their building stock for all future owners to ensure a resilient future.

WHY CODES ARE JUST THE FOUNDATION

Building codes primarily focus on providing immediate life safety—allowing occupants enough time to evacuate a compromised building. However, community resilience relies on the long-term social and economic viability of a community. If codes do their job perfectly, there will likely still be buildings within a community postdisaster that will no longer be suitable for their purpose. Without the ability for reoccupation postdisaster, educational pathways may be disrupted, residents may need to find new housing, and businesses may close. Therefore, communities must have additional policies and programs in place to supplement the minimum life-safety provisions contained in building codes to support "bouncing back."

Discussion on the establishment of "immediate occupancy" codes that would facilitate ongoing use of buildings postdisaster are emerging. Congress asked the National Institute of Standards and Technology (NIST) to identify the research necessary to develop such a code [26], and California is exploring legislation to require exploration of a "functional recovery standard" to address seismic risk [27].

Even outside of the code environment, communities need mechanisms to engage building owners once the code department has issued a certificate of occupancy. This is particularly important as risks change. The level or type of risk present when a building was constructed may evolve over the life of the building. A combination of incentives and mandates is likely necessary to encourage ongoing building upgrades that contribute to the resilience of the building occupants and the community as a whole.

REFERENCES

[1] Insurance Institute for Business and Home Safety, Hurricane Charley: Nature's Force vs. Structural Strength, 2004. http://disastersafety.org/wp-content/uploads/hurricane-charley-report.pdf.

[2] Louisiana State University Hurricane Center, Residential Wind Damage in Hurricane Katrina: Preliminary Estimates and Potential Loss Reduction through Improved Building Codes and Construction Practices, 2003.

[3] National Institute of Building Sciences, Natural Hazard Mitigation Saves 2017 Interim Report: An Independent Study—Summary of Findings. K. Principal Investigator Porter, C. Co-Principal Investigators Scawthorn, N. Dash, J. Santos, P. Schneider, Director MMC. National Institute of Building Sciences, Washington, D.C., 2017.

[4] National Institute of Building Sciences, Natural Hazard Mitigation Saves: 2018 Interim Report. K. Principal Investigator Porter, C. Co-Principal Investigators Scawthorn, C. Huyck, Investigators: R. Eguchi, Z. Hu, A. Reeder, P. Schneider, Director, MMC. National Institute of Building Sciences, Washington, D.C., 2018.

[5] Yale Law School, The Avalon Project. Hammurabi Page, Lillian Goldman Law Library, New Haven, CT, 2008. http://avalon.law.yale.edu/ancient/hamframe.asp.

[6] The Bible – King James Version – Book of Genesis Chapter 6 Verses 14-16.

[7] Vaughan & Turner, The Value and Impact of Building Codes, Environmental and Energy Study Institute, 2013. http://www.eesi.org/papers/view/the-value-and-impact-of-building-codes#7.
[8] U.S. Green Building Council, Greening the Codes – A White Paper. https://www.iccsafe.org/gr/Documents/GreenToolkit/USGBC_Paper.pdf.
[9] Whole Building Design Guide, Codes and Standards Development, National Institute of Building Sciences, 2016. http://www.wbdg.org/resources/codes-and-standards-development.
[10] The International Code Council, Code Development Process, 2018. https://www.iccsafe.org/codes-tech-support/codes/code-development/.
[11] American National Standards Institute, United States Standards Strategy, 2015. https://share.ansi.org/shared%20documents/Standards%20Activities/NSSC/USSS_Third_edition/ANSI_USSS_2015.pdf.
[12] Office of Management and Budget, OMB Circular A-119: Federal Participation in the Development and Use of Voluntary Consensus Standards and in Conformity Assessment Activities, 2016. https://www.whitehouse.gov/sites/whitehouse.gov/files/omb/circulars/A119/revised_circular_a-119_as_of_1_22.pdf.
[13] National Technology Transfer and Advancement Act of 1995, P.L. 104-113. https://www.nist.gov/standardsgov/national-technology-transfer-and-advancement-act-1995.
[14] ICC Code Development Process-How it Works. https://www.iccsafe.org/codes-tech-support/codes/code-development/.
[15] National Fire Protection Association, The Standards Development Process, 2018, in: https://www.nfpa.org/Codes-and-Standards/Standards-development-process/How-the-process-works.
[16] Virginia Legislative Information System, Code of Virginia § 36-99.01, 2018. https://law.lis.virginia.gov/vacode/title36/chapter6/section36-99.01/.
[17] Virginia Department of Housing and Community Development. Virginia Code Academy. Rehab Code Training Module. 2018.
[18] Smart Growth America, National Vacant Properties Campaign. Vacant Properties: The True Costs to Communities. https://www.smartgrowthamerica.org/app/legacy/documents/true-costs.pdf.
[19] Smart Growth America, National Vacant Properties Campaign. Creating Opportunity from Abandonment. Vacant Properties: The True Costs to Communities. https://www.smartgrowthamerica.org/app/legacy/documents/true-costs.pdf.
[20] Wichita State University, Effects of Abandoned Housing on Communities Research Report on The City of Topeka, Hugo Wall School of Public Affairs; Public Policy and Management Center, 2016.
[21] National Institute of Building Sciences, The Role of Existing Building Codes in Safely, Cost-Effectively Transforming the Nation's Building Stock, 2016. https://www.nibs.org/resource/resmgr/ncgbcs/NCGBCS_IEBC_WhitePaper_2016.pdf.
[22] The Daily News, County Faces Lawsuit over Permit Fees, 2009. https://tdn.com/news/local/county-faces-lawsuit-over-permit-fees/article_c5a3c476-e60d-11de-a35d-001cc4c03286.html.
[23] National Institute of Building Sciences/International Code Council, The Future of Code Officials: Results and Recommendations from a Demographic Survey, 2014.
[24] National Institute of Building Sciences, Engaging Code Officials Early in the Process to Achieve High-Performance Buildings, 2018. https://www.nibs.org/resource/resmgr/ncgbcs/NCBSC_EarlyCodeOffInvolement.pdf.
[25] Alliance for National & Community Resilience, Community Resiliency Benchmarks: Buildings Benchmark, 2019. http://www.resilientalliance.org.
[26] National Institute of Standards and Technology, Research Needs to Support Immediate Occupancy Building Performance Objective Following Natural Hazard Events, NIST Special Publication 1224, 2018, https://doi.org/10.6028/NIST.SP.1224.
[27] California Assembly Bill 1857, 2018. http://leginfo.legislature.ca.gov/faces/billNavClient.xhtml?bill_id=201720180AB1857.

Introduction

Thus far, this book has outlined the physical, financial, and analytical actions underway within infrastructure systems to support enhanced resilience. Although such efforts are a vital piece of the overall resilience effort, community-level action does not occur or is ineffective without political and social will and the tools that establish a direction and demonstrate progress. Additionally, the implementer of resilience measures need metrics and tools to set their direction and monitor progress. Ideally, these metrics and tools should allow examination across systems and across geographies.

The chapters in this section outline some of the important components of a comprehensive strategy that supports resilient infrastructure. The policies and practices deployed within the infrastructure sector and by the community leaders citizens rely on must recognize the need for a holistic reframing of disaster planning, response, and recovery. As recognized by the United Nations, achieving resilience requires focus on preventing the creation of risk, the reduction of existing risk, and the strengthening of economic, social, health, and environmental resilience through the adoption and implementation of national and local disaster risk reduction strategies and plans across different timescales with targets, indicators, and time frames [1].

As introduced elsewhere in this book, the concept of incentivization provides a mechanism to align various priorities, costs, and benefits around a singular goal [2,3]. It tries to capture the externalities that often deter stakeholder action by bringing together actors that either bear the cost or reap the benefits of mitigation investments. Incentivization recognizes that a single-discipline solution to optimizing resilience investments is impractical, not cost-effective, and only marginally effective. "The most cost-effective manner to achieve resilience is through a holistic and integrated set of public, private, and hybrid programs based on capturing opportunities available through mortgages and loans; insurance; finance; tax incentives and credits; grants; regulations; and enhanced building codes and their application" [2]. Policy frameworks should be developed and deployed with this holistic approach.

The National Institute of Building Sciences (NIBS) identified seven categories of activities to support effective development of an incentivization strategy. As illustrated throughout this book, many of these activities are underway, but coordinating outcomes to support a holistic incentivization strategy will be key. Categories of activities to support incentivization are as follows [2]:

- Develop the technical support for incentivization, by
  - evaluating the financial impact that resilience strategies will have on regulatory and business processes;
  - mapping community levels of risk as a basis for implementing levels of resilience;
  - designing software tools to support incentivization processes; and
  - improving the flow of resilience information to stakeholders.
- Develop supporting insurance, loan, and investment processes (e.g., assessments and underwriting).
- Develop community-based incentives (e.g., permit acceleration).
- Develop programs tailored to utilities (e.g., small rate increases to help fund resilience efforts).
- Enhance building codes based on political will or the willingness of the public to pay for a higher level of security.
- Develop incentive programs by
  - enhancing and expanding existing programs (e.g., the South Carolina Safe Home Program and FORTIFIED);
  - modifying existing related programs (e.g., the Federal Housing Administration (FHA) 203k and Fannie Mae rehabilitation programs);
  - modifying green programs (e.g., Property Assessed Capital Expenditures, Small Business

Administration loans, and Fannie Mae lower rate mortgages) for resilience;
    - forming new programs (e.g., commercial loans); and
    - initiating public-private partnerships.
- Support incentivizing activities with tax incentives, tax credits, and grants.

In its addendum to the initial incentivization white paper, NIBS identified multiple programs that could be developed or enhanced to serve as components of an incentivization strategy [3]. Some of those measures are captured in Fig. 1.

The NIBS Multihazard Mitigation Council (MMC) and Council on Finance, Insurance and Real Estate (CFIRE) are undertaking an effort to use incentivization in the context of residential mortgages. The concept parallels some of the offerings available for energy-efficient and sustainable properties. A resilience mortgage is so called because it builds upon the existing mortgage structure to achieve the advantages of resilience against natural disasters by incorporating financing for hazard mitigation into the primary mortgage. The resilience mortgage can be designed to be used when a property is purchased or upon refinancing and can be used to

Incentive Strategy	Target (Decision Maker)	Provider (Offerer)	Incentive
Insurance-Based			
Business interruption insurance	Business	Insurance company	Premium reduction
Utility insurance	Utility	Insurance companies	Premium reduction
Mortgage-Based			
Resilience mortgage	Homeowner and business	Bank and loan agency	Increased loan amount, increased appraised value, reduced interest rate and down payment, and reduced associated insurance premium
Secondary mortgage market	Bank, loan agency, and (indirectly) homeowner and business	Financial institution	Credit quality of security-backed mortgages
Finance-Based			
PACE and SBA mitigation support	Small business	Public-private partnership	Loan
Corporate bond	Large business and utility	Bond rating agency	Favorable bond rating
Municipal and revenue bonds	Utility and community	Bond rating agency	Favorable bond rating
REIT	Large business	Financial institution	Capital provided for asset risk reduction
Private equity real estate fund	Large business	Financial institution	Capital provided for asset risk reduction
Tax & Grants-Based			
Taxes	Homeowner, business, utility, and community	Federal, state and local government	Tax deduction, tax credit and fee waiver
Grants	Homeowner, business and utility in insurance and mortgage programs	Federal, state and local government	Allocation and low-interest loan program (revolving fund)
General			
Code and zoning adoption and enhancements	Community	Community and state	Desirable community; favorable BCEGS and CRS ratings
Developer agreements	Developer	Community and economic development corporation	Tax deduction; design standard modifications planning; zoning and fee waivers; accelerated permitting process
Contractor-based financing	Small business and homeowner	Contractor	Loan
Utility rates	Utility	Public utility commission	Rate increase

FIG. 1 Potential incentivization programs. (Source: Reproduced from, Multihazard Mitigation Council and Council on Finance, Insurance and Real Estate, An Addendum to the White Paper for Developing Pre-disaster Resilience Based on Public and Private Sector Incentivization, National Institute of Building Sciences, September 2016.)

make homes more resilient to many local hazards including riverine and coastal flooding, hurricanes and wind storms, earthquakes, and wildfires [4].

The residential resilience mortgage takes advantage of two suppositions: (1) the benefits of reducing certain natural hazards can greatly exceed the costs under certain conditions and (2) many stakeholders enjoy those benefits besides the property owner, who would normally have to bear all of the costs for the mitigation. The resilience mortgage works by aligning the financial interests of several parties to a potentially costly but cost-effective resilience decision.

When a residential property owner mitigates natural hazard risk to a property purchased with a residential resilience mortgage, the lender also benefits through a lower default risk, which has monetary value. A mechanism to fund natural hazard mitigation is introduced here that voluntarily redirects some of the lender's benefits back to the borrower to help pay for resilient construction. The mechanism benefits the borrower through lower cost and the lender through lower risk, without necessarily involving public funds. The rest of society still enjoys various cobenefits, such as lower risk to the regional economy, less demand for local emergency services, less risk of debris and ecological impact, greater employment, and generally greater stability and efficiency.

In addition to a mortgage interest rate reduction, benefits to the borrower can be further increased by a set of complementary or layered incentivization strategies including an insurance premium reduction and local property tax incentives, all of which will reduce the borrower's PITI (principal, interest, taxes, and insurance) and improve the home expense ratio and debt-to-income ratio used to qualify the borrower for the loan. Such incentives reduce the borrower's monthly payment and increase the likelihood that the borrower will build a more resilient home, thus increasing cobenefits to the lender and to other stakeholders. Additionally, the borrower enjoys these benefits even if the home is not exposed to a significant natural hazard. A resilience mortgage in turn provides decreased risk to the insurer that offers a premium reduction and contributes to the resilience of a community that offers a property tax incentive.

Incentivization relies on the institution of a common approach to quantifying benefits and a mutual agreement on methods for measurement and verification of the effectiveness of mitigation activities. The rigorous, peer-reviewed process for calculating benefit-cost ratios developed by NIBS in its *Natural Hazard Mitigation Saves* study [5] could be the basis for such a common approach.

As a concept, incentivization is simple, but the implementation is no doubt complex. Getting multiple segments of the finance and insurance markets to align around a single process is no easy feat, but one that must occur to cost-effectively address our resilience challenges.

Although the resilience mortgage is still taking shape, many states have begun to expand existing energy efficiency programs to include resilience measures. Property Assessed Capital Expenditures (PACE) (formerly called Property Assessed Clean Energy based on its almost exclusive focus on energy efficiency measures) provide a mechanism where approved measures can be funded by either public or private lenders with repayment through the property tax bill. Such an approach provides multiple benefits both to the community and to the building owner. By tying repayment to the property via the tax bill, the likelihood of default is reduced and thus interest on the loan can be reduced. This also allows the cost of the improvement to run with the property, so the initial owner only bears the cost as long as they own the property. Costs after that point shift to the new owner. Nirmul and Scott provide examples of PACE projects in Chapter 10.

Essential to the success of any policy is the ability to set a desired end state and monitor progress toward that end. All elements of a policy or program should have clear connections to achievement of the established goals. For long-term initiatives in particular, benchmarks with key milestones are important to maintain momentum and assure that resources are directed toward the desired outcomes. For an issue as multifaceted as resilience, metrics should be applicable across as many systems as possible (as clearly illustrated by Brashear in Chapter 3). Plodinec in Chapter 14 outlines the importance of conducting regular evaluations and lays out community-level strategies to support such a process.

Plodinec methodically identifies the contributions of efforts to date by the United Nations, the National Institute of Standards and Technology (NIST), and the Community and Regional Resilience Institute (CARRI) and how current initiatives such as the Alliance for National and Community Resilience (ANCR) are leveraging that knowledge to deliver solutions that are useful, usable, and used.

Hopefully, benchmarking and planning at the community scale triggers multiple stakeholders to act. In many cases, community leadership comes from some nontraditional sources. Philanthropic organizations are on the ground in communities everyday—they understand the social needs of the community and are excellent community organizers. As Ottenhoff discusses in Chapter 15, philanthropy is beginning to recognize the importance of investing in community resilience in advance of a disaster event. Such efforts support social, economic, and environmental resilience.

Just as policymakers, practitioners and communities as a whole are beginning to understand the importance of implementing resilience policies and practices, a wrinkle is thrown into the mix—the changing nature of risks. Reilly and Ayyub in Chapter 13 examine how designers, planners, and managers of infrastructure can incorporate strategies that address new risks—particularly those associated with a changing climate and rapid urbanization. Through equipping professionals with these tools and techniques, they can better engage infrastructure owners and policymakers in making informed decisions that address the entire life cycle of a project (which in some cases can be 50 to 100 years).

Achieving resilient communities relies heavily on political decision-making and assuring that the right levers are pulled and the right signals are given. This part of the book and the conclusion that follows in Chapter 16 outline strategies for community leaders and the disciplines and systems that contribute to a community's resilience.

REFERENCES

[1] United Nations International Strategy for Disaster Reduction, Global Assessment Report on Disaster Risk Reduction, 2015.

[2] Multihazard Mitigation Council and Council on Finance, Insurance and Real Estate, Developing Pre-disaster Resilience Based on Public and Private Sector Incentivization, National Institute of Building Sciences, October 2015. https://www.nibs.org/resource/resmgr/MMC/MMC_ResilienceIncentivesWP.pdf.

[3] Multihazard Mitigation Council and Council on Finance, Insurance and Real Estate, An Addendum to the White Paper for Developing Pre-disaster Resilience Based on Public and Private Sector Incentivization, National Institute of Building Sciences, September 2016.

[4] Multihazard Mitigation Council and Council on Finance, Insurance and Real Estate, Residential Resilience Advantage Mortgage: Summary Concept, Unpublished.

[5] Multihazard Mitigation Council, Natural Hazard Mitigation Saves: 2018 Interim Report. K. Principal Investigator Porter, C. co-Principal Investigators Scawthorn, C. Huyck, Investigators: R. Eguchi, Z. Hu, A. Reeder, P. Schneider, Director, MMC, National Institute of Building Sciences, Washington, D.C. https://www.nibs.org/resource/resmgr/mmc/NIBS_MSv2-2018_Interim-Repor.pdf.

CHAPTER 13

Designing for Resilient Systems Under Emerging Risks

ALLISON C. REILLY, PHD • BILAL M. AYYUB, PHD

INTRODUCTION

The number of billion-dollar climate-related disaster events is rapidly increasing. In the United States, the top 6 years for billion-dollar disasters, by count, were in the past 10 years [1]. By far, 2017 was the costliest, with more than $306.2 billion (2017 USD) in cumulative damage costs, dwarfing the previous record of $214.8 billion (2017 USD) set in 2005. Damaging forest fires in the West, Hurricanes Harvey, Irma, and Maria in the South, and tornados in the Midwest all contributed to the 2017 record. Similar loss trends are present throughout the world. The number of international climate-related catastrophes averaged 650 per year in recent years compared with 225 in the early 1980s [2]. Moreover, although the majority (by count) of disasters are climate related, worldwide losses from all-natural disasters, including earthquakes and tsunamis, are also on the rise. Global losses piqued in 2011, which includes the Tōhoku Earthquake and Tsunami, with cumulative costs of more than $366 billion (2011 USD) and 29,700 fatalities [3]. The costs of hazards to local communities are extreme and are expected rise [4–6]. For example, Ayyub et al. [7] examined how climate change could affect hurricane storm surge in Washington, D.C., and estimated potential losses in excess of $20.0 billion (2012 USD).

Although once rare, many emerging factors have been attributed to the rise of catastrophic losses. Population growth is forcing development in areas previously considered "too risky" to develop [8]. In the United States, this behavior is being reinforced by permissive government policies, such as the National Flood Insurance Program (NFIP), which effectively subsidizes high-risk development [9]. Rapid urbanization is simultaneously reducing pervious surface cover, which contributes to flooding. Flooding caused nearly a quarter of all losses from natural disasters since 1980 [2]. Furthermore, deferred maintenance of critical infrastructure is making these lifeline systems more vulnerable to damage when disasters do occur [10]. This creates additional exposure for disaster victims in recovery and slows the restoration process. Finally, climate change is intensifying many climate hazards.

In this chapter, we explore the pressing need for integrating resilience concepts in the planning, design, and management of infrastructure systems. The rationale being that by doing so, it can help to curb losses and speed recovery in a more comprehensive manner than traditional risk management methods. Resilience concepts are well suited for dynamic environments where the events are infrequent and the risk continuously evolves. Infrequent events provide few opportunities for study and to build knowledge; we will refer to these as black swan events. Risks may evolve due to a confluence of factors, including climate change, rapid urbanization, and technological innovation. We refer to these as emerging risks for both their newness and potential for significant consequences [11]. Resilience concepts in engineering contexts have been extensively refined over the previous 15 years and, in their current state, provide a useful framework for conceptualizing hazard loss reduction in the presence of black swans and emerging risks. Present and ongoing research efforts are exploring how resilience concepts can be integrated with current engineering practice to help protect communities [12].

Resilience concepts are particularly powerful in mitigating hazardous situations that possess significant uncertainty and a potential for extreme loss. These types of events, termed black swans, are notable in that they are often unforeseen or surprising when they do happen, even if, perhaps, they should not be. Thus, many aspects of black swans are poorly characterized or even unknown. Many billion-dollar disasters can be considered black swan events. Further exposition on black swans is provided in Section On Black Swans. Traditional risk

Optimizing Community Infrastructure. https://doi.org/10.1016/B978-0-12-816240-8.00013-6

management is not as adept for handling black swan events. These tools tend to be more useful when the risks are well defined and understood [13]. Unfortunately, "well-characterized" is typically not the case for most extreme hazards.

The factors that were briefly mentioned above (e.g., climate change, rapid urbanization, and population growth) each contribute to why many extreme hazards are poorly characterized. They continually change the context in which our understanding of hazards is based. Climate change, for example, is changing the processes through which hurricanes are formed, but significant debate remains in how these changes will downscale to affect different regions [14,15]. All these factors are examples of emerging risks. Emerging risks are risks that are simply new, and because of this, the ways that these factors amplify the negative consequences of disaster are poorly understood [16]. This is forcing some hazardous events that, under different contexts, would have been comparatively mild into the black swan category.

The increased presence of emerging risks in the hazards realm points to the need for a radical paradigm shift in how people, communities, and operators of infrastructure systems prepare and plan. Integrating resilience concepts into the planning and design phases for engineering systems can play a significant role in this paradigm shift. The remainder of this chapter explores how resilience concepts are typically described in engineering and then compares this with more traditional approaches where precision regarding the risk is assumed. It is organized as follows: this introductory section is continued with an exposition on black swan events. Section Risk and Resilience: Terminology and Quantification explores definitional aspects of concepts of risk and resilience and is complemented by a discussion of how risk and resilience is measured. Section Risk and Resilience Analyses for Emerging Risks then compares and contrasts these concepts. Section Engineering for Resilience explores emerging risks and how they complicate planning and design of engineered systems. Conclusions are provided in the final section.

On Black Swans

When viewed individually, the size and scale of many billion-dollar disasters would likely have been considered improbable (or even inconceivable) before their occurrence. Hurricane Harvey was considered to be a 500-year precipitation event [17], despite having had two 500-year precipitation events in the two prior years in the Houston area. The Fukushima Daiichi nuclear disaster in 2011, a nuclear-core-melt accident precipitated by the Tōhoku Earthquake and Tsunami, was considered next to impossible [18] despite a subsequent official report stating the plant operator failed to conduct proper risk assessments [19]. Other "unforeseen" events follow similar patterns, including the destruction that followed Hurricane Katrina and the 2017 California wildfires (in terms of number, size, intensity, and duration). A commonality is that many of these events are considered black swans [20]. Black swans can succinctly be summarized as rare events with significant consequence that are surprising, regardless of whether or not they should be. Significant debate remains over whether a black swan is an event with an extremely low likelihood or an event whose existence was not previously considered (i.e., an "unknown unknown") [21]. A slightly more nuanced perspective could say it is an event with an extremely low likelihood, although significant uncertainty exists in the probabilistic estimate. Nothing prevents a black swan from being a situation stemming from any of the above. However, for the sake of this chapter, we will treat a black swan as being a surprising and rare event about which significant uncertainty exists.

Current risk management practices are well adept for situations where risk is well defined. This includes both a solid understanding of the uncertainty and knowledge about the relationship among entities, be they physical or something less tangible. Because of these strong requirements, traditional risk management frequently places little emphasis on black swan events [22]. This occurs despite a potential for extreme losses absent persistent management of the risk.

Additional factors contribute to traditional risk management practices failing to place emphasis on extreme events. Part of this reflects human behavior, whereby the threat *feels* insurmountable so preparation is simply avoided [11]. For similar reason, even consideration of the threat's existence may be eschewed, forcing the extreme threat into the "unknown unknown" category, regardless of how ripe the system is for disaster. Another reason reflects a perception that costs required to address the hazard may be too significant given the extreme rarity of the event. These causes underlie and reinforce wide-scale practices that exacerbate risk from natural disasters (e.g., permissive land-use planning and deferred maintenance of critical assets). Furthermore, these causes have led to risk management plans that consistently prove insufficient, as evidenced by mounting losses and more catastrophic disasters.

RISK AND RESILIENCE: TERMINOLOGY AND QUANTIFICATION

The concepts of risk and resilience are foundational to discussion of extreme events, emerging risks, and black swans. Without these concepts, it is impossible to adequately understand the impacts of potentially extreme events and to measure the effectiveness of mitigation strategies. Some recent work has argued that much of the terminology in the risk and resilience space has been used synonymously [23], and it likely has; however, a critical review of definitional terminology reveals distinct differences. In this section, we demonstrate how these terms are commonly used. For these terms to be useful in practical applications, what is described must also be measurable. Without measurability, one cannot determine whether interventions are effective. Thus, the second and fourth subsection reviews metrics, measurement, and quantification for risk and resilience, respectively.

Risk: Terminology and Definition

Understanding risk is a key component when preparing for natural hazards. Risk can broadly be defined as a potential for a consequence [13]. Natural hazards are risky because they create a potential for loss. In the context of natural hazards, the risk to a system (be the system a community, infrastructure, cybersystem, or something else) is ostensibly extrinsic; the hazard's forces threaten the system. However, a deeper investigation reveals that risk to systems from natural disasters is similarly intrinsic. There are factors from within a system that can augment risk from external factors. These include permissive land-use practices, poor maintenance practices, and rapid urbanization.

Risk is commonly described by a triplet [24]: "What can happen? How likely is it that that will happen? If it does happen, what are the consequences?" This framework has been used to define and quantify in many contexts, from hazard mitigation [25], to movement of hazardous materials [26], to risks at nuclear power generators [27]. In the first part of the triplet, "what can happen?" possible scenarios are defined. In the context of hazards, this could include, for example, the failure of a structure or breach of a levee. There is often value in dividing scenarios into two components: an incident that is external to the system (i.e., a threat), and a reaction that is internal to the system, conditioned on the initiating incident (i.e., a vulnerability). For example, a threat could be a hurricane or, more specifically, high winds from that hurricane. A vulnerability reflects a system's (e.g., a house) intrinsic ability to withstand the high winds.

For reasons that will be apparent, we skip the second question ("How likely is it that that will happen?") momentarily, to focus on the third question, "If it does happen, what are the consequences?" On the surface, this question is straightforward. Consequences could include financial losses, physical losses, and human injuries and deaths. To allow for comparison, there can be value in reducing the consequence categories to a single economic loss or other form of utility (e.g., Refs. [28,29]). The problem becomes more complicated when uncertainties regarding consequences exist. That is, consequences can be probabilistic estimates.

The question "How likely is it that that will happen?" requires the analyst to assess the frequency of the event. This assessment can be either quantitative, whereby a probabilistic estimate is determined, or qualitative, whereby a relative frequency is estimated (e.g., often, seldom, infrequent). Ideally, the uncertainty is well characterized, meaning that the estimate is accurate and precise. This type of uncertainty is called "aleatory" and is defined by uncertainty that is inherent to the system and well understood [30]. For example, the uncertainty of rolling a "4" on a fair six-sided die is precisely known to be one-sixth. In the hazards realm, if a location has troves of climate data, it may be possible to obtain precise estimates of specific events. However, data on past extreme events are rare, meaning probabilistic estimates of future extreme events may be poor, or at least rife with uncertainty. This type of uncertainty is referred to as "epistemic" uncertainty. It is uncertainty that exists due to a lack of knowledge [30]. In the case of hazards, this could be due to lack of data on past events or lack of a complete understanding of the physical processes that are occurring.

Black swans lack past data for analysts. Thus, analysts might not expect these events to occur or, at least, know very little about what could occur. To reduce this significant epistemic uncertainty, analysts often combine knowledge from subject matter experts and sophisticated computational models to better understand what is possible (e.g., Ref. [31]). The implications of this have been discussed extensively [32]. The strength of this approach is typically conditioned on the level of knowledge on the part of the experts and the quantity/quality of the data for the computer model. To complicate matters, emerging risks change the context in which disasters occur. This implies that the knowledge that experts have about the system may be dated, insufficient, or inaccurate, further adding to the epistemic uncertainty. For example, the rate that hurricanes make landfall at a particular location was once considered a stationary process. However, climate change is changing the rate at which hurricanes make landfall, but how is still being debated [15].

Risk: Measurements and Metrics

To assess whether interventions are needed to reduce harm from hazards and to evaluate whether the interventions are effective, measurements of risk must be taken. Risk has been assessed in numerous different manners. Reilly et al. [26] measure risk based on expected population exposure should a train carrying hazardous material derail. Ayyub et al. [6] measure risk based on potential flooding inundation to structures from riverine and coastal flooding in Washington, DC, under many sea-level rise scenarios.

The commonality among these studies is their approach to quantifying risk. First, a set of scenarios is created. This process may be aided using subject matter experts [33]. Second, likelihoods of the scenarios are generated. These could be derived based on physical processes, historical data, subject matter expertise, or behavioral models. They may be quantitative or qualitative. Third, consequences are evaluated. It is important to note that, unlike resilience, which will be discussed momentarily, no substantive introspection of system capacity is required in risk analysis. Consequences analysis does require knowledge of how the system reacts to a hazard. However, the path to failure and recovery is generally ignored, as is the duration by which service is halted.

The cumulative risk faced by a system is the weighted accumulation of the risk from all the scenarios identified. Significant debate exists over how this practically should be done [34]. A typical approach in quantitative risk assessment is to calculate expected risk for a scenario via multiplication of the likelihood that it is to occur and the consequence. This identifies an "expected risk" for a scenario (e.g., Ref. [25]). These expected risks can then be tallied based on the (relative) likelihood that the scenario is to occur. A typical approach in quantitative risk assessment is similar, but rather than assigning specific numbers, an arbitrary scale—possibly "low, medium, high"—is used and terms are "averaged." For example, a hazard that is unlikely to occur, but could be associated with maximal damage, might be assigned a "medium" risk. These risk outcomes are sometimes color-coded to help effectively communicate the risk.

Recent work has argued that this approach is reductionist and could be extremely problematic for decision-makers planning for extreme events [30,35]. Reducing risk to a single metric obfuscates the degree of knowledge that supports the risk assessment [22]. More specifically, a single metric cannot communicate whether the risk is high because of something specific about the system or because little is understood about the system and the estimates are imprecise. Additionally, this method hides valuable information. For example, it would be hard to believe that a decision-maker would treat and plan for an extremely rare yet consequential event similar to the way they would treat an event occurring with moderate frequency and exhibiting moderate consequence. Yet, the moderate risk that these scenarios pose, on average, would suggest that they should be treated equally [36].

Resilience: Terminology and Definition

The concepts of resilience in the engineering community are born from risk principles. However, the concept tends to deal less with factors extrinsic to the system and rather intrinsic properties of the system. At a high level, and without credence to the nearly two decades of research on the topic, resilience reflects the ability of a system to withstand and recover from some precipitating events. That event could be a black swan. The following discussion reviews prior work defining and scoping resilience, for the purpose of demonstrating the range covered by the topic.

The concept of resilience was initially introduced by Holling [37] to describe an ecosystem's ability to persist over time despite external stressors. The ecosystem may evolve or adapt to the stressor, and the goal is not to return to the original state but rather to a state better suited for the conditions. The concept of resilience has been adopted by other fields including engineering and childhood psychology. The focus of this discussion is limited to infrastructure resilience to extreme events. Broadly, resilience has been defined as:

- "The ability of a system or organization to withstand and recover from adversity."—The Civil Contingencies Secretariat of the Cabinet Office [38].
- "The ability of social units, e.g., organizations and communities, to mitigate hazards, contain the effects of hazard-related disasters when they occur, and carry out recovery activities in ways that minimize social disruption and mitigate the effects of future hazards."—The Multidisciplinary Center for Earthquake Engineering Research [39]. This definition continues by examining intrinsic properties possessed by the system that allow for it to be

resilient. They are robustness, redundancy, resourcefulness, and rapidity.

- "The ability to prepare and plan for, absorb, recover from, or more successfully adapt to actual or potential adverse events."—National Research Council [40].
- "The ability to prepare for and adapt to changing conditions and withstand and recover rapidly from disruptions. Resilience includes the ability to withstand and recover from deliberate attacks, accidents, or naturally occurring threats or incidents."—Presidential Policy Directive (PPD-21, 2013) on *Critical Infrastructure Security and Resilience* [41].
- "The ability to prepare for and adapt to changing conditions and withstand and recover rapidly from disruptions."—Ayyub [28,29]. This definition continues to identify what could threaten a system. They include deliberate attacks, accidents, and naturally occurring hazards.
- "The capability to mitigate against significant all-hazards risks and incidents, and to expeditiously recover and reconstitute critical services with minimum damage to public safety and health, the economy, and national security."—American Society of Civil Engineers, Committee on Critical Infrastructure [42];
- "The ability to reduce the magnitude and/or duration of disruptive events. The effectiveness of a resilient system depends upon its ability to anticipate, absorb, adapt to, and/or rapidly recover from a potentially disruptive event."—The National Infrastructure Advisory Council [43].
- The "capacity to withstand or absorb the impact of a hazard through resistance or adaptation, which enable it to maintain certain basic functions and structures during a crisis, and bounce back or recover from an event."—The United Nations Office for Disaster Risk Reduction [3].

There is significant convergence in these definitions. Resilience in the engineering and infrastructure field is generally viewed as a system's ability to withstand the impact of some extrinsic event and to recover from it. The intension of the word "resilience" includes intrinsic properties, including robustness, redundancy, resourcefulness, adaptability, and rapidity [44]. Each influences a system's ability to be resilient. These properties are discussed momentarily.

The definition of resilience adopted for this work comes from Ayyub [45] and is "the persistence of [a system's] functions and performances under uncertainty in the face of disturbances." This definition harkens Holling's [37] original definition of resilience relating to the persistence of ecological systems and does not provide strict requirements for a system to return to its original state before the disaster but rather possibly to a better or evolved state. This definition is relevant to a wide array of systems affected by hazards, including both infrastructure and communities.

The three keywords in this definition are *performance, uncertainty, and persistence.* Performance implies some objective measures of system requirements, such as structural integrity, lifecycle costs, or even product quality. Uncertainty relates to the uncertainty surrounding the hazards that may transpire. This could include banal episodes or black swans; that which will occur is unknown. Persistence evaluates how the system continues or recovers in the face of the hazard. The one key difference in this definition compared with other definitions is the lack of the word "ability." By excluding this word, it encourages development of metrics that are less process focused and that are more outcome focused [45].

Resilience: Measurements and Metrics

Resilience has been quantified numerous ways by the research community. Garbin and Shortle [46] quantified resilience as the fraction of damaged network links and nodes relative to system performance. Li and Lence [47] leveraged early work by Hashimoto et al. [48] to say resilience is the ratio of performance over two different time periods. Omer et al. [49] measured resilience for the Internet infrastructure systems as the change in information transmitted before and after the event and normalized by the information flow before the event. Unfortunately, these metrics do not capture the outcome of whether and when the system is able to recover or how long it took to recover.

Tierney and Bruneau [50] suggested measuring resilience based on a system's ability to reduce the probability of failure and to hasten recovery. This concept is illustrated using the "resilience triangle," which, in the time since its publication, has become a popular visualization of resilience. A simplistic representation of this concept is provided in Fig. 13.1. In a strictly engineering context, performance (Q) is the metric of concern and reflects one or more dimensions of infrastructure system quality. This is represented on the y-axis. Higher performance of infrastructure is desired, with 100% being the maximum or fully functioning. At an initiating time, t_0, a disruptive event occurs, and performance falls from *A* to *B*. Over time, as repairs are made and systems are restored, quality of infrastructure improves and ultimately returns to

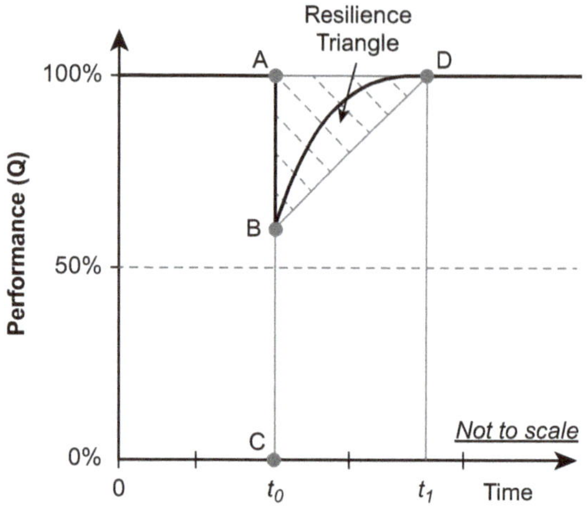

FIG. 13.1 The resilience triangle [29].

D. The shape of this curve abstractly represents system resilience, and the areas above its curve can serve to quantify system resilience. A mathematical representation of this is:

$$\text{Resilience} = \frac{\int_{t_0}^{t_1} Q(t)dt}{100(t_0 - t_1)} \tag{13.1}$$

A more sophisticated representation of the "resilience triangle" for infrastructure is provided in Fig. 13.2. As before, higher performance is desired. Aging and natural deterioration degrade system performance over time. At time t_i, an incident occurs that threatens the system. That incident, as discussed before, possesses uncertainty, and its arrival process could be modeled using a Poisson process, with rate parameter λ. A difficulty when working in environments with emerging risks is that the rate parameter, λ, along with other potentially useful model parameters, may be inaccurate or contain significant epistemic uncertainty. Once the incident occurs, performance drops. The way in which the performance falls (i.e., f1, f2, or f3), and the level to which performance falls (at time t_f), is governed by internal properties or capacities.

A useful way to decompose internal capacities is by using the Birringer et al. [51] framework. This perspective divides resilience capacity into three types: absorptive, adaptive, and restorative capacity. We describe a fourth capacity, evolutionary capacity, at the end of this section. The impact that the initiating event has on performance, that is, the decline in performance between times t_i and t_f, is dictated by a system's absorptive and adaptive capacity. A system's absorptive capacity

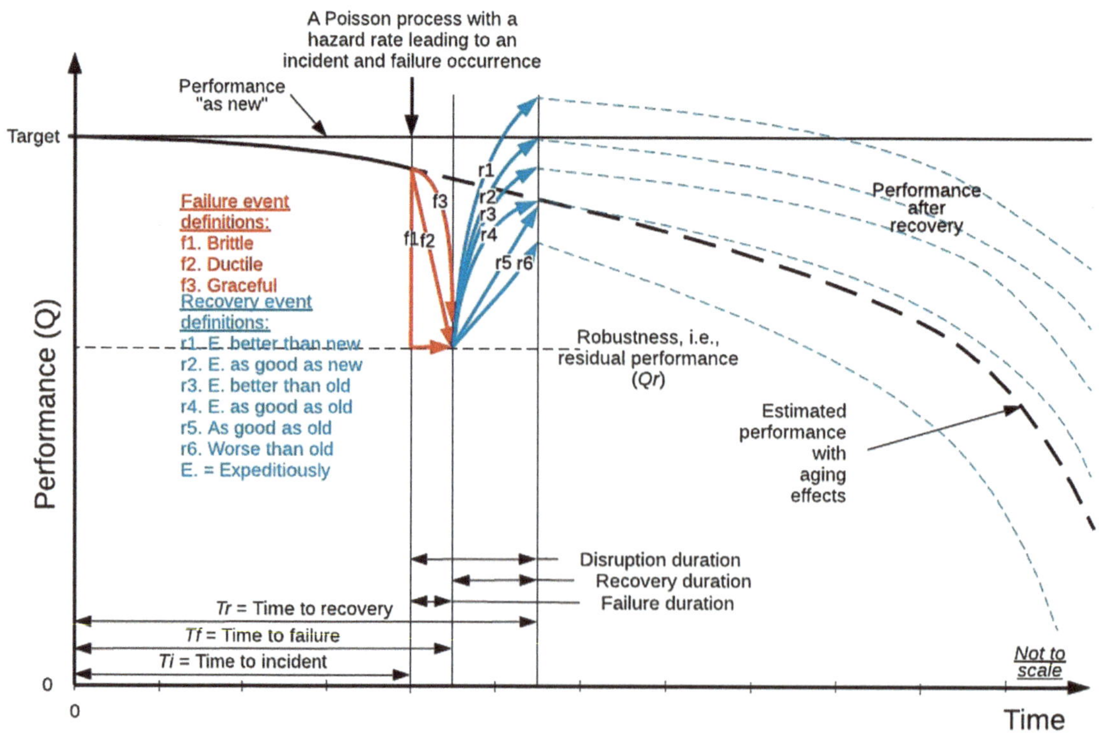

FIG. 13.2 Illustration of resilience [29].

reflects its ability to absorb the force of an event. Enhancing absorptive capacity usually implies building or installing something physical (or engineered), such as an improved roof-to-wall connection or a seawall. Absorptive capacity can be further decomposed into robustness and redundancy [44]. Robustness implies that a system is less likely to fail when exposed to a force, or, in other words, a system is less vulnerable to loss. Actions including hardening, mitigation, and thoughtful land-planning make systems more robust. For example, installing a levee reduces the potential for harm in a community. Redundancy, on the other hand, implies that a "backup" exists, so that if some infrastructure service is degraded, another can stand in its place. For example, Zurich, Switzerland, has two separate water distribution systems, with one being completely gravity fed. Should one fail, the other is available.

The adaptive capacity of a system reflects its ability to self-organize or to be resourceful in the event of disruption, to reduce performance losses. Although, in principle, physical components can be built with adaptive capacity (e.g., smart switching in electric power distribution networks), it is usually human intervention or a combination of human-machine intervention that creates adaptive capacity. For example, evacuating those in harms way before or during the hazard reduces the potential for life loss, and empowering all medical facilities to accept a wide variety of medical traumas reduces pressure on large centralized hospitals that are often overrun following the hazard. Adaptive capacity can be built in physical infrastructure by providing operators with improved situational awareness, so that they can respond in an appropriate manner.

Adaptive capacity and absorptive capacity, two intrinsic properties of a system, define both the degree to which performance will degrade and also the rate by which it degrades. Returning to Fig. 13.2, a system that fails in the same manner as f1—whereby the drop in performance is sudden—is said to be a brittle system [52]. This may be seen as problematic; community members are provided no forewarning about this sudden decline in system performance. The line, f2, represents a ductile system. The rate of performance decline in a ductile system is approximately linear. A system that has a performance decline similar to f3 is called a graceful failure event [52]. This type of performance decline provides time to evacuate individuals, if needed, or to allow individuals to exert their own adaptive capacity and not be harmed by the hazard. For example, predictable brownouts during heatwaves provide time and knowledge for community members to go to a cooling center. The failure duration is measured by ΔT_f. Ayyub [45] proposes measuring the failure profile, F, by:

$$F = \frac{\int_{t_i}^{t_f} f dt}{\int_{t_i}^{t_f} Q dt} \tag{13.2}$$

After the system fails at time t_f, the system can begin to be restored. The rapidity by which a system can be restored, namely the recovery duration, ΔT_r, reflects the system's restorative capacity. Access to capital, situational knowledge, personnel, and replacement components improve a system's restorative capacity. When the system is recovering, the trajectory to recovery could take many paths. The system could be restored to a state that is identical or superior to its original state (r2 and r1, respectively). Similarly, it could be restored to a state that is superior or on par with what is expected after natural deterioration but absent a hazard (r3 and r4, respectively). It is possible that the system recovers, but to an inferior level relative to what would be expected (r6). The system experiences disruption for a duration of ΔT_d. Ayyub [45] proposes measuring the recovery profile, R, by:

$$R = \frac{\int_{t_f}^{t_r} r dt}{\int_{t_i}^{t_f} Q dt} \tag{13.3}$$

The failure and recovery profile can then be integrated to form a resilience measure for a system.

$$\text{Resilience, } R_e = \frac{T_i + F\Delta T_f + R\Delta T_r}{T_i + \Delta T_f + \Delta T_r} \tag{13.4}$$

This equation is relevant for quantifying resilience in that it incorporated both duration measures and measures of the failure and recovery profiles. These profiles are synonymous with system-level performance (i.e., outcome focused) and reflect intrinsic system properties that allow recovery in certain manners when facing external threats.

Often, measuring system-level resilience in complex systems proves challenging. This is particularly true when quantifying resilience at community levels. In these instances, system segregation that logically maps the resilience of system components to overall system performance is required. Renschler et al. [53] provide a logical basis for accomplishing this task, although an assumption of component independence is required. This is atypically the case for hazards. The loading generated by hazards possesses strong spatial dependence. Also, complicating matters is that performance measures that are relevant in one operational context may not be relevant in another. This creates difficulty for determining which and when various metrics are relevant. For example, operators of buildings may be concerned about space availability, and maintainers of bridges may be concerned about throughput; however, after a disaster, they both may primarily be focused on life safety. Cox [34] provides a deeper discussion of the challenges associated with selecting and aggregating resilience measure in complex systems. Table 13.1 provides example performance measure that can elicit resilience for five complex systems. Ayyub [45] provides three illustrative case studies for a community, a building, and a bridge. Each example shows different failure and recovery profiles, and relevant metrics are provided.

Although difficult to represent in Fig. 13.2, it is possible that the system could recover to an evolved state that is better suited for the new environment. This is particularly relevant in environments with emerging risks. This capacity is titled "evolutionary capacity," and the concept, as applied to infrastructure, is introduced in this chapter. A system that possesses evolutionary capacity is able to evolve its form after an impact (or in anticipation of) so that its functionality better reflects the realities of its threats. Evolutionary capacity could be built by adding flexibility within systems and is more likely to exist in systems with access to capital and buy-in from community members, utility services providers, government officials, and others. System performance in systems that have evolved is typically better captured by a whole new performance measure(s). For example, a major hurricane could destroy electric power distribution infrastructure on an island nation. In a traditional sense, the metric of concern could be the number of customers who have been reconnected to the grid as a function of time. However, in the view that hurricanes may become more severe (and thus may repeatedly damage the energy infrastructure to a significant extent) and that household photovoltaic systems are rapidly lowering in cost, an evolved state may be an island nation dependent on household solar energy or microgrids. This is an example of an evolved state and a new metric could be number of households with solar power.

TABLE 13.1
Sample Performance Metric for Quantifying System-Level Resilience [29].

Systems	Performance Metric	Unit
Building	Occupancy	Percentage
Highway structure (e.g., tunnel)	Vehicular throughput	Volume per unit time
Water pumping station	Hours of continuous operation (reliability)	Time
Electric power	Customers without power	Count
Communities	Economic productivity	Dollars

RISK AND RESILIENCE ANALYSES FOR EMERGING RISKS

Risk and resilience theory share many commonalities. Their uses are both rooted in a desire to better understand how systems behave and how to protect (or add capacity) to new or existing systems. Combined, they provide a deeper understanding of internal and external threats that could harm the system and how intrinsic system properties interact to withstand and react to those threats. Differences exist too. One key difference relates to their primary objectives. Resilience homes in on system properties (e.g., capacities), but conversation surrounding what might threaten that system can be vague. More precisely, accurate threat estimates, perhaps ones that are probabilistic, are useful but generally not required as a prerequisite for resilience assessments. Risk, on the other hand, requires careful scenario creation, along with assessing the likelihood of these scenarios. Here, accurate threat and vulnerability estimates are needed. But the accuracy of these estimates is unclear.

First, it is worthwhile pointing out that even accurate estimates are probabilistic and not deterministic. They are best described using a probabilistic exceedance plot [54]. As an example, a particular building may have a 30% chance of collapse when wind speeds exceed 45 m per second. The building system possesses inherent uncertainty, and additional study of the building system will not change (or "improve") this chance of collapse. Only structural modifications to the building can do that. This leads to the second point regarding accuracy of probabilistic estimates. Many estimates are based on (limited) sampling of the system, perhaps using past data or using simulation. This implies that some information about the system is known, but not *all* information. That is, there is knowledge left to be gathered that could improve accuracy. This is the case with emerging risks, such as climate change, and the level of knowledge can vary dramatically.

In this section, we compare risk and resilience analyses specifically through their ability to address emerging risks. To accomplish this, we first review knowledge, ignorance, and uncertainty. This then leads to a discussion on emerging risks and how risk and resilience theory treat uncertainty.

Knowledge, Information, Ignorance, and Uncertainty

Knowledge includes know-how and propositions that are based on justified true beliefs [55]. A key word in this statement is "justified"; this implies the existence of evidence in support of the knowledge. This could be observational evidence that supports a conclusion. A stronger degree of knowledge is provided using the scientific method, where all other explanations that could explain a hypothesis are discredited by the evidence.

Gaps in knowledge may occur. This creates numerous potential problems when preparing for hazardous climatic events, and this lack of knowledge could create severe consequences, from ineffective use of funds to additional causalities. These knowledge deficiencies or "ignorances" consist of two primary categories: (1) blind ignorance and (2) conscious ignorance [56]. The first relates to the discussion above relating to insufficient data or information. This ignorance can be unintentional or deliberate, but the core reason for the ignorance is a lack of information. It is remedied by more fundamental research and other scientific pursuits. Of course, this presumes the evidence, and the methods by which the evidence is collected lack error. Although important, we save that discussion for future research. The latter ignorance, conscious ignorance, refers to purposeful avoidance of information, perhaps due to biases and taboos, which makes a knowledge set incomplete [57]. This can be a recognized self-ignorance through reflection [56]. The theory of near-misses is an example of a bias [58]. For individuals who experience a hazard and are "lucky," meaning that they do not experience harm, they tend to believe their likelihood of future hazard-afflicted harm is less than a control group and, thus, are less likely to adequately prepare. Essentially, these individuals misinterpreted the signal and thus become ignorant to the true risk due to bias. A deeper discourse on the topic can be found in Smithson [57] and Ayyub [56], which include complementary taxonomies of ignorance.

The strength of a risk assessment is conditioned on the degree of knowledge possessed by the assessor [59]. If the assessor is ignorant to knowledge or possesses biases, uncertainty is generated. Unfortunately, without self-reflection, this uncertainty could go undetected. Regardless of the cause of the lack of knowledge, this type of uncertainty can be classified as epistemic uncertainty. Epistemic uncertainty is the knowledge-based, subjective uncertainty. It can be reduced by gathering additional data to form new knowledge. Aleatory uncertainty, on the other hand, is the inherent, random, or nonreducible uncertainty. This includes concepts such as material strength randomness [29]. It is not created by ignorance, but rather underlying stochastic phenomena.

Finally, uncertainty can be further decomposed among its aleatory and epistemic uncertainties, depending on the level of knowledge that can be created and how well the uncertainty can be (or is) characterized [55,60]:

1. Uncertainty that is identified and is probabilistically well characterized (e.g., materials properties)
2. Uncertainty that is identified and is moderately well characterized (e.g., future global precipitation, temperature)
3. Uncertainty that is identified and is poorly characterized (e.g., future demands on water supply by growing populations)
4. Uncertainty that is identified and cannot be characterized (e.g., global governance trends)
5. Uncertainty sources that are unknown to exist

This decomposition is useful for both determining which analytical approaches are best to provide additional insight into the problem and for reporting the level of knowledge that supports the recommendations that are generated as a result of the inquiry.

Emerging Risk and Uncertainty

Emerging risks, such as population growth and climate change, change the context in which disasters occur. They generate new uncertainties, some of which may be aleatory in principle. However, given our current limited understanding of their influence and impact, much of the uncertainty, at present, is epistemic.

Although the precise definition of an emerging risk is being debated, it is broadly considered to be either a new risk or a newly identified risk [16], about which relatively little is understood. Two examples include climate change and technological innovation. Climate change is amplifying already perilous climate hazards through hotter and drier summers, stronger and slower-moving hurricanes, and heavier extreme precipitation events. How the global effects of climate change will downscale and affect local communities is still heavily researched and debated. Furthermore, how to best respond is still an open question. It will depend not only on how the threats evolve but also on how individuals react to the changes.

Technological innovations include technologies such as unmanned aerial vehicles (i.e., drones) and "smart" HVAC systems in buildings and have the potential to ease the impact of many of these extreme disasters (such as heat waves) and speed recovery following disasters (such as earthquakes). Although the role of and standards for these technologies during extreme events have not been established, in part because it is not well understood how they may interact with a dynamic hazardscape, which includes both the build environment and individuals. Absent a better understanding of how they may serve to complement existing processes, they could potentially slow or complicate recovery efforts. Some argue that technological innovations could create conditions for an extreme, yet positive event, in what otherwise would be catastrophic. This still makes technological innovations an emerging risk as it is possible to include positive outcomes in the consequence attribute of risk, although this topic is heavily debated.

The concepts of emerging risk and black swans are complementary, yet distinct, concepts. A black swan is an extreme event with significant epistemic uncertainty, which, when occurs, is typically surprising. An emerging risk is a catalyst that disrupts a system's environment or context. This may perturb how likely black swan events are to occur or may create the conditions for wholly new black swan events. The uncertainty for most emerging risks is profound and mainly stems from lack of knowledge. Thus, planning for and adapting to emerging risks is challenging.

The uncertainty generated by emerging risks is often poorly characterized, and in their infancy—when the risks are just being discovered—emerging risks are unknown to exist by many. Treating emerging risks using traditional engineering practices would require a reduction in uncertainty, and thus, many are leveraging insights from new models and frameworks. Some modern approaches are attempting to provide uncertainty estimates or confidence bounds around probabilistic estimates of the aleatory uncertainty. In some cases, the bounds are difficult to validate. Other approaches have been developed to support decision-making in light of the uncertainties. These are discussed in the subsequent section titled "Engineering for Resilience."

Differentiating Risk and Resilience for Addressing Emerging Risks

Risk assessments tend to eschew focus on epistemic uncertainty for reasons including difficulty in quantification and, at times, the assessors simply being blind of its existence [13]. This poses difficulties when trying to make risk-informed decisions. Resilience assessments, on the other hand, focus on intrinsic properties that reduce the severity of potential threats. Focusing on internal properties that go beyond the physical structure to include human and policy capacities allows for greater creativity when it comes to improving resilience. Thus, resilience assessments and planning do not require the same level of knowledge about threats.

The downside to focusing on internal capacity building is its sacrifice of efficiency. Simply, adding capacities can easily prove far costlier than any potential benefits that might come from them. Thus, cost-benefit analyses, which include measure of uncertainty surrounding the uncertainty, are needed. Furthermore, it points to the value of identifying strategies that are multifaceted in their capacity building. For example, policies that encourage development in less vulnerable locations do not only reduce the severity of an impact but also enable faster restoration because the landscape after an impact is more hospitable (e.g., flood waters are likely lower and transit infrastructures are littered with less debris).

ENGINEERING FOR RESILIENCE

Design Philosophies

Uncertainty is an underpinning of the engineering design philosophy. The primary goal of design is to balance safety and economics. Today, the engineering design paradigm primarily accounts for aleatory uncertainty. For example, a 100-year flooding event is based on local historical precipitation data and not a scenario where the rainfall process may be heavier and flashier (perhaps due to climate change). This current approach is suitable in the context of stationarity but is less suitable in environments where the change is occurring. More recently, in some cases, engineering design approaches have evolved based on a broader understanding of uncertainty.

It is imperative that engineers adopt a paradigm that allows their systems to adapt to emerging risks. Obviously, accomplishing this task is challenging at best given the uncertainty and lack of knowledge associated with emerging risks. Principles in resilience analysis could help; designing systems based on multifaceted capacities could help them to withstand or adapt to changing environments. Furthermore, it is possible that when designing for capacities, nonstructural interventions can be considered as well, including the possibility for human interventions.

We return to the five differentiations of uncertainty that is divided based on whether the uncertainty is identified and the degree to which it is, or could be, quantified. In the classic case where uncertainty is well characterized, as is the case of aleatory uncertainty, traditional factors of safety suffice. This is an example of designing to have sufficient absorptive capacity. In some systems, there is value in ensuring that the system possesses resilience capacities beyond absorptive. For example, a power distribution system may be designed to meet minimum safety thresholds, although there is still value in evaluating its restorative capacity in the event of a system failure. For the cases where uncertainty is both recognized and poorly or moderately characterized, the best practice is to use reliability-based or risk-informed designs when creating absorptive capacity. This implies that a system can withstand a probable load or is sufficiently redundant so that few, if any, failures occur. However, possessing greater epistemic uncertainty places more emphasis on ensuring the system has sufficient adaptive and restorative capacities. For example, if the system is a levee and the uncertainty surrounding the future rainfall distribution is uncertain, having an evacuation plan should the levee overtop must be part of the resilience planning. In some cases, this need points to adopting more of a systems approach.

In situations where the uncertainty cannot be defined or identified, both adaptive design principles and constant risk management are required [60]. This could include flexibility in designs, so that the structure or system can be significantly altered in the future, based on new information or a changing environment. An example of this is including flexibility for raising bridge piers to combat sea level rise, as is the case in an Amtrak bridge in California, or overdesigning a foundation so that a building can be raised, as was the case of the Blue Cross Blue Shield Building in Chicago, IL, and Four Season Building in Baltimore, MD. Adaptive management has been successfully applied to water resource management [61].

When uncertainties are well characterized, scenario modeling can provide tremendous insights into how the built environment may interact with hazards. This includes Monte Carlo simulation and can include conditional probabilistic data such as random phasing of ground motions, local soil conditions, and structural fragilities. Recently, alternative methods have been developed for problems with "deep uncertainty," meaning poorly quantified or unknown uncertainty. Robust decision-making is a popular method, and instead of asking "what can happen," it instead asks "what has to occur for something to go wrong" [61,62]. Its goal is to identify strategies that are robust against the wide range of threats.

Performance-based design is becoming more common among the seismic engineering community. Here, design options are evaluated by monetizing costs and benefits for various scenarios of differing intensities. This helps to inform decision-makers by reducing the problem space to common units for comparison. Regardless, these approaches should be conducted in tandem with other resilience assessments to ensure that when the system is overtaken by a load, there is a plan for both human and system recovery. Adaptive risk management should be integrated within performance-based design to account for nonstationary processes as opposed to focusing on mean recurrence intervals.

Economics of Resilience

Strategies for improving resilience are typically evaluated in economic terms within a structured decision-making framework. Improvements and harms generated by each alternative strategy are typically reduced to a single economic metric to simplify future comparisons. This process thus reflects the values of those doing the conversion between potential outcomes and the monetary equivalent of this potential outcome. Ultimately, these conversions should reflect the priorities and goals of the society at large as opposed to those of a single decision-maker. To aid in this process, methods that quantify the total economic value of resilience are needed. Concepts from risk analysis and management may be useful for this purpose [28,29]. Valuation of outcomes relies heavily on ethics. It is useful to distinguish among types of values, including (1) instrumental and intrinsic values, (2) anthropocentric and biocentric (or ecocentric) values, (3) existence value, and (4) utilitarian and deontological values [28].

Two common economic measures that help to translate outcomes to values are willingness to pay and willingness to accept. The former measure reflects how much someone is willing to pay for a good or a service. It is useful for placing cost caps on a project, and ultimately, as a comparative metric among alternatives. However, in an absolute sense, it simply reflects one's resources more than one's need for hazard risk or other harm reduction. Willingness to accept, on the other hand, is the minimum amount one is willing to accept to tolerate something negative in perpetuity, such as risks associated with a hazard.

Benefit-cost analyses provide insight into which alternative strategies provide more benefit than cost, a minimal metric for project viability. Benefits, in this context, refer to the value of the potential loss avoidance. This typically includes both direct and indirect loss avoidance. Costs refer to the costs, typically financial, of the strategy. Given that hazards are stochastic, the benefits and costs themselves should be treated as random variables, likely with benefits receiving a larger distribution spread given more uncertainty [28,29]. Thus, there is potential for costs to outweigh the benefits.

Given the potential for extreme epistemic uncertainty, especially for potential benefits of various strategies, a strict "racking and stacking" of alternatives—meaning prioritizing strategies using the cost-benefit ratio alone—is likely less productive than in other contexts. When evaluating benefits provided by alternatives, credence should be given to alternatives that emphasize resilience principles. Ultimately, experts and community members must decide which alternative is superior both in terms of potential outcomes and in terms of resilience principles.

CONCLUSION

Incorporating risk reduction and management strategies in engineered systems has repeatedly been demonstrated to reduce long-term costs, injuries, and mortality. Risk principles require knowledge about the system and what threatens it, along with how likely the threats are to occur. Many infrastructure systems are being designed today for future environments about which we know little. Emerging risks, including climate change and population growth, may amplify the consequences associated with known hazards or even cause the hazard itself to change.

Incorporating resilience principles into engineered systems by increasing their internal capacities to withstand, repair, and adapt is an effective way to address many negative emerging risks when more precise information is unavailable. The four capacities discussed in this chapter are absorptive, adaptive, restorative, and evolutionary. Capacity building can be accomplished via a combination of structural and nonstructural interventions. The former tends to be more effective for absorbing the force of a disaster, whereas the latter, which includes fiscal resources, training, communication, and effective policy interventions, has been shown effective along all resilience dimensions. The science and practice of resilience are still in their infancy, with much foundational work still needed. For example, measuring resilience is still imprecise and little is understood on how to quantitatively evaluate internal capacities. Furthermore, the topic of how dependencies and interdependencies among physical and social systems influence resilience outcomes is ripe for discovery. Immediate work should identify novel, effective, and implementation-ready strategies that build resilience capacity across a variety of communities.

REFERENCES

[1] National Oceanic and Atmospheric Administration, National Centers for Environmental Information, "Billion-Dollar Weather and Climate Disasters: Overview,", 2018 [Online]. Available: https://www.ncdc.noaa.gov/billions/.

[2] Insurance information Institute, Facts + Statistics: Global Catastrophes, Insurance Information Institute, 2017 [Online]. Available: https://www.iii.org/fact-statistic/facts-statistics-global-catastrophes.

[3] UNISDR, Making Cities Resilient: My City Is Getting Ready! A Global Snapshot of How Local Governments Reduce Disaster Risk , United Nations Office for Disaster Risk Reduction, 2012 [Online]. Available: www.unisdr.org/campaign.

[4] B.M. Ayyub, M. Kearney, Towards the development of regional risk profiles and adaptation measures for sea level rise, Journal of Risk Analysis and Crisis Response 1 (1) (2011) 75–89.

[5] B.M. Ayyub, M.S. Kearney, Sea Level Rise and Coastal Infrastructure: Prediction, Risks, and Solutions, vol. 6, ASCE Publications, 2012.

[6] B.M. Ayyub, H.G. Braileanu, N. Qureshi, Prediction and impact of sea level rise on properties and infrastructure of Washington, DC, Risk Analysis 32 (11) (2012) 1901–1918.

[7] B.M. Ayyub, J. Ramirez, K. Markham, P. Broqueres, Quantifying regional risk profiles attributable to sea level rise, in: Sea Level Rise and Coastal Infrastructure: Prediction, Risks, and Solution, 2012, pp. 1–19.

[8] P.R. Berke, J. Kartez, D. Wenger, Recovery after disaster: achieving sustainable development, mitigation and equity, Disasters 17 (2) (1993) 93–109.

[9] C. Kousky, E. Michel-Kerjan, Examining flood insurance claims in the United States: six key findings, Journal of Risk & Insurance 84 (3) (2017) 819–850.

[10] American Society of Civil Engineers, Report Card for America's Infrastructure, 2017 [Online]. Available: https://www.infrastructurereportcard.org.

[11] N. Taleb, The Black Swan: Why Don't We Learn that We Don't Learn, NY Random House, 2005.

[12] D.O. Prevatt, et al., Making the case for improved structural design: Tornado outbreaks of 2011, Leadership and Management in Engineering 12 (4) (2012) 254–270.

[13] M.E. Paté-Cornell, Uncertainties in risk analysis: six levels of treatment, Reliability Engineering & System Safety 54 (2–3) (1996) 95–111.
[14] P.S. Dailey, G. Ljung, G. Zuba, J. Guin, Probability of hurricane intensification and United States hurricane landfall under conditions of elevated Atlantic sea surface temperatures, in: Hurricanes and Climate Change, Springer, 2009, pp. 121–138.
[15] A. Staid, S.D. Guikema, R. Nateghi, S.M. Quiring, M.Z. Gao, Simulation of tropical cyclone impacts to the US power system under climate change scenarios, Climatic Change 127 (3–4) (2014) 535–546.
[16] R. Flage, T. Aven, Emerging risk–Conceptual definition and a relation to black swan type of events, Reliability Engineering & System Safety 144 (2015) 61–67.
[17] M. Davlasheridze, K.O. Atoba, S. Brody, W. Highfield, W. Merrell, B. Ebersole, A. Purdue, R.W. Gilmer, Economic impacts of storm surge and the cost-benefit analysis of a coastal spine as the surge mitigation strategy in Houston-Galveston area in the USA, Mitigation and Adaptation Strategies for Global Change (2018) 1–26.
[18] Safety of Nuclear Power Reactors, World Nuclear Association, 2018 [Online]. Available: www.world-nuclear.org/info/inf06.html.
[19] The Official Report of the Fukushima Nuclear Accident Independent Investigation Commission, The National Diet of Japan, 2012.
[20] N.N. Taleb, The Black Swan: The Impact of the Highly Improbable, vol. 2, Random House, 2007.
[21] T. Aven, On the meaning of a black swan in a risk context, Safety Science 57 (2013) 44–51.
[22] E. Paté-Cornell, On 'Black Swans' and 'Perfect Storms': risk analysis and management when statistics are not enough, Risk Analysis 32 (11) (2012) 1823–1833.
[23] R. Francis, B. Bekera, A metric and frameworks for resilience analysis of engineered and infrastructure systems, Reliability Engineering & System Safety 121 (2014) 90–103.
[24] S. Kaplan, B.J. Garrick, On the quantitative definition of risk, Risk Analysis 1 (1) (1981) 11–27.
[25] M. Legg, R.A. Davidson, L.K. Nozick, Optimization-based regional hurricane mitigation planning, Journal of Infrastructure Systems 19 (1) (2012) 1–11.
[26] A. Reilly, L. Nozick, N. Xu, D. Jones, Game theory-based identification of facility use restrictions for the movement of hazardous materials under terrorist threat, Transportation Research Part E Logist. Transp. Rev. 48 (1) (2012) 115–131.
[27] S. Schroer, M. Modarres, An event classification schema for evaluating site risk in a multi-unit nuclear power plant probabilistic risk assessment, Reliability Engineering & System Safety 117 (2013) 40–51.
[28] B.M. Ayyub, Risk Analysis in Engineering and Economics, Chapman and Hall/CRC, 2014.
[29] B.M. Ayyub, Systems resilience for multihazard environments: definition, metrics, and valuation for decision making, Risk Analysis 34 (2) (2014) 340–355.
[30] T. Aven, E. Zio, Some considerations on the treatment of uncertainties in risk assessment for practical decision making, Reliability Engineering & System Safety 96 (1) (2011) 64–74.
[31] C.L. Berner, A. Staid, R. Flage, S.D. Guikema, The use of simulation to reduce the domain of 'black swans' with application to hurricane impacts to power systems, Risk Analysis 37 (10) (2017) 1879–1897.
[32] W.P. Aspinall, R.M. Cooke, Quantifying scientific uncertainty from expert judgement elicitation, in: Risk and Uncertainty Assessment for Natural Hazards, Cambridge University Press, Cambridge, UK, 2013, p. 64.
[33] B.M. Ayyub, Elicitation of Expert Opinions for Uncertainty and Risks, CRC press, 2001.
[34] J. Cox, Community resilience and decision theory challenges for catastrophic events, Risk Analysis 32 (11) (2012) 1919–1934.
[35] J. Cox, Some limitations of 'Risk= Threat× Vulnerability× Consequence' for risk analysis of terrorist attacks, Risk Analysis 28 (6) (2008) 1749–1761.
[36] L.A.(T.) Cox Jr., What's wrong with risk matrices? Risk Analysis 28 (2) (2008) 497–512.
[37] C.S. Holling, Resilience and stability of ecological systems, Annual Review of Ecology and Systematics 4 (1) (1973) 1–23.
[38] Civil Contingencies Secretariat of the Cabinet Office, 2018 [Online]. Available: www.cabinetoffice.gov.uk.
[39] M. Bruneau, A. Filiatrault, G. Lee, T.D. O'Rourke, A. Reinhorn, M. Shinozuka, K.J. Tierney, White Paper on the SDR Grand Challenges for Disaster Reduction, in: Multidisciplinary Center for Earthquake Engineering Research, 2006.
[40] National Research Council, Disaster Resilience: A National Imperative, The National Academies Press, Washington, DC, 2012.
[41] Critical Infrastructure Security and Resilience, 2013. Washington, DC.
[42] Policy Statement 518 - Unified Definitions for Critical Infrastructure Resilience, American Society of Civil Engineers, 2013 [Online]. Available: http://www.asce.org/issues-and-advocacy/public-policy/policy-statement-518—unified-definitions-for-critical-infrastructure-resilience/.
[43] N. I. A. Council (US), Critical Infrastructure Resilience: Final Report and Recommendations, National Infrastructure Advisory Council, 2009.
[44] M. Bruneau, S.E. Chang, R.T. Eguchi, G.C. Lee, T.D. O'Rourke, A.M. Reinhorn, M. Shinozuka, K.J. Tierney, W.A. Wallace, D. Von Winterfeldt, A framework to quantitatively assess and enhance the seismic resilience of communities, Earthquake Spectra 19 (4) (2003) 733–752.
[45] B.M. Ayyub, Practical resilience metrics for planning, design, and decision making, ASCE-ASME Journal of Risk and Uncertainty in Engineering Systems 1 (3) (2015) 04015008.
[46] D.A. Garbin, J.F. Shortle, Measuring resilience in network-based infrastructures, in: Crit. Think. Mov. Infrastruct. Prot. Infrastruct. Resil., 2007.

[47] Y. Li, B.J. Lence, Estimating resilience for water resources systems, Water Resources Research 43 (7) (2007).
[48] T. Hashimoto, J.R. Stedinger, D.P. Loucks, Reliability, resiliency, and vulnerability criteria for water resource system performance evaluation, Water Resources Research 18 (1) (1982) 14–20.
[49] M. Omer, R. Nilchiani, A. Mostashari, Measuring the resilience of the global internet infrastructure system, in: Proc. of the 2009 IEEE International Systems Conference, 2009. Vancouver, Canada.
[50] K. Tierney, M. Bruneau, Conceptualizing and measuring resilience: a key to disaster loss reduction, TR News (250) (2007).
[51] B. Biringer, E. Vugrin, D. Warren, Critical Infrastructure System Security and Resiliency, CRC press, 2016.
[52] S. Gilbert, B.M. Ayyub, Models for the economics of resilience, ASCE-ASME Journal of Risk and Uncertainty in Engineering Systems 2 (4) (2016) 04016003.
[53] C.S. Renschler, A.E. Frazier, L.A. Arendt, G.P. Cimellaro, A.M. Reinhorn, M. Bruneau, A Framework for Defining and Measuring Resilience at the Community Scale: The PEOPLES Resilience Framework, MCEER Buffalo, 2010.
[54] G.P. Cimellaro, A.M. Reinhorn, M. Bruneau, Framework for analytical quantification of disaster resilience, Engineering Structures 32 (11) (2010) 3639–3649.
[55] B.M. Ayyub, G.J. Klir, Uncertainty Modeling and Analysis in Engineering and the Sciences, Chapman and Hall/CRC, 2006.
[56] B.M. Ayyub, From dissecting ignorance to solving algebraic problems, Reliability Engineering & System Safety 85 (1–3) (2004) 223–238.
[57] M. Smithson, Ignorance and Uncertainty: Emerging Paradigms, Springer Science & Business Media, 2012.
[58] R.L. Dillon, C.H. Tinsley, M. Cronin, Why near-miss events can decrease an individual's protective response to hurricanes, Risk Analysis 31 (3) (2011) 440–449.
[59] T. Aven, Risk Analysis, John Wiley & Sons, 2015.
[60] B.M. Ayyub, et al., Climate-Resilient Infrastructure: A Manual of Practice on Adaptive Design and Risk Management, vol. 140, American Society of Civil Engineers, Reston, VA, 2018.
[61] D.G. Groves, R.J. Lempert, A new analytic method for finding policy-relevant scenarios, Global Environmental Change 17 (1) (2007) 73–85.
[62] R.J. Lempert, Shaping the Next One Hundred Years: New Methods for Quantitative, Long-Term Policy Analysis, Rand Corporation, 2003.

CHAPTER 14

Where Are We? Why Community-Wide Benchmarking Is Important

M. JOHN PLODINEC, PHD

INTRODUCTION

American communities are faced with a spectrum of potential disruptions and disasters unique in our nation's history. They are threatened by the specter of natural and human-induced disasters, while trying to recover from an historic economic disaster. Growing community complexity and globalization have resulted in networks of interdependencies, creating new vulnerabilities for communities. The rise of global terrorism has also brought new threats. Many communities are plagued by crime and social unrest. Global climate change may bring its own panoply of hazards.

At the same time, communities are constrained in ways they never have been before. Recovery from the global recession has severely depleted community coffers, thus limiting the resources available for recovery from disasters. The retirement of the "Baby Boomers" (and the promises made to them) constitutes a lien that amounts to trillions of dollars [1]. As the Boomers retire, they also take with them their accumulated experience and their knowledge of why institutions work the way they do. In the past, communities often could adapt to emerging trends and new hazards at their own pace. In today's technocentric world, the accelerating rate of change means that communities are almost continually faced with the need to adapt or reinvent themselves. The combination of the new spectrum of risk and the constraints on resources means that communities must decide how to address current needs while preparing for inevitable future challenges.

This is complicated by the public's expectation that disruptions of the community's normal functions—water, electricity, healthcare, business—will be rare. If essential functions are disrupted, the public expects that they will be quickly restored with little disruption of normal life. In short, the public expect their communities to be resilient to the significant risks they face.

It falls to community leaders to make the decisions expected by the public, to make investments, and to take action to enhance their communities' resilience. Community resilience benchmarks can be important tools informing community leaders what actions are needed to make their communities more resilient [2].

DIFFICULTY OF ASSESSING COMMUNITY RESILIENCE

Globally, the assessment of community resilience has become a major focus of activity especially as it relates to natural disasters [3]. The number of measurement approaches for community resilience speaks both to the subject's importance and to the difficulty of actually doing it. In considering various types of approaches, Plodinec [4] identified the following factors that make it difficult to assess community resilience:

- *Resilience is a fuzzy concept.* Some efforts, particularly when focused on a community's infrastructure, seem to equate *resilience* with *resistance*, thus focusing on the ability to resist damage (for example, Park et al. [5]). Others, particularly in social and economic contexts, see resilience as the ability of a community to adapt to adverse circumstances.
- *A community's resilience is only revealed through response to and recovery from a disruption.* Although the goal of measurement is to enhance community resilience in some way, a community's true resilience can only be known after it recovers from a crisis.
- *A community's resilience depends on how it is stressed.* As pointed out by Carpenter et al. [6], a community's resilience will depend on the type and magnitude of the crisis as well as the nature of the community (its structure and its governance), or "disasters have direction"—different types of crises will attack different parts of a community [7]. Thus, a

Optimizing Community Infrastructure. https://doi.org/10.1016/B978-0-12-816240-8.00014-8

community may follow a different recovery path after a natural disaster than it does from a pandemic or an economic crisis. For example, Butler and Sayre [8,9] found very different recovery paths for US Gulf Coast communities after the Deepwater Horizon oil spill (an economic disaster for the affected communities) than after Hurricane Katrina (a natural disaster).

- *Different parts of a community will have different levels of resilience.* New Orleans in the wake of Katrina provides many examples to illustrate the point. Some neighborhoods, such as Broadmoor, have not only recovered from the devastation of Hurricane Katrina but have built many new structures (such as a neighborhood center) that are significantly improving residents' quality of life. Conversely, the Lower Ninth Ward, with so many houses still boarded up even a decade after Katrina, has not, and may never be, recovered. Although the city's population is only about three-quarters of what it was before Katrina, residents' median household incomes and the achievements of their children in the city's schools have both significantly improved above pre-Katrina levels.
- *Resilience is a manifestation of a community's strengths, but it is often difficult to measure those strengths directly.* Aldrich [10], Weil [11], and many others have pointed out the importance of a community's social capital—the connections that bind the community together and those that extend beyond the community to sources of external resources—to the community's resilience. Although these might be determinable by social network analysis, that science is still in its infancy.
- *The decision (and decision-maker) determines the data needed.* Too often, the developers of approaches to assess community resilience forget that measurement is not the goal; action is. The purpose of the assessment is to provide the information decision-makers need to make good decisions. At the community level, community leaders wanting to invest to improve their community's resilience certainly need to know the strengths and weaknesses of each part of their community. However, they also need to know their community's risk profile so that they can prioritize their investments.
- *A community's resilience is not constant; it changes over time.* It is almost a cliché—in a world of kaleidoscopic change, both communities and the contexts they find themselves in are changing. Former areas of strength may atrophy; the increasing complexity of the community's economic and social supply chains may introduce new risks; economic shifts may create new weaknesses.

DESIGN PRINCIPLES FOR APPROACHES TO ASSESS COMMUNITY RESILIENCE

Plodinec [4] also developed principles to guide the development of practical approaches to inform decisions by community leaders relating to community resilience.

- *Purpose.* The specific purpose for which the assessment is needed—the kinds of decisions to be made—must be clearly identified. Doing this will provide a less fuzzy definition of resilience for decision-makers. This chapter is focused on supporting decisions to enhance the community's resilience, primarily prior to a disaster or a disruptive event.
- *Decision-maker(s).* It is imperative to ascertain who the decision-makers are before developing an approach. Although the decisions to be made will in part determine what data should be collected, the value of the data collected (and, in fact, whether it is used at all) will also depend on the capability and perceptions of the decision-makers: their level of understanding of community resilience; the time they can devote to analysis of data; and the amount of data they believe is necessary for decisions. Ultimately, community leaders will be responsible for decisions about how to enhance the community's resilience.
- *Data collection.* The data collection process—who will collect data, how (and how frequently) they will collect data, in what domains—must be clearly specified. Decision-makers may easily become frustrated if the data collection process has not had this scrutiny: the wrong data may be collected, or the data may not be provided in a timely manner. Some decision-makers (e.g., a transportation department) may only be interested in data on one part of the community; inundation by irrelevant (to them) data will also cause frustration.
- *Understanding and trust.* If the message—data—is to inform decisions, then decision-makers must trust the messenger and understand the message. The better acquainted the decision-makers are with the data providers, the more likely they will be to trust the data. By using the community's own subject matter experts (SMEs) as the data providers, community leaders' distrust of those "not from around here" is minimized. Local SMEs also can communicate data to community leaders in the

context of the community itself, increasing understanding.

Case Study: Alliance for National and Community Resilience

The Alliance for National & Community Resilience, or ANCR (www.resilientalliance.org), is a coalition of more than 40 organizations working together to increase the resilience of America's communities. Members include business corporations, trade associations, standards setting organizations, nonprofits interested in community resilience, and academia. ANCR has undertaken the task of developing action-oriented community resilience benchmarks for community leaders. ANCR began this work recognizing the lack of this kind of information. Comparisons of the community's policies and systems against proven standards and practices were a significant barrier to communities seeking to become more resilient. Simply put, it is hard to develop a path to greater resilience if a community does not know where its starting point is.

The ANCR system is being developed based on many of the key fundamentals identified throughout this chapter. ANCR's goal is a system of community resilience benchmarks that are truly *useful* to communities, informing them where they are now; *usable* by communities, requiring only knowledge available within their own community; and thus, most importantly, *used* by community leaders to decide what actions are needed to become more resilient.

Because each community segment has its own resilience, it is essential that benchmarks associated with that segment reflect the knowledge and nuances associated with it. Therefore, ANCR has assembled teams of subject matter experts (SMEs) to revise and enhance the content of an initial strawman benchmark. The SMEs will ensure that existing codes and standards are used to the maximum possible extent and that the benchmarks are consistent with resilience-building processes such as that laid out in the National Institute of Standards and Technology's *Community Resilience Planning Guide* [12]. SMEs will also identify acceptable evidence that the requirements have been met.

It is important to note that ANCR's benchmarks are designed to elicit the knowledge of the community's own experts, that is, self-assessment by those who should know best.

DEFINING COMMUNITY RESILIENCE

The development and implementation of a system of benchmarks requires a well-settled definition of what resilience means in the context of the benchmark. There is a plethora of definitions all centered around the concept of "bouncing back," but each with a distinct flavor reflecting the domain for which it is intended [13]. Rather than adding to the welter of words, ANCR, for example, adopted an operational definition based on the experience of actual communities that embodies the concept common to these definitions of "bouncing back." This provides a clear formulation of resilience. It is shown in Fig. 14.1.

The line represents some measure of a community's capacity from before a disruptive event through recovery. The capacity might be the amount of water or electrical power provided to the community or the number of habitable homes or perhaps the number of businesses able to open.

As indicated, the functional capacity of the community (e.g., its ability to deliver clean water) prior to a disruption may be changing (its **trajectory**)—increasing (as shown), decreasing, or staying constant. When a community is "shocked"—faces a disaster or a disruptive event—it experiences a **loss of capacity** (the first arrow in the figure). This may be due to downed power lines or roads washed away or homes damaged or destroyed. Over time the community recovers (second arrow) toward a "new normal." Recovery requires **resources** and the ability to use them effectively (**competence**).

Experience indicates that each part of a community—its neighborhoods, its economy, each of its infrastructural systems, and its natural and built environment—will have a similar curve to that in the figure if negatively impacted by a disaster [14,15]. **However**, the trajectory, the pre-event capacity and condition, the amount of loss, the amount of capacity recovered, and the time scale for loss and recovery will likely differ depending on the type and magnitude of the shock.

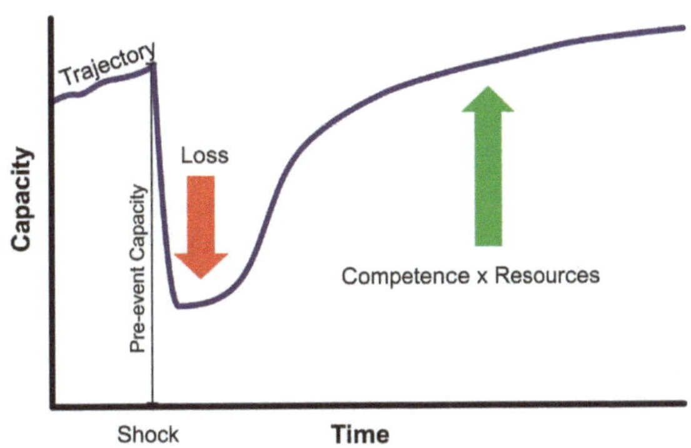

Adapted from UK Department for International Development

FIG. 14.1 Community resilience.

Experience has shown that two preshock attributes can help predict how well the community will recover (i.e., the community's resilience): its **trajectory** [16] and its **pre-event capacity** [17]. There is a sort of conservation of "community momentum" that makes the community's trajectory important: a community growing or gaining capacity prior to a shock is likely to recover faster than one that is contracting. A positive trajectory often means that the community has invested in itself—for example, replacing more fragile with more robust infrastructure. Conversely, a community that is having difficulty providing adequate service to all of its members prior to a shock is unlikely to recover very rapidly or to achieve a "new normal" that is better than before the disruption or disaster.

Thus, an assessment of a community's resilience needs to answer questions like the following about each of these essential elements:

- *Pre-event capacity.* Are all members of the community being adequately served? Are there geographic "pockets" where residents are underserved?
- *Trajectory.* Is the community's capacity growing or contracting? Is the community doing this with its own or external resources?
- *Loss of capacity.* What are the significant risks (potential shocks) facing the community? What services may be affected?
- *Resources.* What internal resources does the community have to reach its "new normal?" What external resources can it access?
- *Competence.* Does the community have a plan for dealing with a shock? Does the community have the connections and the know-how to access external resources? Does the community have experience in recovering from a shock?

RESILIENT TO WHAT?

Experience on the Gulf Coast has shown that a community's resilience to one type of risk (e.g., Hurricane Katrina) is not necessarily the same as its resilience to other risks (e.g., the Great Recession or the BP oil spill). Furthermore, every community has its own distinctive risk profile. Chicago provides an interesting example. Summer temperatures are rising, bringing with them increased risk to the poor, especially the elderly indigent. At the same time, the city is facing an explosion of violent crime, a financial crisis brought on by poor pension planning and a decaying educational system. It would be of little use to the city if a benchmarking system were to provide communities with benchmarks relating only to natural disasters or only to climate change or financial risk. Just as the community is facing all of these, community resilience benchmarks need to consider all of these. Thus, a community's resilience only has meaning in terms of its risk profile, which should include all of the significant risks the community faces.

PARSING THE COMMUNITY

All communities carry out the same essential functions; what differentiates communities is the manner and effectiveness with which these functions are performed. In every community, there is a system that carries out the function (usually a combination of several subsystems). The system often includes agents both within the community and outside the community. For example, a community may receive its electric power from a generating facility in another state or country. As another example, a regional hospital may be an important player in providing healthcare to a rural community.

The advantages of this type of approach are that "functions" are a fairly intuitive concept for most stakeholders (community leaders and residents) to grasp and accept. In addition, the boundaries between the functions can be readily delineated, which makes the assessment process more straightforward.

A community's resilience depends on every segment of the community; each segment has its own resilience. This assumption implies that the benchmarks should reflect a whole community concept. This requires a consistent framework for parsing a community into its component parts. It also requires that the benchmarks indicate the degree of resilience of each of those parts.

The ANCR development team elected to parse communities into 19 functions as identified in Fig. 14.2. For each function, there are identifiable stakeholders—those responsible for carrying out that function—to facilitate use of the benchmarks by community leaders.

Approaches based on self-assessment appear to have advantages in terms of understanding and trust. They are not constrained by the availability of consistent publicly available data sets. They can be designed to collect and use the community's own data. Because the data come directly from the community itself, the data are more likely to be trusted [4].

COMMUNITY RESILIENCE BENCHMARKS BUILT ON STRONG FUNDAMENTALS

Although the ability to benchmark a community's resilience across all its essential functions is just emerging,

FIG. 14.2 ANCR's community functions.

significant groundwork has been laid through the development of sound fundamentals. Any resilience benchmarking system that a community uses should reflect these fundamentals.

- The United Nations International Strategy for Disaster Reduction (UNISDR) assessment tools ("10 Essentials") developed to support the Sendai framework [18]. These provide a consistent and defensible basis for benchmarks. They help to resolve the tension between the public's focus on service and continuity, whereas those providing services are more focused on protecting their assets. However, the government-centric Sendai framework is not easily applied in the American context of private ownership and shared responsibility for infrastructural services. It requires significant adaptation before it is used in this context.
- The practical experience of the Community and Regional Resilience Institute (CARRI). CARRI found that resilience is a manifestation of the strengths of a community; thus, benchmarks need to focus on determining the community's strengths and weaknesses. CARRI's experience in applying its Community Resilience System to communities across the nation indicated that performance-based, outcome-oriented approaches tend to be most useful to communities looking to become more resilient [19]. Other similar approaches lead to the same conclusion [20,21].
- Existing standards. In many cases, existing standards—especially those used for accreditation—include one or more of the essential elements of resilience. For example, a healthcare facility meeting the standards of the Joint Commission will have demonstrated risk awareness, essential capacity, and competence to deal with crises. Similarly, the codes published by the International Code Council, National Fire Protection Association and others establish conditions that increase the resilience of individual facilities.
- The *Community Resilience Planning Guide for Building and Infrastructure Systems*, developed by the National Institute of Standards and Technology (NIST). This document provides a process for communities to follow to increase their resilience, as well as detailed guidance relevant to the resilience of some of a community's functions.

BENCHMARKING TODAY, ACTIONS TO IMPROVE

Benchmarking provides a community with a point-in-time assessment of its current trajectory, the likely loss of capacity following a shock, and its competence and available resources to recover from that shock. In some cases, a community may only have taken minimal steps to improve their resilience, whereas others have made significant progress. Assuring that the benchmarking system can differentiate between these investments is important—leaders should be acknowledged and laggards must be given an ideal to strive for. Therefore, benchmarks must exist at multiple levels. ANCR has selected three levels—essential, enhanced, and exceptional—to capture the continuum.

Tier 1 (Essential). These requirements, if met, should ensure that the community function can be restored after a disruption. The questions are aimed at the entities that carry out the specific community function. Taken together, achieving these benchmarks establishes the following:

- The community has identified the risks it faces and their potential impacts. This means it recognizes the potential for **loss of capacity.**
- The community has prepared to deal with the potential shocks it faces (**competence**). This implies that the community either has developed plans for dealing with disruptive shocks or has otherwise shown that it has the human capital to manage the effort (e.g., an upward **trajectory**).
- The community can meet the needs of all of its members even before a shock (**pre-event capacity**). Otherwise, it is unlikely to be able to meet the public's needs after a shock.

Tier 2 (Enhanced). Many communities have enhanced the resilience of one or more of the community functions in some way, thus increasing confidence that the community can recover more rapidly. Some communities have strengthened building and fire codes and their enforcement or have hardened their infrastructure (reducing **loss of capacity**). Some local governments have set up reserve funds (providing **resources** for recovery). Many utilities have established mutual assistance agreements (increasing both **competence** and **resources**). The requirements in this category, therefore, go beyond the Essential. In general, they will require a greater commitment on the community's part to be met.

Tier 3 (Exceptional). A few communities have taken additional actions to bolster their resilience that are highly innovative or very rare. The success, the novelty, and/or rarity of these actions boost them above the well-recognized and common approaches to enhance resilience and make communities that employ them exceptional. The "catastrophe bonds" sold by New York's Metropolitan Transportation Authority focused on storm surge are a good example. These could become a very important new tool as communities hedge their risk, providing a valuable source of new **resources** for recovery, in this case of the transportation function. The requirements in this tier thus are indicative of the types of actions that an exceptionally resilient community might take.

Each requirement must also include acceptable evidence that the requirement has been met. In general, each requirement addresses risk awareness, capacity or functionality, community competence, or resources for recovery, as shown in Table 14.1.

The value of the acceptable evidence, however, extends beyond demonstrating that the requirements are met. The acceptable evidence for each requirement also indicates action that the community can take to become more resilient. For example, a requirement for the buildings functional area might be "*Have owners of privately owned critical facilities threatened by the significant risks facing the community been encouraged to mitigate the threats to their facilities?*" The acceptable evidence would point to actual examples of ordinances, policies, and programs developed by communities.

TABLE 14.1
Examples of Acceptable Evidence for Each Community Resilience Benchmark Category.

Resilience Element	Essential Resilience	Enhanced Resilience	Exceptional Resilience
Risk awareness	Demonstrate most significant risks and their impacts have been identified	Demonstrate risks due to interdependencies have been identified	Demonstrate all risks have been considered and prioritized based on impact and frequency, including those arising from interdependencies
Capacity or functionality	Demonstrate sufficient capacity for the entire community and have fully functioning systems in place	Demonstrate development of redundant or additional sources or mitigated loss of capacity	Demonstrate community demand has been met through highly innovative means
Competence	Demonstrate development of a credible plan for dealing with each risk	Demonstrate exercising of each plan and implemented recommendations from after action report(s) Demonstrate development of a credible plan for risks arising from interdependencies	Demonstrate implemented plans now include actions for risks arising from interdependencies
Resources	Demonstrate resources required by plan(s) have been identified	Demonstrate actions have been taken to secure necessary resources (e.g., mutual aid agreements, training of grant writers, agreements with funders, insurance)	Demonstrate that resources have been secured in a highly innovative manner

CONCLUSION

Resilience is about building capacity, adaptive strength, and opportunity across the entire community—its economy, its infrastructure, and its social functions. It is about unleashing the expertise and innovative spirit within the business sector, building strong ties, and exceptional quality of life within neighborhoods and creating forward-looking, supportive community organizations. It is about making the whole community better, more competitive, more robust, more productive, and better able to cope with change.

Assessment will not make a community more resilient. However, a well-designed benchmarking approach can illuminate the community so that its strengths and weaknesses stand out in bold relief. The approach should provide the data needed by decision-makers to inform their decisions and guide their actions. The data must be provided in a timely manner, in a form that decision-makers can understand and trust. As resilience-enhancing decisions will impact the entire community, the resilience of the whole community must be assessed. An assessment must point to concrete actions the community can take to boost its resilience. Usable and useful assessments are important tools that can be used to boost a community's resilience.

REFERENCES

[1] American Legislative Exchange Council, Unaccountable and Unaffordable, American Legislative Exchange Council, Arlington, VA, 2017.

[2] A. Datnow, V. Park, Data-driven Leadership, John Wiley & Sons, New York, 2014.

[3] A. Ostadtaghizadeh, A. Ardalan, D. Paton, H. Jabbari, H. Khankeh, Community disaster resilience: a systematic review on assessment models and tools, PLOS Curr (2015), https://doi.org/10.1371/currents.dis.f224ef8ef bdfcf1d508dd0de4d8210ed.

[4] M.J. Plodinec, Measuring resilience: "why" is as important as "how, in: Proceedings of the 6th International Conference on Building Resilience, 2016. Retrieved from: https://www.scribd.com/document/344718981/ Conference-Proceedings-BUILDING-RESILIENCE.

[5] J. Park, T.P. Seager, P.S.C. Rao, M. Convertino, I. Linkov, Integrating risk and resilience approaches to catastrophe management in engineering systems, Risk Analysis 33 (3) (2013) 356–367.

[6] S. Carpenter, B. Walker, J.M. Anderies, N. Abel, From metaphor to measurement: resilience of what to what? Ecosystems 4 (8) (2001) 765–781.

[7] M.J. Plodinec, Disasters have direction: understanding cascading events, in: Proceedings of the 5th International Conference on Building Resilience, 2015. Retrieved from: https://www.newcastle.edu.au/__data/assets/pdf_file/0008/202967/Final-5th-BRC-Proceedings-23-07-15.pdf.

[8] D.L. Butler, E.A. Sayre, Modeling Micro-economic Resilience and Restoration after a Large Scale Catastrophe: An Analysis of the Gulf Coast after Hurricane Katrina, SERRI Report 80041-01, Oak Ridge National Laboratory, Oak Ridge, TN, USA, 2012.

[9] D.L. Butler, E.A. Sayre, Resilience and Restoration after a Large-Scale Disaster: An Analysis of the BP Deep Water Horizon Oil Spill, SERRI Report 80040-02, Oak Ridge National Laboratory, Oak Ridge, TN, USA, 2012.

[10] D.P. Aldrich, Building Resilience: Social Capital in Post-disaster Recovery, University of Chicago Press, 2012.

[11] F.D. Weil, Rise of community organizations, citizen engagement, and new institutions, in: Resilience and Opportunity: Lessons from the U.S. Gulf Coast after Katrina and Rita, Brookings Institution Press, Washington, DC, 2011, pp. 201–219.

[12] NIST, Community Resilience Planning Guide for Buildings and Infrastructure Systems – Volume II, NIST Special Publication 1190, Gaithersburg, 2015. Retrieved from: http://www.nist.gov/el/building_materials/resilience/upload/NIST_Guide_Volume_2_042515_For-Web-2.pdf.

[13] M.J. Plodinec, Definitions of Community Resilience, 2008. Retrieved from: http://www.resilientus.org/wp-content/uploads/2013/08/definitions-of-community-resilience.pdf.

[14] M. Bruneau, et al., A framework to quantitatively assess and enhance the seismic resilience of communities, Earthquake Spectra 19 (4) (2003) 733–752.

[15] D. Wallace, R. Wallace, Urban systems during disasters: factors for resilience, Ecology and Society 13 (1) (2008). Article 18, http://www.ecologyandsociety.org/vol13/iss1/art18/.

[16] F.H. Norris, S.P. Stevens, B. Pfefferbaum, K.F. Wyche, R.L. Pfefferbaum, Community resilience as a metaphor, theory, set of capacities, and strategy for disaster readiness, American Journal of Community Psychology 41 (1–2) (2007) 127–150.

[17] K. Sherrieb, F.H. Norris, S. Galea, Measuring capacities for community resilience, Social Indicators Research 99 (2010) 227–247.

[18] United Nations Office for Disaster Risk Reduction, Disaster Resilience Scorecard for Cities, United Nations Office for Disaster Risk Reduction, New York, 2017.

[19] R.K. White, W.C. Edwards, A. Farrar, M.J. Plodinec, "A practical approach to building resilience in America's communities, American Behavioral Scientist 59 (2) (2015) 200–219.

[20] R.L. Pfefferbaum, B. Pfefferbaum, R.L. Van Horn, R.W. Klomp, F.H. Norris, D.B. Reissman, The communities advancing resilience toolkit (CART): an intervention to build community resilience to disasters, Journal of Public Health Management and Practice 19 (3) (2013) 250–258.

[21] A. Haines, Asset-based community development, in: An Introduction To Community Development, 2009, pp. 38–48.

CHAPTER 15

How Philanthropy Is Transforming Resilience Theory Into Practical Applications at the Local Level

ROBERT G. OTTENHOFF, MCRP, BA

WORKING TO TRANSFORM THE FIELD OF DISASTER PHILANTHROPY

When disasters strike communities, American individuals, foundations, and corporations typically respond with an incredible outpouring of donations of cash, volunteer time, products, and services. It is estimated that over $1 billion was donated in the aftermath of 2017's Hurricane Harvey alone. In a typical response, hundreds of millions of dollars will be donated to disaster-related activities. But with the frequency and intensity of natural disasters increasing, pressing questions arise. When and how should donors respond? What type of activities should be supported? With billions of dollars coming from government response to disasters, how can a donor truly make a difference?

Promoting resilience as part of a larger strategy to respond to disasters seems to make a lot of sense in theory, but how does one actually put it into practice? This chapter explores how the Center for Disaster Philanthropy (CDP) and a small number of corporations and foundations are beginning to use grantmaking activities to support community resilience.

CDP was founded to help institutional donors and individuals become more strategic and intentional in their philanthropic approaches to disasters. CDP is motivated by an ambitious mission: "to transform the field of disaster philanthropy in order to increase donor effectiveness." CDP advises donors on their giving strategies and manages millions of dollars on behalf of donors for medium- and long-term disaster recovery work. Over a decade of organizational experience has helped CDP develop some approaches to effective disaster giving that can help donors be engaged with the full life cycle of disasters, from preparedness and mitigation to recovery and increasingly, the inclusion of resilience.

Funding projects that promote disaster-related resiliency is a relatively new phenomenon among philanthropists and is certainly still at modest levels compared with other approaches that receive donor support. To gain a better understanding of the philanthropic landscape, for the last 5 years, CDP has partnered with Foundation Center to produce a survey of disaster-related giving from sources worldwide and how those contributions are being used, Measuring the State of Disaster Philanthropy. This research consistently finds that the vast majority of philanthropic contributions are made for immediate relief within 60 days after a disaster occurs. Relatively small amounts—10% or so—are given to resilience, risk reduction, and mitigation. Another 10% is given to long-term recovery work. On average, only about 20% of the largest foundations support disaster-related activities and their contributions average nearly $200 million per year. Corporate support varies from year to year and includes cash contributions as well as donations of products and services [1] (Fig. 15.1).

In some ways, these results emphasizing immediate relief are not surprising. Very few foundations or corporations have disasters as a grantmaking subject area, and only a handful of foundations have a professional program officer specializing in disasters. Media attention, as well, emphasizes emergency relief and all but ignores planning and long-term recovery. Media coverage, in particular, affects individual giving to disasters. Funding for immediate relief is important because it provides much needed food, clothing, and temporary housing to people badly in need. However, as disasters increase

Optimizing Community Infrastructure. https://doi.org/10.1016/B978-0-12-816240-8.00015-X

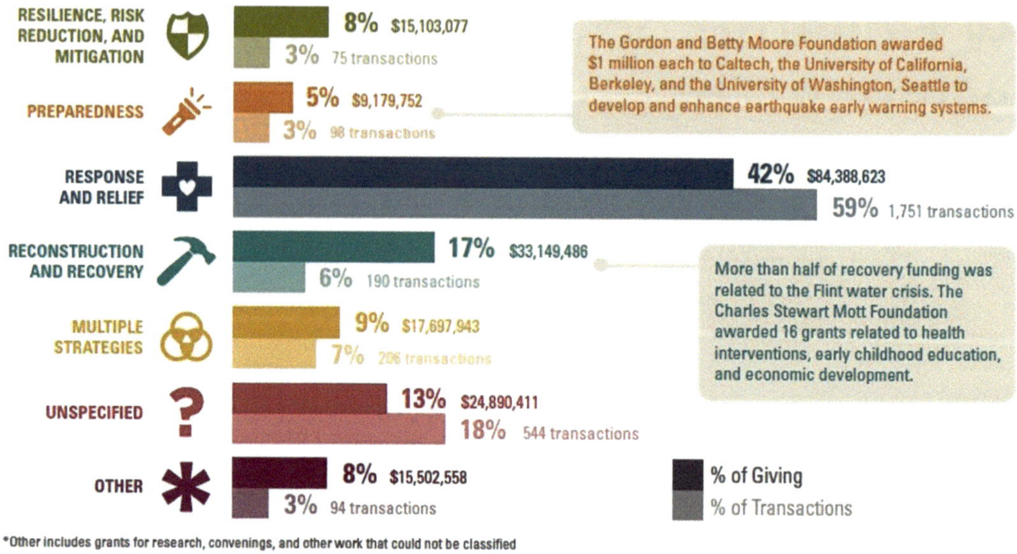

FIG. 15.1 Philanthropic funding by disaster assistance strategy, 2016. (*Reproduced from*: Center for Disaster Philanthropy, Foundation Center. Measuring the State of Disaster Philanthropy 2018: Data to Drive Decisions, November 2018. http://disasterphilanthropy.foundationcenter.org/key-findings/.)

in quantity and intensity, a growing number of institutional donors are realizing that such a reactive, episodic philanthropic approach is neither sufficient nor sustainable. It does not allow communities to be better prepared for the next disaster or take advantage of lessons learned. Responding to disasters effectively will require becoming much more strategic, with a combination of approaches before and after disasters, including better planning and preparation, additional mitigation projects, and more efforts to create adaptable, resilient communities. For the rest of this chapter, we report on how donors are not just reacting but also beginning to think ahead about creating resilient communities—from small towns to big cities.

CENTER FOR DISASTER PHILANTHROPY, MIDWEST EARLY RECOVERY FUND

Low-attention disasters are the norm in the Midwest. With 20% of the land mass and only 10% of the population, it is a region with hundreds of small and rural communities that do not attract much public attention when a disaster strikes. Without news coverage, recovery resources are scarce. Federal and state dollars are rarely allocated to address the needs of those most vulnerable to the impact of flooding, fires, tornadoes, or severe storms. With little expectation of receiving outside support, building more resilient communities is an essential and key focus of the Midwest Early Recovery Fund.

In July 2014, CDP was awarded a $2.1 million grant from the Margaret A. Cargill Foundation (MACF) Midwest Disaster Relief and Resilience program to develop and implement the Midwest Early Recovery Fund (the Fund). With that grant, CDP developed a replicable process to efficiently and effectively allocate monies to organizations supporting the needs of vulnerable populations within communities affected by low-attention disasters in the 10 state MACF area of interest. In 2017, a new 3-year grant was awarded for $3.1 million.

The Fund is utilized to address needs 2 weeks to 18 months after natural disasters, including tornadoes, flooding, earthquakes, landslides, and wildfires, in a region that includes Arkansas, Iowa, Kansas, Minnesota, Missouri, Montana, Nebraska, North Dakota, Oklahoma, and South Dakota. Increasingly, grants are being used to help communities become better prepared and more resilient to withstand future disasters. In these communities, developing strong recovery protocols is essential to their ability to consider more future-looking approaches to build resiliency.

The Fund has identified five key challenges faced by communities affected by low-attention disasters [2]. During early recovery, communities often lack the capacity to do the following (Fig. 15.2):

1. Identify and develop sufficient resources to meet the needs of those affected by the disaster.
2. Develop robust long-term recovery efforts without additional support from national, regional, or state disaster organizations and/or other partners.
3. Coordinate client information and resources from multiple agencies.
4. Identify affected vulnerable populations and develop appropriate resources.
5. Meet the unique needs of children postdisaster.

Because they lack civic and philanthropic homegrown resources, communities in the Midwest impacted by low-attention disasters have difficulty recovering. This creates an expanding array of challenges. For example, they consistently need more funding and resources for outreach and support services, especially for children, veterans, those living in poverty and/or inadequate housing, people with physical and mental health disabilities, and the elderly. They also need more disaster recovery trainers, educators, and resources for unmet needs. There is a gap between response and recovery that all too often results in a homeless population for those without the resources to secure temporary housing. Furthermore, essential recovery workers often are not provided sufficient training, ongoing support, and financial compensation, although they are key to a community's recovery. More disaster-trained mental health workers are needed for survivors and caregivers along with the resources to deploy them to communities in need. More investments are needed in disaster recovery technology and data collection. In addition, response practitioners often lack the resources to document best practices when a grant results in a process that can and should be replicated in a subsequent disaster. Because of all of these response and recovery implications, the Fund has begun to embed disaster recovery best practices into the fabric of these communities in an effort to make them more resilient in future disasters.

FIG. 15.2 After a tornado struck Morristown, Minnesota, a family lost not only their home but also their childcare business. (*Courtesy of: Child Care Aware of Minnesota is a grantee of the Midwest Early Recovery Fund.*)

The first Midwest Early Recovery Fund grant was awarded to Pilger, Nebraska, on March 24, 2015. Since then, the Fund investments have provided much needed hope and invaluable resources to thousands of disaster survivors across the Midwest. The grants have helped build the capacity of local organizations and empowered them to serve their neighbors at a time of great need and suffering. As Nancy Beers, the Fund's director observes, "We have intentionally sought out disaster-impacted communities that few others have noticed, let alone invested their time, effort, energy or financial resources in."

The Fund and its focus on low-attention disasters have underscored the need for creating resiliency. But how is that done through the recovery process? Here are three examples of how resiliency is built at the local level.

Tulsa County, Oklahoma

An EF-2 tornado tore through Tulsa County, Oklahoma, on March 25, 2015, and damaged nearly 200 homes and businesses; more than 60 of those homes were destroyed. The River Oaks Mobile Home Park was one of the hardest hit areas with all but 4 of the 56 homes destroyed, leaving many of the most vulnerable families in this area homeless.

Although Moore, Oklahoma, only 100 miles away, had suffered one of the most devastating tornadoes in US history just 2 years earlier, the Tulsa County community was ill-prepared to develop the resources needed to recover. With the support of a handful of national partners who had been involved in nearby Moore's recovery, along with funding and technical assistance from CDP's Midwest Early Recovery Fund, a few committed community members took a leap of faith and developed a local long-term recovery committee. The committee struggled to get community support, but eventually, with guidance from CDP and other NGO partners, the last case was closed almost a year to the date after the tornado.

Then the unthinkable happened. On March 30, 2016, just 5 days after the 1-year anniversary, a tornado once again hit Tulsa and neighboring Rogers County. More than 390 homes and businesses were affected [3].

But this time, the long-term recovery committee was fully prepared. They knew what to do and how to do it. They immediately contacted local officials and leaders, foundations, and nonprofits. They raised significant funding to address client needs and complete recovery for all those in need of support. There was even enough to support their own cadre of disaster case managers as well as other part-time and full-time employees to manage donations. Because of the assistance, both technical and financial, offered after the first tornado, the community's leadership was incredibly resilient, understanding immediately how to help their community. This time, the long-term recovery committee quickly developed into a regional coordinating committee or Community Organizations Active in Disasters (COAD), more than ready to help their community in its time of need, no matter what.

Rural Nebraska

A small community in rural Nebraska lost a major employer just a few years before a flood severely impacted more than 500 homes. Families were already struggling financially as were many of the local nonprofits. CDP's Midwest Early Recovery Fund worked with Nebraska's Volunteer Organizations Active in Disasters (VOAD) leadership and partners to develop and support the community's efforts to recover. CDP funded disaster case management, a critical asset for the long-term recovery committee. The committee worked long and hard to help everyone repair and rebuild their homes.

But the committee did not stop working, even when the last family finally returned home 16 months later. They asked if CDP would help support a postdisaster economic recovery effort. With the support of the Society of St. Vincent de Paul and the Midwest Early Recovery Fund, a successful economic recovery summit was hosted at the University of Lincoln's Rural Futures Institute where leaders and partners worked with the community's leaders to begin development of an economic revitalization program. Today, this community is more resilient than ever.

The Northern Plains Indian Reservations

The Northern Plains Indian Reservations are among the poorest counties in the United States. The life expectancy of a male on the Crow Creek reservation in South Dakota is 47; over 50% of the population suffers from diabetes; less than 10% of the residents have full-time employment; and their high school has a 0% graduation rate. It is a very hard place to live and yet the Dakota people who call Crow Creek home generously treat everyone like family.

In 2016, a severe storm with over 100 mile per hour winds tore through the reservation, damaging hundreds of poorly constructed homes and flimsy trailers. CDP's Midwest Early Recovery Fund worked with a local nonprofit, Diamond Willow Ministries, to develop local recovery assets. CDP provided funding to Diamond Willow to hire staff to work with individuals and families to repair and rebuild their homes, with the support of South Dakota VOAD partners and Lutheran Social Service of South Dakota.

However, it was evident that the homes on Crow Creek, although repaired and livable, were well below the Federal Emergency Management Agency (FEMA) guidelines for safety, sanitation, and security. In partnership with Diamond Willow, CDP funded the nonprofit Sustainable Native Communities Collaborative (SNCC) to develop a "resilient housing model" that would withstand the harsh weather conditions of the South Dakota plains, while being culturally appropriate and affordable, as well. The community worked closely with SNCC, and plans were developed for a new resilient neighborhood development project. Although that plan is still in the planning stages, another exciting project came from this work. Diamond Willow Ministries and the Crow Creek Tribal leadership have now developed a plan to build the first-ever permanent Tokata Youth and Family Center facility. With over 50% of the residents of Crow Creek under the age of 18 years, this new center, designed by SNCC, will be a place of "renewing hope and strengthening our future." SNCC recently completed the schematic design, and Native American–owned First American Design is in the process of completing the design development and construction drawing phase. A capital campaign is under way, and they hope to break ground soon (Fig. 15.3).

RESILIENCY PROJECTS IN TEXAS—CDP HURRICANE HARVEY RECOVERY FUND

Hurricane Harvey made its initial landfall on August 26, 2017, near Rockport, Texas, as a Category 4 storm. It circled back out to the Gulf Coast and made a second landfall on August 27 near Copano Bay and moved slowly toward the Houston area. Harvey then moved back to the Gulf of Mexico and again made landfall on August 30 just west of Cameron, Louisiana.

Harvey was the strongest hurricane to hit Texas in more than 50 years. It caused an estimated $180 billion

FIG. 15.3 Habitat for Humanity and Sustainable Native Communities Collaborative (SNCC) staff visit a site slated for a new community-designed home. (*Courtesy of*: Sustainable Native Communities Collaborative (SNCC).)

in damage and will require years to make a full recovery along the Gulf Coast. A significant portion of the damage came from flooding in the Houston and Beaumont areas. As a result of the storm, Cedar Bayou near Houston set a North American rainfall record with 51.88 in.

The storm killed 77 people and displaced more than 1 million in the immediate aftermath. It damaged an estimated 200,000 homes across a 300-mile stretch of the Gulf Coast. Thousands of individuals, corporations, and foundations contributed to CDP's Hurricane Harvey Recovery Fund, and a total of $14.2 million was raised. In its first year, the Fund awarded approximately $5.5 million to 19 different organizations. The first full round of grant funding provided capacity building for long-term recovery to nine groups working in several counties that were affected by Hurricane Harvey.

The first challenge leaders faced in these communities was how to build back. Unquestionably, there is a sense of urgency to build back quickly—to restore order and keep the community together. However, it is not only impractical but also unwise to build back and experience the same traumas after the next disaster. Is it possible to not just build back but also build back better in such a way to be in an improved position to endure future challenges?

In Fayette County, the Disaster Recovery Team in La Grange, Texas, used their grant funds to pay staff to focus on raising additional funds, oversee housing rebuild and repair, and disaster case management. As a result of their expanded capacity, the team was able to coordinate with national nonprofit organizations, local government, and the local community to identify an entire neighborhood along the river that should be relocated. "We're not building houses. We're not building streets. We're building a community," said Kenny Couch, the executive director of the Fayette County Disaster Recovery Team (FCDRT) at a ribbon-cutting ceremony held at the site of a new neighborhood being developed in La Grange. The aptly named "Hope Hill" is a neighborhood development formed through a collaborative effort among the FCDRT, Mennonite Disaster Services, Samaritan's Purse, city and county leaders, and many other community partners. On 23 acres, these partners will work together, with soon-to-be-residents of the neighborhood providing sweat equity, to build their new community (Fig. 15.4).

The residents of Hope Hill are being relocated, mostly from mobile homes in the floodplain along the river where they have been flooded before and will likely flood again if they stay. The group plans to

FIG. 15.4 Hope Hill ribbon-cutting ceremony, May 17, 2018. (*Courtesy of*: Sally Ray.)

construct 64 homes on what was once lovely elevated farm land. Along with the homes, FCDRT and partners will also create a beautiful community park, nestled among trees and bordered by a fishing pond.

In Victoria County, the Long-Term Recovery Group (LTRG) collaborated with the Victoria County United Way to use funds to hire staff to oversee disaster case management, housing rebuild/repair, and community outreach. Because they were quick to get working on the ground and provided regular communications, this LTRG has been welcomed and accepted by their political and community leaders. They brought in other resources (including a significant gift from the Rebuild Texas Fund and additional funds from the American Red Cross) to support their efforts. They are collaborating with nonprofit organizations to relocate and build more resiliently, not just "back to where they were before."

The Wharton County Recovery Team (WCRT) received CDP Hurricane Harvey Recovery Funds for adding staff. This group has worked collaboratively with city and county leaders, as well as local funders and the school district, to develop an overall plan for recovery that is connected to city plans for resiliency. One of the most exciting partnerships is with the Wharton West-End Initiative of Wharton County (WWEI) whose mission is to rebuild the West-End in Wharton, an area of town that has flooded for the past 3 years. The WWEI has a vision and plans for new multifamily housing; the elevation of existing structures; and rebuilding to revitalize this devastated neighborhood. The WCRT will provide case management for West-End clients and partner in the rebuilding efforts. This project is enhanced and reinforced by the announcement that the US Army Corps of Engineers will construct a federally funded levee system on the Colorado River, which will be future protection against flooding in Wharton and the West End. Construction of the levee will begin in 2019. Groundbreaking for the first rebuilds in the West End is scheduled for September 2019.

The Fund also provided a grant to a new nonprofit organization called Attack Poverty. The purpose of the grant was to build the capacity of this organization and expand its recovery model to other areas. Their model is a holistic one, designed to support a more resilient future for their clients. As a result of this capacity building grant, they have already been able to add two staff members and purchase additional organizational materials to support their work. In addition, they are collaborating with CDP, the Rebuild Texas Fund and BC Workshop to identify creative and bold ways to rebuild homes that were affected. This is a new and developing collaboration, but it will be part of a bigger story that will come about in the next round of funding. "By building the capacity and resiliency of this organization," says Fund Director Sally Ray, "they are also building the resiliency of the communities where they work and potentially developing a model for resiliency to be used beyond this disaster."

The CDP Hurricane Harvey Recovery Fund also worked with the Rebuild Texas Fund to provide $1.5 million (an increase after starting with $1 million) in preparedness and resiliency grants throughout the 41 affected counties. As the team entered the 2018 hurricane season, they realized recovery from Hurricane Harvey was just beginning, but they knew they needed to begin the discussion about how to prepare for the next big storm. The Fund received 161 grant applications from cities, counties, neighborhoods, municipalities, schools, libraries, and nonprofit organizations. Grants were awarded to 29 organizations with the purpose of those grants ranging from distribution of disaster kits and generators to creating a shelter and more [4] (Fig. 15.5).

One example of the type of grants awarded is the two grants provided to the City of Port Arthur to support their response efforts during and after a disaster. Port Arthur was one of the most affected communities and is at high risk for future disasters because of its location and its high concentration of petrochemical industry companies. During Harvey, Port Arthur had only one search and rescue boat to cover 143 square miles. This hampered their ability to access those most in need. One grant provided the city with funding to purchase

FIG. 15.5 A crew repairing and rebuilding a home in Texas to restore hope to those who were affected by the hurricane to help make families and communities more resilient for the future. (*Courtesy of: Attack Poverty is a grantee of the Hurricane Harvey Recovery Fund and mentioned in the copy.*)

four additional boats and trailers. The second grant provided for equipment needed to have a fully functioning, quickly activated emergency operations center (laptops, tablets, projectors, etc.).

MOVING FROM REACTIVE TO RESILIENT IN LOUISIANA

Over the past two decades, citizens of Louisiana have experienced a series of catastrophic disasters—Hurricanes Katrina, Rita, Gustav, Ike, Isaac, and the Deepwater Horizon Gulf Oil Spill. But the floods of 2016 were like no other disaster.

The Great Floods of 2016 were triggered by a complicated, slow-moving low-pressure weather system that dumped as much as 2 feet of rain on parts of East Baton Rouge, Livingston and St. Helena parishes in 48 hours. Accumulations peaked at 31.39 in. in Watson, just northeast of Baton Rouge. The record 2-day rainfall had a 0.1% chance of occurring in any year because it was the equivalent of a 1000-year rain [5]. The Washington Post noted that this "no-name storm" dumped three times as much rain on Louisiana as Hurricane Katrina [6].

Governor John Bel Edwards called the disaster a "historic, unprecedented flooding event" and declared a state of emergency. Because the floods hit areas that previously had not flooded, most homeowners affected were without flood insurance. FEMA declared at least 20 parishes as federal disaster areas. Thirteen deaths were reported as a result of the flooding, and tens of thousands of people were stranded in their homes and vehicles. Damages were anticipated to reach $10 to $15 billion, projections that would rank this storm as the seventh most expensive natural disaster in the United States since 1978.

In the aftermath of the floods, a consortium of Louisiana-based private, community, and corporate foundations and donors founded the Louisiana Disaster Recovery Alliance (LDRA) to address issues of disaster recovery and resilience to reduce levels of risk and vulnerability across the state in the face of repetitive disaster events. In a novel approach, the consortium brought Louisiana's foundation community together with federal and state government partners in a first-of-its-kind public-private partnership that takes a long-term approach to building resilience at every phase of the disaster life cycle from preparedness to recovery. The vision of this group is to examine the threats facing Louisiana with a long-term lens and to produce a set of innovative, local-specific solutions that will rebuild healthy, sustainable, and resilient communities. A Joint Fund for Recovery and Resilience was created to support the work of LDRA [7].

In October 2017, the founding members of LDRA awarded inaugural grants totaling $175,000 to meet pressing recovery needs and rebuild healthy, resilient communities throughout the state [8]. LDRA Board President Lori J. Bertman, CEO and President of the Baton Rouge-based Irene W. and C.B. Pennington Foundation observed "the grants will provide deserving nonprofit organizations the resources they need to continue to help Louisiana become healthier, economically stronger and ultimately, more resilient." The grants included the following:

- Catholic Services of Acadiana to expand and extend the development of a Disaster Data Management System as established by the Acadiana Long-Term Recovery Committee following the August 2016 floods.
- The Center for Planning and Excellence (CPEX) to expand programming based on two community workshops in a Lafayette-based pilot that shifts water management from a drainage focus to a comprehensive water management approach. The programming includes green infrastructure, low-impact development, and smart growth solutions.
- The Food Bank of Northwest Louisiana to pilot a sustainable Community Food Hub that prioritizes a long-term, disaster recovery model that addresses the needs of communities in Shreveport and Northwest Louisiana.
- Save the Children in support of their Journey of Hope program that addresses the psychosocial needs of children affected by natural disasters in the State of Louisiana.
- Southeast Louisiana Legal Services to support the Flood Proof Project that continues efforts to clear property titles—a hurdle that often prevents individuals from receiving benefits through FEMA, Community Development Block Grants (CDBG), or other programs, as well as insurance benefits or loans. The program serves 22 parishes in Southeast Louisiana, 13 of which were involved in the federal declaration for the August 2016 floods.

Years of catastrophic events have tested the philanthropic sector in Louisiana but have also allowed local foundations to accumulate experience with disaster grantmaking, function as engines of community recovery, take risks, forge best practices, as well as innovate and learn from the challenges of ongoing processes of disaster relief and recovery. The entrepreneurial thinking the philanthropic sector demonstrated during Hurricane Katrina and other disasters has informed recovery programs in other parts of the United States and represents an asset that can be used in support of their communities facing disasters in the future.

The formation of LDRA was part of a realization that efforts to tap into the knowledge across different foundations and share best practices had not been highly successful in the past given the fragmented nature of the philanthropic sector in the state. The Great Floods of 2016 shifted this dynamic by creating a desire to collaborate across institutional boundaries and become a collective force for change. The Louisiana Disaster Recovery Alliance was designed to support cross-sector collaborative efforts and communication, in addition to sharing knowledge and resources—all with the goal of improving recovery outcomes and increasing the resilience of Louisiana residents and communities.

THE ROCKEFELLER FOUNDATION ADOPTS CONCEPT OF RESILIENCE

Community foundations are becoming centers of information and leadership before, during, and after disasters. At a recent conference of community foundations, an attendee wisely noted, "We have learned that there should no longer be any single-purpose infrastructure." It was a sophisticated answer for a group of foundation leaders still learning about disaster philanthropy. It may have been influenced by the work over the last decade of the Rockefeller Foundation.

Although the word "resilience" has been used for many years in both disaster and nondisaster settings (with many different definitions), probably no single philanthropic entity has done more to promote and define the concept of resilience than the Rockefeller Foundation. Over the last decade, it has awarded more than a half billion dollars in grants to advance the concept of resilience and support various activities.

The Foundation and its former president, Judith Rodin, aggressively promoted its resilience work through its website, media connections, and a 2014 book authored by Rodin, *The Resilience Dividend* [9].

For decades, the Rockefeller Foundation was mostly noted for its contributions to the discipline of public health, supporting advancements in the eradication of diseases such as yellow fever and malaria and leading the Green Revolution development of new high-yield crops.

Judith Rodin became president of the Foundation just before Hurricane Katrina hit the New Orleans area, killing nearly 2000 people and causing extensive damage, which is one of the biggest disasters to ever strike the United States, which captured the attention of the country for months. As Rodin observed in a newspaper article, "We intervened heavily in Katrina, and I realized this was a much bigger issue and a much bigger domain, and that we could do something transformational if we invested heavily, worked in a really creative way, took some risks" [10].

In another interview, Rodin observed, "While we did what we could to help address the human suffering in the immediate aftermath of the storm, the Rockefeller Foundation understood that the scale of this disaster was long in the making, due to a compendium of social, economic, and environmental challenges that had been simmering for decades, and because strategies that were designed to address crisis and natural shocks were narrow. In the rebuilding and recovery, we saw an opportunity for the city to take a longer, more holistic view of the capacity of individuals, communities and systems to survive, adapt, and transform in the face of shocks and stresses. We wanted the city—and thousands of other communities—to develop that capacity to recover quickly and effectively when crises arise" [9].

Soon thereafter, "resilience" became a new focus of the Foundation.

But what is resilience? Rockefeller defined it as "the ability to survive and thrive in the face of increasingly unpredictable natural or manmade disasters, often spurred by climatic change or hiccups in the global economy" [9].

Rodin stated, "Crisis is becoming the new normal... Globally, in 2011, we spent $320 billion just on recovery from natural disasters. That's not sustainable given all the other types of disasters we have to confront... But hasn't the world always been faced with one crisis or another? They are becoming far more expensive-the data show that" [10].

At the root of much of this is climate change and urbanization. "It has dramatically upped the stakes in terms of the number of crises," Rodin said. "Then there's rapid urbanization: it's estimated that at the rate at which urbanization is occurring, particularly in the developing world, 40% of the infrastructure that will exist in 35 years is not yet built. We can get that right, or we can get it horribly wrong" [10].

Superstorm Sandy sharpened Rodin's and the Foundation's approach because it underscored just how complacent cities could be. "New York has viewed itself as quite resilient, and there was a great deal of real and significant attention and investment after 9/11 to becoming more resilient-and yet Sandy was catastrophic" [10].

Unless resilience is built into the culture of cities, Rodin asserted there is always the prospect of disruption turning into disaster. Disruptions, to her way of thinking, are inevitable; disasters are not. Indeed, with an integrated approach to resilience, disruption can be turned to advantage. Every crisis is an opportunity.

As resilience gathered attention and adherents, it also created new challenges. It introduced seemingly new costs for local and state governments, identified new complexities, and added new tasks. It was seen as one new complication to add to the enormous problems already faced. Rockefeller addressed these concerns in effect by expanding the concept of resilience to include both current and future benefits. In her book, *The Resilience Dividend*, this is how Rodin described it: "Building resilience creates two aspects of benefits: it enables individuals, communities, and organizations to better withstand a disruption more effectively, and it enables them to improve their current systems and situations. But it also enables them to build new relationships, take on new endeavors and initiatives, and reach out for new opportunities, ones that may never have been imagined before. This is the resilience dividend. The resilience dividend not only enables people and communities to rebound faster from disasters or deal with stresses; it spurs economic development, job creation, environmental sustainability, and social cohesion. It brings benefit to people, organizations, and communities when things are going right as well as when they go wrong" [11].

Yet, to ensure that cities can imagine, finance, and plan the infrastructure needs of tomorrow, Rodin pointed out that we need to change the mind-set around infrastructure from "keeping all bad things out" to creating new kinds of capacity to respond to the challenges that will inevitably come.

The first step is to move away from responding only to the last disaster and instead anticipate future threats and changes. As an example, in 2010, designers in Portland, Oregon, revisited their plans for a light-rail bridge spanning the Willamette River to ensure it could handle higher and more rapid waters. These were costly changes, but now the bridge can withstand a broader range of potential impacts and is the first transportation project in Oregon's history to be conceived and planned with future storms and weather-related incidents in mind.

Another way to build resilience is to better integrate infrastructure projects for public good with the needs of the private sector. To this end, the Rockefeller Foundation teamed up with the White House, the US Conference of Mayors, and innovators in the private sector to fund an initiative called RE-invest, which is supporting eight cities to establish a new form of public-private partnership that will help them package portfolios of investments aimed at building more resilient infrastructure [12].

With the assistance of leading engineering, law, and finance firms, the cities will be able to use public resources more efficiently to leverage private investments—for example, in the development of better stormwater infrastructure.

In this way, infrastructure investments can achieve multiple wins, or what Rockefeller calls the "resilience dividend." Simply stated, this means financing, planning, and implementing solutions that help cities, systems, institutions, and people rebound more quickly from disaster if and when it hits while helping spur economic development, job creation, environmental sustainability, and social cohesion between shocks. For example, the effort to create and maintain green infrastructure will necessarily spur the expansion of education and employment opportunities for a new generation of highly skilled workers.

No matter how much planning for and predicting of major disruptions occurs, infrastructure failure is sometimes unavoidable given the increasing severity of shocks and stresses to our systems. Thus, the second step is to build in mechanisms for infrastructure to fail safely, minimizing the disruption that can ripple across systems. We saw this vital need in New York City during Superstorm Sandy. The electrical grid was too networked, so when one part of the system went down in a fantastic explosion, the entire lower half of Manhattan went with it.

The third step to adopting more resilient capacity is to expand the expectation of who pays for infrastructure. Traditionally, this has been viewed as solely the realm of government. But the resilience of businesses—and indeed an entire sector—is intertwined with the resilience of its community. The private sector has a clear interest and responsibility to prioritize this work. A resilient nation requires resilient communities; this calls for multisector commitment and collaboration.

As Rodin concludes: "There is no ultimate or end state of resilience. By working together to build resilience to the greatest degree possible, we can reduce our reliance on crisis as a driver of change and instead, deliberately take the future into our own hands—for the well-being of our families, our communities, our cities and indeed, the planet we all share."

THE RESILIENT COMMUNITIES MOVEMENT

In this final section, we review three examples of how philanthropists are active in ways to promote resilient cities and communities.

100 Resilient Cities

In the United States, cities have experienced a remarkable resurgence in the last decade, with unparalleled building booms, rising populations (particularly among young people), and a renewed sense of energy and confidence. It was not too many years ago that the future of urban America looked bleak. Many big cities hit bottom in the 1970s and 1980s, with falling populations and crumbling infrastructure; more than a few, such as New York City and Cleveland, were even threatened by bankruptcy. As revitalization began toward the end of the 20th century, cities also experienced some tremendous shocks—ones that challenged the way they were governed, financed, and planned for the future. The 2001 attack on the World Trade Center in lower Manhattan, with nearly 3000 deaths, forever changed our nation's approach to terrorism and personal safety. Since then, tens of billions of dollars have been spent on hard and soft security systems and increased employment of large numbers of people with security responsibilities. We also became more aware of infrastructure failures in 2005 when the levees broke in New Orleans after Hurricane Katrina, causing the deaths of nearly 2000 people and widespread, catastrophic destruction... all televised live. It was a traumatic national experience and another turning point for the country as we reconsidered our responses to natural disasters and the role of FEMA. The destruction wrought by Superstorm Sandy to parts of New Jersey and New York in 2012 reminded us once again of the vulnerabilities of cities to disasters and the imperative to change our approach to water and our concept of resiliency.

The resurgence of cities has not immediately solved all their problems. They still face a myriad of challenges including persistent pockets of poverty, lack of affordable housing, inadequate public transportation, and inconsistent funding. Building a more resilient city sounds appealing but rarely rises to the level of a high priority, given the never-ending competition for attention and money. And, in most cases, the concept of resilience is still elusive and rarely the kind of bread and butter issues to win elections.

Looking for ways to increase its impact and leverage its investments in resilience, in 2013, the Rockefeller Foundation launched 100 Resilient Cities (100RC) to inspire planning for more resilient cities. The Foundation has had a long history with grantmaking programs dealing with urban poverty and design, but its new focus on resilience placed renewed attention on cities, as populations around the world continue to migrate to urban life.

100RC takes a broad and comprehensive approach to resiliency. As it states in its literature: "It isn't solely about responding to an occasional hurricane or "100-year floods." That approach would not capture people's imaginations." And looking from a political perspective, it also would not capture the countless economic and political challenges faced by cities. Thinking broadly by including natural disasters, civil disturbances, economic disruptions, and other potentially disruptive events builds constituencies. 100RC defines urban resilience as "the capacity of individuals, communities, institutions, businesses, and systems within a city to survive, adapt, and grow no matter what kinds of chronic stresses and acute shocks they experience" [13].

This requires looking at cities holistically: "Understanding the systems that make up the city and the interdependencies and risks they may face ... strengthening the underlying fabric of a city and better understanding the potential shocks and stresses it may face, a city can improve its development trajectory and the well-being of its citizens." This fits well with other efforts underway by foundations, including civic engagement, community organizations, and nonprofit vitality. It is a people-first focus that foundations embrace.

The 100RC approach to infrastructure will also appeal to cash-strapped cities. Rather than solely focusing on building a new highway, for example, the 100RC approach to resiliency suggests looking for multiple benefits, such as considering a new highway project that doubles as a water barrier along a river that also provides new recreational opportunities. Resiliency underscores the new mantra that "single uses of infrastructure are over."

In writing about these efforts, Michael Berkowitz, now president of 100 Resilient Cities, described the program in a Living Cities' Medium article in 2018: "In creating 100RC, the Foundation understood the need to work directly with municipal governments to upend the old structures that stymie this potential. With seed funding and training for a new position in city government, the Chief Resilience Officer (CRO), Rockefeller forged a new model for interceding on this systemic scale" [14].

The cities received four types of support:

1. To hire and empower a city CRO as a central point of contact within each city to coordinate, oversee, and prioritize resilience activities.

2. To develop a resilience strategy that analyzes and mitigates their vulnerabilities and builds on their unique strengths.
3. Access to a platform of services leveraging resources significantly beyond the city to support solutions that integrate Big Data analytics, technology, resilience land use planning, infrastructure design, and new financing and insurance products.
4. Membership in the 100 Resilient Cities Network, a peer-to-peer network that shares new knowledge and resilience best practices and fosters new connections and partnerships.

Even the most visionary mayors preside over a governance structure created in the 20th century that has solidified over generations and administrations. The agencies they oversee have been optimized for efficiency, but those silos often hinder the kind of integration cities need to successfully address their challenges. Transport people talk to transport people, as do economic development, housing, and immigration. Those conversations are more efficient at least in the short term. But that approach has risks and misses significant intersectional benefits. How many times in the 20th century has road construction damaged existing development (or missed upside) because it was planned and built by people using a single metric—moving cars? Similarly, how often do social services operate in silos, where a coordinated approach would provide better impact, more cheaply?

Taking a more comprehensive systems view has allowed cities to begin to address this issue. Hiring a CRO is a rare opportunity to create the disruption city leaders seek in order to to change the status quo. A CRO is a senior city official that often reports directly to the city's chief executive and works across departments and city sectors to help a city address its complexities systemically.

Will this approach make a difference for cities in the long run? Or will this end up only as a one-time grant that allowed cities for a year or two to do something they otherwise could not? Will "resilience officers" make a sufficient enough difference to earn ongoing funding within a city budget or will they be pushed aside by more urgent priorities? For philanthropy, another big question will be whether this movement can migrate from a Rockefeller Foundation–inspired effort into one with broad support within the philanthropic community. Only time will tell whether these and other issues are sufficiently addressed, so that building urban resilience becomes a movement or just a passing fancy.

In his year-end report in December 2018, Berkowitz noted that more than 80 CROs around the world continue to lead pioneering work to build resilience in their cities. Nearly half of all cities in the global network have released visionary, actionable Resilience Strategies, which contain almost 2600 tangible initiatives to build resilience in cities, regions, and beyond. More than $3.35 billion has been leveraged by member cities to implement these resilience solutions [15].

In late 2018, the Urban Institute released what it termed "an independent assessment of 100RC" (funded by the Rockefeller Foundation) that found member cities are widely adopting holistic resilience planning practices and "desiloing" city operations to tackle social, economic, and physical challenges of the 21st century. Moreover, the analysis found 100RC is among the first global urban initiatives to employ a consistent set of tools, supports, and resources across so many diverse cities for which no alternative exists. Findings suggest 100RC is contributing positively to six key areas of interest in its member cities by embedding resilience principles in city planning and operations. These six areas of positive change include the explication of resilience in city planning; the internal consistency across cities' various planning documents; the establishment of a CRO or similar cross-sectoral coordinator; a reduction in the strength of the government silos that promote ineffective solutions, duplication, and inefficiency; better collaboration across city, state, and national levels of government; and changes to budgetary review procedures or leveraged funds for resilience-building efforts, which may ultimately lead to more efficient and effective use of city funds [16].

The current list of member cities can be found at www.100resilientcities.org/cities. [Editor's note: In April 2019 as this book was being finalized, 100RC announced that it was ceasing operations and that the Rockefeller Foundation was funding new resilience initiatives. It is unclear at this time how the announcement will impact efforts within the member cities and the resilience movement as a whole.]

Rebuild by Design

A second major nationwide initiative worth our attention is Rebuild by Design. It began as a design competition launched by the US Department of Housing and Urban Development (HUD) in partnership with nonprofit organizations and the philanthropic sector, in response to Superstorm Sandy's devastating impact on the eastern United States. In June 2013, it launched a multistage planning and design competition to promote resilience in the Sandy-affected region. HUD conducted the competition under the authority of the America COMPETES Reauthorization Act of 2010 and administered the competition in partnership with philanthropic, academic, and nonprofit organizations [17]. The lead funder was the Rockefeller Foundation. Additional funding was provided by JPB Foundation, Surdna Foundation, The Deutsche Bank Americas Foundation, The Hearst Foundation, and The New Jersey Recovery Fund.

In June 2014, then-HUD Secretary Shaun Donovan announced the award of $930 million to seven winning ideas. Each one comprised multiple phases, which collectively represented a master plan for resiliency in the New York/New Jersey metropolitan area, particularly in water management [18].

Looking back, researchers have identified 10 takeaways from their work in the Sandy design competition:

- Design for and encourage projects that provide multiple benefits.
- Achieving comprehensive resilience will require a long-term approach.
- Align multiple streams of funding and administrative requirements.
- Create more flexibility in disaster recovery.
- Identify additional funding sources to support long-term monitoring and maintenance.
- Use lessons learned from project implementation to reform permitting.
- Encourage coordination across agencies and levels of government.
- In addition to infrastructure projects, pursue legal and policy mechanisms.
- Encourage better predisaster planning and mitigation.
- Encourage robust public engagement and partnerships.

Today, Rebuild by Design is working in 10 locations, including six American cities: San Francisco, Atlanta, Boston, Oakland, Boulder, and Los Angeles. Projects include studying coastal areas, housing, transportation, and climate change. In all these programs, Rebuild by Design looks for a multidimensional approach, with collaborations between designers, researchers, community members, government officials, and subject matter experts.

In June 2018, the organizers of the original Rebuild by Design competition published a report reflecting on the impact of the competition 5 years later. Top on their list was the fact that it had helped to create a community of resilience practitioners. Nearly every respondent (31/33 or 94%) has worked on resilience projects following the conclusion of the design competition and many report a deeper understanding of resilience in their subsequent work. One respondent, Stephen Whitehouse, Principal, Starr Whitehouse Landscape Architecture, observed: "I think it has helped redefine the scope of what we do as a firm and the projects which we are considered for." Another respondent observed that a "resilience perspective" is now being included in projects that previously might not have talked about resilience. "Resilience has shown up in projects that aren't necessarily 'resilience projects'—for example, master planning or residential development where clients aren't being informed about what decision making should take place in order to be resilient" [19].

The Philanthropic Preparedness, Resiliency, and Emergency Partnership

Every year, communities in the United States are challenged by natural disasters including floods, tornadoes, and earthquakes. And every time a major disaster strikes a community, members are forced to figure out how to respond, often with little preparation. With support from individuals, nonprofit organizations, faith-based groups, funders, and local/state/federal agencies, incredible relief and recovery work gets done, but perhaps not as effectively or as efficiently as it could.

How can the concept of resiliency be promoted beyond big cities? One great example is a program created by The Funders' Network for Smart Growth and Livable Communities (TFN) called the Philanthropic Preparedness, Resiliency, and Emergency Partnership (PPREP). CDP is a partner to TFN and a major consultant to the project; funding comes from Margaret A. Cargill Philanthropies. Currently, 20 community foundations and 3 regional associations are participating.

The program is intended to build community foundation leadership and capacity that will help their institutions and communities be better prepared for, respond to, and recover from natural disasters. Working with partners in the private, nonprofit, and public sectors, community foundations have a unique ability to convene service providers and community-based organizations (CBOs)—including CBOs working with low-income communities, immigrant communities, vulnerable populations, and others who are often neglected after a disaster strikes—as well as corporations and businesses, government actors, and others who all have an interest in contributing to the betterment of their communities. Community foundations also have a unique capacity to pool monies from donors to distribute funds efficiently to the most effective actors and through the most effective means, ensuring the communities most in need get access to resources in a timely manner. Moreover, as communities begin the long process of recovering after a disaster—when the media are gone and attention has shifted to other pressing stories—these foundations are part of the community in which they serve.

CDP develops and implements a curriculum for the group, which regularly convenes online for educational webinars and in-person for site visits, presentations, and discussions on the role of philanthropy in disaster response and recovery. Their work has focused on providing practical resources to increase the internal capacity of foundations to lead in the face of disaster and expand their knowledge of the government and nongovernment actors in a disaster.

The PPREP program provides the following:

- Strategies for building community resilience,
- Tips for enhancing relationships with local partners,
- Tactics for identifying and attending to socially vulnerable populations, and
- Examples for designing disaster grants programs.

Participants also complete a Disaster Preparedness Workbook that is updated over time to provide a central source of institutional information and processes in the event of a disaster. The workbook, and subsequent cocreated curriculum, will help these community foundations:

1. Enhance their readiness to respond before, during, and after a disaster, in partnership with other community organizations.
2. Understand the range of resources and vulnerabilities in the community foundation's service area.
3. Develop operational capacity, disaster grantmaking policies, and processes.
4. Increase the knowledge of community-wide disaster management and the role of philanthropy in that work.

The work of PPREP puts community foundations as the centerpiece in an effort to plan and prepare for disasters. Community foundations can act as "neutral" parties bringing together disparate segments of a community around a common cause. These efforts have the potential to result in stronger communities through relationship building, sharing of best practices, and putting resilience at the forefront of both disaster response and philanthropy.

CONCLUSION

Hurricanes Katrina and Sandy were traumatic events for the United States with thousands of people killed and billions of dollars of buildings and infrastructure damaged. These disasters prompted a widespread reconsideration of the readiness to respond to disasters and the fundamental approaches to protecting communities. These storms exposed how vulnerable any town or city can be in a disaster. Meanwhile, across the country, smaller-scale disasters were challenging states, rural communities, and small towns to rethink their own approaches.

There is growing philanthropic interest in and understanding of the concept of exploring and funding the full life cycle of disasters. Some are moving beyond traditional philanthropic approaches of focusing on the few days immediately after a disaster. Interest is growing in supporting planning and preparation, as evidenced by the Midwest Early Recovery Fund. Recognizing that we can never fully eliminate death and destruction from disasters, others such as the Louisiana Disaster Recovery Alliance, the CDP Hurricane Harvey Recovery Fund, and the Rebuild by Design programs are looking at ways to help communities create approaches that build more resilient communities—ones better able to come back quickly when disasters occur.

As with any large-scale commitment of tax dollars and philanthropic support, applying resources in an equitable and just manner is essential. Whose community are we making resilient? And at what cost? If not carefully and fairly undertaken, the resiliency movement could experience some of the pitfalls of the urban planning movement of the 1950s and 1960s where decades later we can more fully understand some of the negative consequences of large-scale destruction and rebuilding of neighborhoods and more clearly see those who benefitted from the movement and those that lost out.

The hopeful signs of the resilience movement are only a beginning. The relatively few philanthropic commitments to resilience do not yet represent a philanthropic trend. However, they indicate a growing recognition that philanthropic contributions to disaster-related activities need to be aware of the full life cycle and can make a significant difference in creating more resilient communities better able to withstand crises that disrupt normal life.

REFERENCES

[1] Center for Disaster Philanthropy, Foundation Center, Measuring the State of Disaster Philanthropy 2018: Data to Drive Decisions, November 2018. http://disasterphilanthropy.foundationcenter.org/key-findings/.

[2] N. Beers, Midwest Early Recovery Fund, Center for Disaster Philanthropy, 2018. https://disasterphilanthropy.org/midwest-early-recovery-fund/.

[3] S. Ryburn, 390+ Homes and Businesses Affected by Tornado in Tulsa, Tulsa World, April 2, 2016.
[4] S. Ray, Hurricane Harvey Recovery Fund, Center for Disaster Philanthropy, 2018. https://disasterphilanthropy.org/cdp-fund/cdp-hurricane-harvey-recovery-fund/.
[5] Hydrometeroligical Design Studies Center Office of Water Prediction, Exceedance of Probability Analysis for the Louisiana Rainfall Event, 11-13, August 2016, National Weather Service, August 16, 2016.
[6] J. Samenow, No name storm dumped three times as much rain in Louisiana as hurricane Katrina, Washington Post (August 19, 2016).
[7] Center for Disaster Recovery, Louisiana Recovery Alliance, 2017. http://louisianarecoveryalliance.org/.
[8] Center for Disaster Philanthropy, Louisiana Disaster Recovery Alliance Announces Inaugural Grant Awards, October 3, 2017. http://disasterphilanthropy.org/press-release/ldra-inaugural-grant-awards/.
[9] J. Rodin, The Resilience Dividend: Being Strong in a World where Things Go Wrong, Public Affairs, a member of the Perseus Books Group, New York, 2014.
[10] S. Moss, Judith Rodin's warning for the world: crisis is becoming the new normal, The Guardian 27 (2015).
[11] J. Rodin, Valuing the Resilience Dividend, February 27, 2017. https://www.rockefellerfoundation.org/blog/valuing-resilience-dividend/.
[12] J. Paynter, How Judith Rodin Created A New Model for Philanthropic Funding at the Rockefeller Foundation, Fast Company, December 6, 2016. https://www.fastcompany.com/3065904/how-judith-rodin-created-a-new-model-for-philanthropic-funding-at-the.
[13] J. Rodin, Infrastructure and the Resilience Dividend, May 2014. Available at: https://assets-prod.mckinsey.com/featured-insights/urbanization/infrastructure-and-the-resilience-dividend.
[14] 100 Resilient Cities. http://100resilientcities.org.
[15] M. Berkowitz, Leading the Way to a More Resilient Future, September 25, 2018. https://www.100resilientcities.org/leading-way-resilient-future/.
[16] C.M. McTarnaghan, Institutionalizing Urban Resilience: A Midterm Monitoring and Evaluation Report of 100 Resilient Cities, December 6, 2018. https://www.urban.org/urban-resilience.
[17] Georgetown Climate Center, Rebuilding with Resilience: Lessons from the Rebuild by Design Competition after Hurricane Sandy, November 14, 2016. https://www.georgetownclimate.org/reports/rebuilding-with-resilience-lessons-from-the-rebuild-by-design-competition-after-hurricane-sandy.html.
[18] U.S. Department of Housing and Urban Development, Hurricane Sandy Rebuilding Strategy, Fall, 2014. https://www.hud.gov/sites/documents/HURRSANDREBSTRATPRF2014.PDF.
[19] R. Basalaev-Binder, D. Wachsmuth, Rebuild by Design Five Years Later: Reflections from the Designers, November 20, 2018. http://www.rebuildbydesign.org/data/files/1083.pdf.

PART VII

Conclusion

CHAPTER 16

A Vision for Resilient Infrastructure

RYAN M. COLKER, JD, CAE

True community resilience is built at the intersection of disciplines and systems. Hopefully, through this book, the importance of focus within and between individual infrastructure systems and leveraging multiple strategies including natural solutions and emerging financing models is clear. However, it is also clear that the fragmented or single-discipline optimization of infrastructure will have limited impact on addressing complex communities.

To effectively achieve resilience at a national and community level, a new strategy is needed, one built on multiple actions that cut across social, economic, and environmental realms and across disciplines and systems. The challenge moving forward is to bring together some of the great resilience-building activities underway within disciplines and systems (some of which are examined in this book). Educating practitioners across infrastructure systems to recognize their contributions to community resilience and their interdependencies across systems is essential. Developing the next generation of design, planning, and operations professionals with competence and comfort in interdisciplinary and cross-disciplinary approaches to addressing the wicked problems of today is necessary. Shaping the economy to value resilience investments and leveraging technology advancements to help address complexities are also required.

A NEW POLICY APPROACH

The concept of a holistic approach to achieving resilience at both a national and community scale is not exclusive to the content of this book. Organizations like the United Nations (UN) and the World Bank at a global level and the Alliance for National and Community Resilience (ANCR) at the national scale are raising awareness and advancing policies around the importance of a systems-based approach to achieving resilience.

The Sendai Framework for Disaster Risk Reduction 2015–30 was adopted by UN Member States on March 18, 2015 at the Third UN World Conference on Disaster Risk Reduction in Sendai City, Miyagi Prefecture, Japan. It is the first major agreement of the 2030 development agenda and aims for "the substantial reduction of disaster risk and losses in lives, livelihoods and health and in the economic, physical, social, cultural and environmental assets of persons, businesses, communities and countries" [1].

The Sendai Framework identified Seven Global Targets essential to achieving resilience at a global scale (Fig. 16.1):

1. Substantially reduce global disaster mortality by 2030, aiming to lower the average per 100,000 global mortality rate in the decade 2020–30 compared with the period 2005–15.
2. Substantially reduce the number of affected people globally by 2030, aiming to lower the average global figure per 100,000 in the decade 2020–30 compared with the period 2005–15.
3. Reduce direct disaster economic loss in relation to global gross domestic product (GDP) by 2030.
4. Substantially reduce disaster damage to critical infrastructure and disruption of basic services, among them health and educational facilities, including through developing their resilience by 2030.
5. Substantially increase the number of countries with national and local disaster risk reduction strategies by 2020.
6. Substantially enhance international cooperation to developing countries through adequate and sustainable support to complement their national actions for implementation of this Framework by 2030.
7. Substantially increase the availability of and access to multihazard early warning systems and disaster risk information and assessments to the people by 2030.

Optimizing Community Infrastructure. https://doi.org/10.1016/B978-0-12-816240-8.00016-1

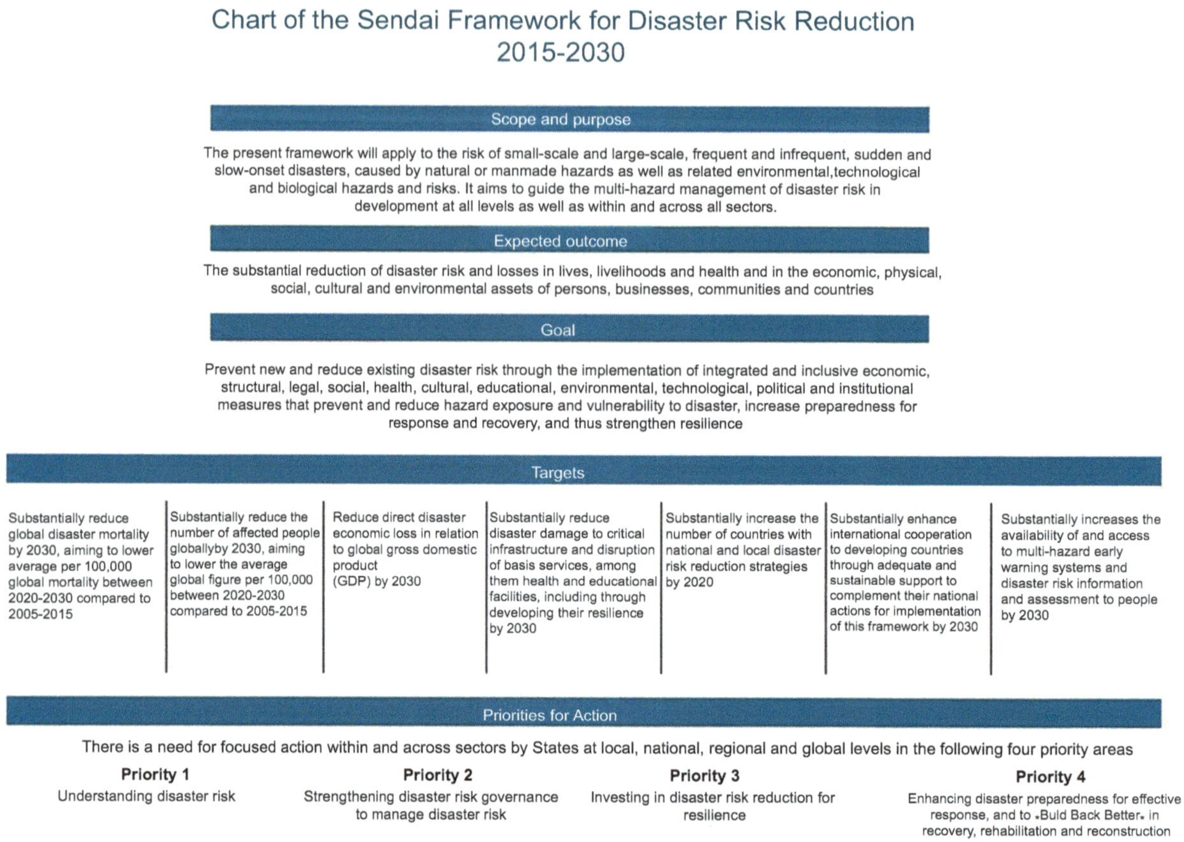

FIG. 16.1 Components of the Sendai Framework. (Source: Courtesy United Nations Office for Disaster Risk Reduction, Sendai Framework for Disaster Risk Reduction, 2015. https://www.unisdr.org/we/coordinate/sendai-framework.)

To achieve these global targets, the framework identifies Four Priorities for Action:

Priority 1. Understanding disaster risk

Disaster risk management should be based on an understanding of disaster risk in all its dimensions of vulnerability, capacity, exposure of persons and assets, hazard characteristics, and the environment. Such knowledge can be used for risk assessment, prevention, mitigation, preparedness, and response.

Priority 2. Strengthening disaster risk governance to manage disaster risk

Disaster risk governance at the national, regional, and global levels is very important for prevention, mitigation, preparedness, response, recovery, and rehabilitation. It fosters collaboration and partnership.

Priority 3. Investing in disaster risk reduction for resilience

Public and private investment in disaster risk prevention and reduction through structural and nonstructural measures are essential to enhance the economic, social, health, and cultural resilience of persons, communities, countries, and their assets, as well as the environment.

Priority 4. Enhancing disaster preparedness for effective response and to "build back better" in recovery, rehabilitation, and reconstruction

The growth of disaster risk means that there is a need to strengthen disaster preparedness for response, act in the anticipation of events, and ensure capacities are in place for effective response and recovery at all levels. The recovery, rehabilitation, and reconstruction phase is a critical opportunity to build back better, including through the integration of disaster risk reduction into development measures [1].

The United Nations went further to translate the targets and priorities outlined in the Sendai Framework to essential actions for implementation at the local level. These actions look across governance, planning, and response to support a community's holistic strategy for enhancing its resilience. The "Ten Essentials for Making Cities Resilient" are as follows [2] (Fig. 16.2):

1. *Organize for disaster resilience*. Put in place an organizational structure with strong leadership and clarity of coordination and responsibilities. Establish disaster risk reduction as a key consideration throughout the city vision or strategic plan.
2. *Identify, understand, and use current and future risk scenarios*. Maintain up-to-date data on hazards and vulnerabilities. Prepare risk assessments based on participatory processes, and use these as the basis for urban development of the city and its long-term planning goals.
3. *Strengthen financial capacity for resilience*. Prepare a financial plan by understanding and assessing the significant economic impacts of disasters. Identify and develop financial mechanisms to support resilience activities.
4. *Pursue resilient urban development and design*. Carry out risk-informed urban planning and development based on up-to-date risk assessments with particular focus on vulnerable populations. Apply and enforce realistic, risk compliant building regulations.
5. *Safeguard natural buffers to enhance the protective functions offered by natural ecosystems*. Identify, protect, and monitor natural ecosystems within and outside the city geography and enhance their use for risk reduction.
6. *Strengthen institutional capacity for resilience*. Understand institutional capacity for risk reduction including those of governmental organizations, private sector, academia, professional and civil society organizations, to help detect and strengthen gaps in resilience capacity.
7. *Understand and strengthen societal capacity for resilience*. Identify and strengthen social connectedness and culture of mutual help through community and government initiatives and multimedia channels of communication.
8. *Increase infrastructure resilience*. Develop a strategy for the protection, update and maintenance of critical infrastructure. Develop risk-mitigating infrastructure where needed.
9. *Ensure effective preparedness and disaster response*. Create and regularly update preparedness plans, connect with early warning systems, and increase emergency and management capacities. After any

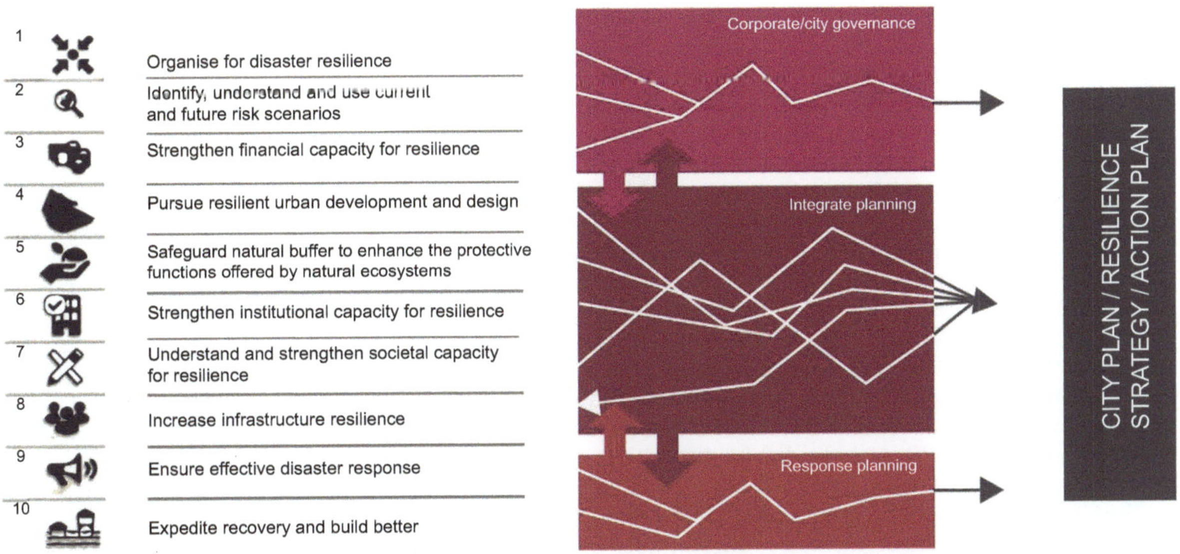

FIG. 16.2 Ten essentials for making cities resilient. (Source: Reprinted from United Nations Office for Disaster Risk Reduction, How to Make Cities More Resilient: A Handbook for Local Government Leaders, 2017. https://www.unisdr.org/campaign/resilientcities/toolkit/article/a-handbook-for-local-government-leaders-2017-edition.)

disaster, ensure that the needs of the affected population are placed at the center of reconstruction, with support for them and their community organizations to design and help implement responses, including rebuilding homes and livelihoods.

10. *Expedite recovery and build back better*. Establish postdisaster recovery, rehabilitation, and reconstruction strategies that are aligned with long-term planning and providing an improved city environment.

The wholistic approach captured within the Sendai Framework provides a valuable lens through which to view resilience policy. With the growing impacts of disasters outlined in the Introduction and clearly documented by the National Oceanic and Atmospheric Administration (NOAA) [3], the current approach to recovery through postdisaster funding and rebuilding is not sustainable. It is also highly disruptive to communities and the economy.

The work of the National Institute of Building Sciences (NIBS) to identify the cost-effectiveness of approaches to predisaster mitigation provides policymakers, businessowners, homeowners, and members of the resilience economy (more on this below) with a compelling approach [4]. Investing in resilience before a disaster strikes can provide considerable benefits. Ideally, this methodology and thinking find their way into the marketplace, serving as the basis for investment decisions and as a mechanism for identifying value.

These findings should also be used to support a new approach to benefit-cost analysis in the case of policy and business decisions, one that provides greater value on the avoidance of future costs. Today, many decisions are made solely on first costs or with a very limited time frame for capturing future benefits (which are often highly discounted themselves). A total cost of ownership approach, capturing benefits and costs across the life of a project, will provide a more realistic picture of the benefits associated with particular investment decisions. This lack of focus on total cost is one reason why the infrastructure investment gap is as large as it is—funding is approved for the initial investment, but funding for operations and maintenance is not approved at necessary levels to maintain the initial levels of service.

The US Congress has taken some steps in this direction through the Disaster Recovery Reform Act, providing incentives for state and local governments to adopt and enforce up-to-date building codes. However, Congress is often hamstrung in making long-term investments in mitigation due to evaluation rules that limit their ability to capture long-term savings as part of the consideration in the cost of a project. Changing Congressional Budget Office (CBO) rules to allow the capture of a greater proportion of the benefits of investments should allow additional Congressional support for mitigation investments.

In most cases, building in infrastructure improvements earlier in the design and construction of a project is significantly cheaper than having to redesign and reconstruct an existing piece of infrastructure in the face of changed risk. As climate change brings a new set of risks, designing to the reasonably foreseeable risk and allowing for future adaptability to address further changes in risk over the life of the project is a wise strategy.

Infrastructure owners should not have to solely bear the costs of these forward-looking benefits—particularly when others reap some of these benefits. The concept of incentivization provides an initial strategy for assuring that all stakeholders who benefit from a resilience investment actually contribute to that investment [5,6]. It aims to address the classic market failure presented by externalities. Optimizing investment in resilience requires capturing these externalities. Once these externalities are captured, the most cost-effective strategy, benefiting multiple stakeholders, will reveal itself.

A NEW RESILIENCE ECONOMY AND SUPPORTING WORKFORCE

To truly achieve a marketplace that values resilience and supports resilience decision-making, the establishment of a resilience economy is essential. A resilience economy is its own system of functions that feeds upon itself to create an ecosystem of interdependent systems and disciplines that ultimately drive the achievement of resilience. NIBS identified some of the offerings that would feed consumer interest in resilience and deliver solutions. Components of the resilience economy are outlined in Fig. 16.3.

In addition to a supportive policy structure coordinated across finance, planning, and government, a resilience economy relies on the existence of a workforce prepared to deliver resilience services. Today's practitioners should become increasingly aware of how their projects and decisions impact adjacent systems and the community as a whole. This book is intended to help start that journey, but additional resources are needed including continuing education requirements covering resilience. Certifications and certificate programs will

Operation	Organization	Activity	Output
Resilience support	Federal agencies, universities, non-profits	Resilience technique and mitigation strategies	Studies, journal articles, website articles, and software tools
		Performance indices and metrices	
		Resilience and mitigation strategy valuations	
		Risk assessment models	
		Data gathering and analysis	
	Codes and standards bodies	Codes and standards development	Codes and standards publications, training, and certifications
Resilience planning	Communities and states	Resilience planning, code and Zoning adoption, and regulatory modifications	Resilience plans, data, codes, ordinances, and regulations
Resilience incentivization	Insurance and mortgage companies, secondary insurance and mortgage organizations, and industry organizations	Insurance underwriting, mortgage origination	Policies, mortgages, and company and industry guidelines and models
	Communities	Zoning-based incentives, developer agreements, and bonds	Ordinances, agency guidelines, and offerings
	Utilities	Rate increases	Guidelines
	Communities, state legislatures, U.S. Congress	Tax incentives, new programs, and regulatory modifications	Legislation, ordinances, agency guidelines
	Financial organizations	Loans, loan support, bond investment and ratings, and financial management	Prospectuses, industry guidelines, offerings, and models
	Contractors	Financing, one-stop shopping bundling financing and mitigation strategy implementation	Brochures, contracts, and consumer education
Resilience implementation	Communities, states, foundations, and public-private partnerships	Resilience programs, data development and zoning and code enforcement	Resilient buildings, data, and case studies
	Developers, construction companies, non-profits	Construction, retrofits, project management, and building information modeling (BIM)	Resilient buildings,real estate information, and case studies
Resilience evaluation	Private companies, non-profits, and state and federal agencies	Performance evaluations of resilience programs and supporting insurance, mortgage, finance, and tax-based programs	Reports, studies, journal articles, website articles
	Appraises and real estate brokers	Identification and promotion of increased property values based on implementation of resilience measures	Appraisal reports, reporting resilience measures in property descriptions/MLS
Resilience communication	All stakeholders	Peer-to-peer communication on methods for achieving resilience	Conferences, web meetings, webinars, and website articles
	Media	Reporting on resilience activities (particularly in context of pre- and post-hazard reporting) to build a culture of resilience	News stories, public service announcements, community fairs

FIG. 16.3 Elements of the resilience economy. (Source: Reprinted from Multihazard Mitigation Council and Council on Finance, Insurance and Real Estate, An Addendum to the White Paper for Developing Pre-disaster Resilience Based on Public and Private Sector Incentivization, National Institute of Building Sciences, September 2016. https://www.nibs.org/resource/resmgr/MMC/MMC_IncentivizationWB_Add.pdf.)

allow resilience thinkers to highlight their skills and recognition of the complex challenges before us.

A new type of professional is likely to emerge—one that has the ability to look across social, economic, and environmental systems to identify resilience strategies that cut across all of them. These polymaths are beginning to emerge, not through a formal educational pathway but out of necessity. The Chief Resilience

Officers established through the 100 Resilient Cities project represent the first cadre of such professionals, but more are needed to address this global challenge. Ideal candidates will be truly interdisciplinary with a focus on working through complex, multifaceted issues. Universities are beginning to pick up on the need for this new type of professional, developing interdisciplinary departments or even colleges.

NEW TOOLS

Today's professionals and those entering the field have access to a growing toolbox of technologies that can help understand the complexities of developing resilient systems and communicate the results to decision-makers and the public. The availability of data and the ability to collect and process large amounts of data have unlocked multiple new strategies for identifying, communicating, and addressing risks. Light detection and ranging (LIDAR) technology is providing new tools for mapping natural and man-made environments with accuracy, precision, and flexibility [7]. Marrying these new maps with computing power provides access to new models that can pinpoint where roads or other pieces of infrastructure are vulnerable under various disaster scenarios. North Carolina has mapped the entire state using LIDAR and has been leveraging the results to determine the impact of sea level rise [8].

Several start-up companies are using artificial intelligence and machine learning to improve mapping of risk and predicting the likelihood and severity of a hazard—even down to the parcel level. These efforts are a far cry from the static maps relied on today to set flood insurance rates or building code requirements. These localized risk management tools can also inform investment decisions.

In addition to more informed decision-making, these new tools should help communicate risk to homeowners and businessowners. It can more clearly illustrate the impacts of certain decisions. Rather than just seeing if a property is inside or outside a floodplain, prospective homeowners can see their risk today and into the future. New visualization technologies make resilience more accessible. Entities like The Weather Channel are using their reach to bring resilience messages to the public. Through augmented reality and computer gaming technology, they are showing viewers the real-world impacts of hazard events [9].

The integration of existing and emerging data tools from different disciplines can provide planners and emergency response personnel with additional clarity to support decision-making. Integrating building information models (BIM) into geographic information systems (GIS) can allow decision-makers to drill down from the macrolevel of a community to the microlevel of an individual property and some key attributes of that property, which may inform its resilience. Interconnecting additional data points can provide even more context. However, as decisions are being made on who should have access to what data, be mindful of the potential to overwhelm the user.

Unlocking Value Through BIM

Building information modeling (BIM) has emerged as a tool to support design and construction of first buildings and now other forms of infrastructure including bridges and other transportation system components. BIM allows the capture of design information and other important pieces of data in a three-dimensional environment. The National BIM Standard-United States (NIBIMS-US) defines BIM as, "a digital representation of physical and functional characteristics of a facility. A BIM is a shared knowledge resource for information about a facility forming a reliable basis for decisions during its life-cycle; defined as existing from earliest conception to demolition. A basic premise of BIM is collaboration by different stakeholders at different phases of the life cycle of a facility to insert, extract, update or modify information in the BIM to support and reflect the roles of that stakeholder" [10].

As the definition recognizes, BIM has evolved beyond a design tool to be a repository for information across the life cycle of a facility. Information in the model can be valuable in advancing resilience at both the individual building and the community level. If planners have access to building level data, planning and exercising can be based on real-world scenarios rather than assumptions. Incentive programs for mitigation or sustainability improvements can be tailored directly to the community's needs. When tied to LIDAR or GIS data, the utility of BIM data is further enhanced.

In times of emergency response, access to BIM data could be valuable. First responders could enter facilities with knowledge of potential hazardous conditions. Egress points could be easily identified. When matched with real-time sensors, situational awareness can improve dramatically.

BIM data are of greatest utility when it is interoperable—meaning that the data can be accessed by multiple parties (assuming permission is given) in a format that is useful for them. This interoperability is captured within NBIMS-US. Extending interoperability to other data-rich systems will allow the generation of a complete picture of the community.

Smart city initiatives are unlocking new sensor and control-based data that can help inform all aspects of a community's resilience from the current condition of infrastructure to developing issues in public health. The emerging concept of a digital twin can be utilized to enhance modeling and planning [11]. As communities move forward in implementing smart city technologies, it is important to understand what they hope to achieve through such efforts. Hopefully, improved resilience is a key component of their strategy, and the data and knowledge gained through deployment of smart city technology feeds into their goals.

CONCLUSION

Communities and the infrastructure that supports their ongoing functionality are complex. Layer on the growing number and impact of hazards and the uncertain impacts of climate change at a community level and the complexity grow. Where and how we build significantly influences whether a community faces a risk and how it survives. These complexities, the interconnectedness of systems, and limited funding lead to the notion that community resilience is ultimately a 21st century wicked problem. Only through optimizing community infrastructure to avoid and withstand the shocks and stresses they will face today and into the future can we truly deliver societal, economic, and environmental resilience.

In the words of a colleague and resilience thought leader Ann Kosmal, "Let's be resilient together," because, there really is no other approach to achieve community and national resilience than a holistic one.

REFERENCES

[1] United Nations Office for Disaster Risk Reduction, Sendai Framework for Disaster Risk Reduction, 2015. https://www.unisdr.org/we/coordinate/sendai-framework.

[2] United Nations Office for Disaster Reduction, How to Make Cities More Resilient: A Handbook for Local Government Leaders, 2017. https://www.unisdr.org/campaign/resilientcities/toolkit/article/a-handbook-for-local-government-leaders-2017-edition.

[3] NOAA National Centers for Environmental Information (NCEI), U.S. Billion-Dollar Weather and Climate Disasters, 2019. https://www.ncdc.noaa.gov/billions/.

[4] Multihazard Mitigation Council, Natural Hazard Mitigation Saves: 2018 Interim Report, K. Principal Investigator Porter, C. co-Principal Investigators Scawthorn, C. Huyck, Investigators: R. Eguchi, Z. Hu, A. Reeder, P. Schneider, Director, MMC. National Institute of Building Sciences, Washington, D.C. https://www.nibs.org/resource/resmgr/mmc/NIBS_MSv2-2018_Interim-Repor.pdf.

[5] Multihazard Mitigation Council and Council on Finance, Insurance and Real Estate, Developing Pre-disaster Resilience Based on Public and Private Sector Incentivization, National Institute of Building Sciences, October 2015. https://www.nibs.org/resource/resmgr/MMC/MMC_ResilienceIncentivesWP.pdf.

[6] Multihazard Mitigation Council and Council on Finance, Insurance and Real Estate, An Addendum to the White Paper for Developing Pre-disaster Resilience Based on Public and Private Sector Incentivization, National Institute of Building Sciences, September 2016. https://www.nibs.org/resource/resmgr/MMC/MMC_IncentivizationWB_Add.pdf.

[7] National Ocean Service, National Oceanic and Atmospheric Administration. What is LIDAR. https://oceanservice.noaa.gov/facts/lidar.html.

[8] L. Montgomery, This new mapping technology will show whether global warming could drown your town, Washington Post (June 27, 2014). https://www.washingtonpost.com/news/wonk/wp/2014/06/27/this-new-mapping-technology-will-soon-show-you-whether-global-warming-will-drown-your-town/.

[9] R. Withers, The Weather Channel uses video game simulation to convey the severity of the hurricane threat, Slate (September 13, 2018). https://slate.com/technology/2018/09/weather-channel-hurricane-florence-flood-simulation.html.

[10] National Institute of Building Sciences, Building Smart Alliance, National BIM Standard-United States, Version 3, 2015. https://www.nationalbimstandard.org.

[11] B. Marr, What is digital twin technology - and why is it so important? Forbes (March 6, 2017). https://www.forbes.com/sites/bernardmarr/2017/03/06/what-is-digital-twin-technology-and-why-is-it-so-important/#2c1bc9372e2a.

Index

Note: Page numbers followed by "b" indicate text boxes, "f" indicate figures and "t" indicate tables.

S